CONCEPTUAL MODELS
OF **FLOW** AND **TRANSPORT** IN THE
FRACTURED VADOSE ZONE

Panel on Conceptual Models of Flow and Transport
in the Fractured Vadose Zone

U.S. National Committee for Rock Mechanics

Board on Earth Sciences and Resources

Commission on Geosciences, Environment, and Resources

National Research Council

D1166303

.... PRESS
Washington, D.C.

This study was supported by the U.S. Nuclear Regulatory Commission, award number NRC-04-96-078, and the U.S. Department of Energy, DE-FG08-97NV12056. The opinions, findings, conclusions, and recommendations expressed herein are those of the authors and do not necessarily reflect the view of the Nuclear Regulatory Commission or the U.S. Department of Energy.

International Standard Book Number 0-309-07302-2
Library of Congress Control Number: 2001087694

Additional copies of this report are available from:

National Academy Press
2101 Constitution Avenue, N.W.
Box 285
Washington, DC 20055
800-624-6242
202-334-3313 (in the Washington metropolitan area)
http://www.nap.edu

Cover: Illustrations courtesy of David A. Feary (photographs from Yucca Mountain and Busted Butte), and Peters, R. R., and E. A. Klavetter, 1988, a continuum model for water movement in an unsaturated fractured rock mass, Water Resources Research 24(3): 416-430, copyright by American Geophysical Union (for schematic figure showing fluid in fractures).

Printed in the United States of America

THE NATIONAL ACADEMIES

National Academy of Sciences
National Academy of Engineering
Institute of Medicine
National Research Council

The **National Academy of Sciences** is a private, nonprofit, self-perpetuating society of distinguished scholars engaged in scientific and engineering research, dedicated to the furtherance of science and technology and to their use for the general welfare. Upon the authority of the charter granted to it by the Congress in 1863, the Academy has a mandate that requires it to advise the federal government on scientific and technical matters. Dr. Bruce Alberts is president of the National Academy of Sciences.

The **National Academy of Engineering** was established in 1964, under the charter of the National Academy of Sciences, as a parallel organization of outstanding engineers. It is autonomous in its administration and in the selection of its members, sharing with the National Academy of Sciences the responsibility for advising the federal government. The National Academy of Engineering also sponsors engineering programs aimed at meeting national needs, encourages education and research, and recognizes the superior achievements of engineers. Dr. William A. Wulf is president of the National Academy of Engineering.

The **Institute of Medicine** was established in 1970 by the National Academy of Sciences to secure the services of eminent members of appropriate professions in the examination of policy matters pertaining to the health of the public. The Institute acts under the responsibility given to the National Academy of Sciences by its congressional charter to be an adviser to the federal government and, upon its own initiative, to identify issues of medical care, research, and education. Dr. Kenneth I. Shine is president of the Institute of Medicine.

The **National Research Council** was organized by the National Academy of Sciences in 1916 to associate the broad community of science and technology with the Academy's purposes of furthering knowledge and advising the federal government. Functioning in accordance with general policies determined by the Academy, the Council has become the principal operating agency of both the National Academy of Sciences and the National Academy of Engineering in providing services to the government, the public, and the scientific and engineering communities. The Council is administered jointly by both Academies and the Institute of Medicine. Dr. Bruce Alberts and Dr. William A. Wulf are chairman and vice-chairman, respectively, of the National Research Council.

PANEL ON CONCEPTUAL MODELS OF FLOW AND TRANSPORT IN THE FRACTURED VADOSE ZONE

PAUL A. HSIEH, *Chair*, U.S. Geological Survey, Menlo Park, California
JEAN M. BAHR, University of Wisconsin, Madison
THOMAS W. DOE, Golder Associates, Inc., Redmond, Washington
ALAN L. FLINT, U.S. Geological Survey, Sacramento, California
GLENDON GEE, Battelle Pacific Northwest Laboratory, Richland, Washington
LYNN W. GELHAR, Massachusetts Institute of Technology, Cambridge
D. KIP SOLOMON, University of Utah, Salt Lake City
MARTINUS VAN GENUCHTEN, U.S. Salinity Laboratory, Riverside, California
STEPHEN W. WHEATCRAFT, University of Nevada, Reno

NRC Staff

DAVID A. FEARY, Senior Program Officer (beginning 3/2000)
THOMAS M. USSELMAN, Senior Program Officer (through 2/2000)
JENNIFER T. ESTEP, Administrative Associate

vii

Preface

The purpose of this study is to describe the processes through which conceptual models of flow and transport in the fractured vadose zone are developed, tested, refined, and reviewed. The Panel convened a two-day workshop in March 1999, during which a large group of specialists from the hydrogeologic, geochemical, soil science, and related fields discussed the current state of knowledge, lessons learned from field investigations, and needs for future research. A series of invited presentations provided the basis for much of the discussion at this workshop. Individually authored papers based on these presentations are presented as Chapters 2-11 in the second part of this volume.

The Panel was charged with preparing a consensus report on the development and testing of conceptual models for fluid flow and transport in the fractured vadose zone. The Panel's conclusions and recommendations were based in large part on the workshop presentations and discussions. This report is intended to describe the present status of conceptual model building in the fractured vadose zone; to provide guidance to regulatory agencies on the review process for conceptual models developed for site licensing; to compile knowledge and experiences from related disciplines so that technical communities can benefit from advances in related fields; and to identify future research needed to advance the technical basis for developing and evaluating vadose zone conceptual models.

In its consideration of transport in the vadose zone, the Panel focused on the application of environmental tracers (such as tritium and chlorine-36) because they provide integrated responses that are difficult to determine by point measurements of fluid potential or moisture content. The Panel briefly reviewed approaches for modeling transport of conservative solutes, but the scope of the study did not include reactive solutes or water-rock interactions.

The Panel report (Chapter 1) is composed of three main sections. First, we discuss general considerations applicable to the development and testing of conceptual models. Second, we summarize the current state of knowledge of flow and transport processes in the fractured vadose zone. Third, we present our conclusions and recommendation.

This report has been reviewed in draft form by individuals chosen for their diverse perspectives and technical expertise, in accordance with procedures approved by the NRC's Report Review Committee. The purpose of this independent review is to provide candid and critical comments that will assist the institution in making its published report as sound as possible and to ensure that the report meets institutional standards for objectivity, evidence, and responsiveness to the study charge. The review comments and draft manuscript remain confidential to protect the integrity of the deliberative process. We wish to thank the following individuals for their review of this report:

John D. Bredehoeft, The Hydrodynamics Group, Story, Wyoming
June T. Fabryka-Martin, Environmental Technology, Los Alamos National
    Laboratory, New Mexico
Jane C. S. Long, Mackay School of Mines, University of Nevada, Reno
Mark Person, Department of Geology and Geophysics, University of
    Minnesota, Minneapolis
Daniel B. Stephens, Daniel B. Stephens and Associates, Inc., Albuquerque,
    New Mexico

Although the reviewers listed above have provided many constructive comments and suggestions, they were not asked to endorse the conclusions or recommendations nor did they see the final draft of the report before its release. The review of this report was overseen by George Hornberger, University of Virginia, Charlottesville, appointed by the Commission on Geosciences, Environment, and Resources, who was responsible for making certain that an independent examination of this report was carried out in accordance with institutional procedures and that all review comments were carefully considered. Responsibility for the final content of this report rests entirely with the authoring committee and the institution.

In addition, we acknowledge peer reviews provided by the following for the invited papers (Chapters 2-11): S. Bradford, J. D. Bredehoeft, J. Fabryka-Martin, R. Healy, D. L. Hughson, V. Kapoor, K. Karasaki, K. Keller, S. Kung, E. Kwicklis, L. D. McKay, J. W. Mercer, R. L. Michel, B. Mohanty, J-V. Parlange, L. Pyrak-Nolte, E. H. Roseboom Jr., B. R. Scanlon, D. B. Stephens, T. K. Tokunaga, and E. P. Weeks. Although these papers have undergone peer review, their inclusion in this report does not constitute any specific endorsement of their contents, either by the Panel or the National Research Council.

Paul Hsieh
Chair, Panel on Conceptual Models of
    Flow and Transport in the Fractured
    Vadose Zone

Contents

xiii

Note: Color plates are located after p. 184.

CONCEPTUAL MODELS
OF **FLOW** AND **TRANSPORT** IN THE
FRACTURED VADOSE ZONE

Executive Summary

Fluid flow and solute transport within the vadose zone, the unsaturated zone between the land surface and the water table, is the cause of expanded plumes arising from localized contaminant sources, and an understanding of vadose zone processes is an essential prerequisite for cost-effective contaminant remediation efforts. Contamination of the vadose zone can result from many causes, including chemical spills, leaky underground storage tanks, leachate from waste disposal sites and mine tailings, and application of agricultural chemicals. Another major environmental concern is the potential for long-term migration of radionuclides from low- and high-level nuclear waste disposal facilities. Development of flow and transport models for the vadose zone is a key requirement for designing remediation and long-term stewardship strategies. The presence of fractures and other channel-like openings in the vadose zone poses a particularly significant problem, because such features are potential avenues for rapid transport of chemicals from contamination sources to the water table.

The underpinning of any vadose zone fluid transport model is the conceptualization of (1) the relevant processes, (2) the structure of the subsurface, and (3) the potential events or scenarios that impact the behavior of the modeled system. These conceptualizations together form a "conceptual model," and it is such conceptual models that are the focus of this Panel's study. In cases where multiple competing conceptual models could lead to drastically different conclusions, strategies for evaluating these models must be based on sound technical criteria. The need to develop such strategies and criteria was a key reason for the appointment of this Panel (see Box 1).

1

BOX 1
Statement of Task

A panel under the auspices of the U.S. National Committee for Rock Mechanics will organize and conduct a workshop on conceptual models of fluid infiltration in fractured media. The study will focus on the scale, complexity, and site-specific conditions and processes that need to be determined in order to develop an appropriate conceptual infiltration model. Examples of questions that may be addressed are: (1) Does the conceptual model provide an adequate characterization of the system? (2) How well does the model perform in comparison with competing models? (3) Is the data base adequate to estimate model parameters with sufficient reliability that the associated prediction uncertainties are acceptable in light of the intended application of the model? (4) What are the opportunities for field testing and verification of the model? The Panel will produce a consensus report of its findings and conclusions. A series of individually authored papers that were presented at the workshop will be appended to the report.

When carefully developed and supported by field data, models can be effective tools for understanding complex phenomena and for making informed predictions. However, model results are always subject to some degree of uncertainty due to limitations in field data and incomplete knowledge of natural processes. Thus, when models form the basis for decision-making, uncertainty will be an inescapable component of environmental management and regulation. A key consideration in any modeling process is whether the model has undergone sufficient development and testing to address the problem being analyzed in a sufficiently meaningful manner.

After reviewing the process through which conceptual models of flow and transport in the fractured vadose zone are developed, tested, refined, and reviewed, the Panel produced the following conclusions grouped according to the two major topics addressed in this report: (1) general considerations during the development and testing of conceptual models, and (2) flow and transport in the fractured vadose zone. These conclusions are followed by the Panel's recommendations for research activities that will contribute to the conceptual modeling process.

CONCLUSIONS ON DEVELOPMENT AND TESTING OF CONCEPTUAL MODELS

1. Development of the conceptual model is the most important part of the modeling process. The conceptual model is the foundation of the quantitative, mathematical representation of the field site (i.e., the mathematical model), which in turn is the basis for the computer code used for simulation.

2. The context in which a conceptual model is developed constrains the range of its applicability. A conceptual model is by necessity a simplification of the real system, but the degree of simplification must be commensurate with the problem being addressed.

3. It is important to recognize that model predictions require assumptions about future events or scenarios, and are subject to uncertainty. Quantitative assessment of prediction uncertainty should be an essential part of model prediction. A suite of predictions for a range of different assumptions and future scenarios is more useful than a single prediction.

4. Testing and refinement of the conceptual model are critical parts of the modeling process. Reasonable alternative conceptualizations and hypotheses should be developed and evaluated. In some cases, the early part of a study might involve multiple conceptual models until alternatives are eliminated by field results.

5. Although model calibration does provide a certain level of model testing, a good fit to the calibration data does not necessarily prove that the model is adequate to address the issues in question. A model that matches different types of calibration data collected under different field conditions is likely to be more robust than a model that matches a limited range of calibration data. However, if the model cannot be calibrated to match the calibration data, this is an indication that the conceptualization should be re-examined.

6. Checking model simulation results against field data (that were not used for calibration) is one, but not the exclusive, approach to model testing. Where all field data are needed for calibration, and none are left for further testing, this does not mean that the model cannot be used for prediction.

7. From an operational perspective, the goal of model testing is to establish the credibility of the model. In addition to testing and evaluation, the credibility of a model can be enhanced by independent peer review and by maintaining an open flow of information so that the model is available for scrutiny by concerned parties.

CONCLUSIONS ON FLOW AND TRANSPORT IN THE FRACTURED VADOSE ZONE

1. There exists a body of field evidence indicating that infiltration through fractured rocks and structured soils does not always occur as a wetting front advancing at a uniform rate; accordingly, any model simulations that assume a uniform wetting front within a homogeneous medium may provide erroneous estimates of flux and travel times through the vadose zone.

2. The current state of knowledge is not adequate to determine which processes are likely to control unsaturated flow and transport at a given field site. Although laboratory and theoretical analyses demonstrate that film flow in frac-

tures can transport fluid and solute at rates substantially higher than transport by capillary flow, the significance of film flow at the field scale is controversial.

3. Although not identical, structured soils and fractured rocks exhibit many similarities in flow and transport processes. Macropores and aggregates in structured soils are respectively analogous to fractures and matrix blocks in rock, and therefore communication between workers in both soil science and fractured rock fields will be mutually beneficial.

4. Models of varying complexity have been developed for preferential flow, but their adequacy for field-scale application requires further testing. In order to avoid explicitly simulating the mechanisms that cause preferential flow, current models simulate fast and slow flow by use of a composite hydraulic conductivity curve or by dual permeability domains. Further testing is required to determine whether such models are adequate for field-scale application over a broad range of field conditions.

5. The interaction between fracture and matrix exerts a strong control on fluid and solute movement. However, the strength of this interaction in the field is not well known. The simplified representation of this interaction in current models also requires further evaluation.

6. Solute transport in the fractured vadose zone can exhibit complex behavior due to the large variations in fluid velocity and the interplay of advective and diffusive transport between fractures and matrix. Solute transport models are more complex than flow models, and can involve multiple regions to represent the diversity of macropore and micropore sizes.

7. Environmental tracers should be included in field investigation strategies from the very beginning of a site characterization program. Use of geochemical and environmental tracer data in several studies has led to substantial revisions of conceptual models initially based upon hydrodynamic analysis.

RECOMMENDED RESEARCH

Flow and transport in the fractured vadose zone have been, and will continue to be, an active area of research in both the soil science and subsurface hydrology disciplines. The research recommended in this report is not meant to be inclusive. Instead, the list below reflects topics that address the issues identified in this report so that conceptual models of flow and transport can be improved.

1. Fundamental research to understand flow and transport processes in unsaturated fractures should continue. Particular emphasis should be placed on understanding mechanisms that cause non-uniform (preferential) flow, film flow, and intermittent behavior.

2. Research is needed to understand the spatial variability in vadose zone properties, and to develop upscaling methods. Spatial variability is a key cause of model uncertainty, because the subsurface cannot be exhaustively sampled. Up-

scaling methods are needed to derive field-scale flow and transport properties from small-scale laboratory measurements.

3. There is a need for comprehensive field experiments in several fractured vadose zone geologic environments. These experiments should be designed to understand the controlling processes for a broad range of field conditions, to evaluate methods of parameter upscaling, and to test alternative conceptual models.

4. Current models should be evaluated for their adequacy for simulating flow and transport in the presence of fingering, flow instability, and funneling. Of particular importance is the evaluation of transfer coefficients to represent fluid and solute exchange between fracture and matrix.

5. There is a need to develop quantitative assessments of prediction uncertainty for models of flow and transport in the fractured vadose zone. Meaningful quantification of uncertainty should be considered an integral part of any modeling endeavor, as it establishes confidence bands on predictions given the current state of knowledge about the system.

6. Research should be undertaken to develop improved techniques for geochemical sampling of the fractured vadose zone. Improved sampling techniques will facilitate the use of environmental tracers and geochemical data for conceptual model building.

Panel Report

1

Conceptual Models of Flow and Transport in the Fractured Vadose Zone

INTRODUCTION

A significant number of subsurface environmental problems involve fluid flow and solute transport in the fractured vadose zone. In this report, the vadose zone refers to that part of the subsurface from land surface to the lowest seasonal water table elevation. The vadose zone may be composed of consolidated rock and/or unconsolidated granular material, including soils. Contamination of the vadose zone can result from many sources, including chemical spills, leaky underground storage tanks, leachate from waste disposal sites and mine tailings, and application of agricultural chemicals. Another major environmental concern is the potential for long-term migration of radionuclides from low- and high-level nuclear waste disposal facilities. The presence of fractures and other channel-like openings in the vadose zone poses a particularly significant problem, because such features are potential avenues for rapid transport of chemicals from contamination sources to the water table.

A key component in assessing contamination hazards and designing remedial actions is the development of flow and transport models to approximately represent real systems. Present-day models often use computer programs to simulate flow and transport processes, and such models are applied throughout the scientific, engineering, and regulatory arena. When carefully developed and supported by field data, models can be effective tools for understanding complex phenomena and for making informed predictions for a variety of assumed future scenarios. However, model results are always subject to some degree of uncertainty due to limitations in field data and incomplete knowledge of natural processes. Thus, when models form the basis for decision-making, uncertainty will be an inescapable component of environmental management and regulation. A key consideration in the modeling process is whether or not the model has undergone sufficient development and testing to address the problem being analyzed in a sufficiently meaningful manner.

The underpinning of any vadose zone fluid transport model is the conceptualization of (1) the relevant processes, (2) the structure of the subsurface, and (3) the potential events or scenarios that impact the behavior of the modeled system. These conceptualizations together form a "conceptual model," and it is such conceptual models that are the focus of this Panel's study. The evolution of a conceptual model, from the initial formulation through the subsequent revisions and refinements, is the crux of the model development process. Conceptual models of fully saturated flow in granular aquifers have progressed to a relatively mature state due to a long history of model development and application. By contrast, conceptual models of partially saturated flow and transport in fractured vadose zone environments are poorly developed and largely untested. In cases where multiple competing conceptual models could lead to drastically different conclusions, strategies for evaluating these models must be based on sound technical criteria. The need to develop such strategies and criteria was a key reason for the appointment of this Panel.

The purpose of this study is to investigate the processes through which conceptual models of flow and transport in the fractured vadose zone are developed, tested, refined, and reviewed. The Panel convened a two-day workshop in March 1999, during which a large group of specialists (see Appendix A) from the hydrogeologic, geochemical, soil science, and related fields discussed the current state of knowledge, lessons learned from field investigations, and needs for future research. A series of invited presentations provided a basis for much of the discussion at the workshop. Individually authored papers based on these presentations are presented as Chapters 2-11.

The Panel was charged with preparing a consensus report on the development and testing of conceptual models for fluid flow and transport in the fractured vadose zone. The Panel's conclusions and recommendations were based in large part on the workshop presentations and discussions. The report is intended to:

- provide information on contemporary philosophies, approaches, and techniques for conceptual model building;
- provide guidance to regulatory agencies on the review process for conceptual models developed for site licensing;
- bring together knowledge and experiences from related disciplines so that technical communities can benefit from advances in related fields; and
- identify future research needed to further the technical basis for developing and evaluating conceptual models of flow and transport in the fractured vadose zone.

The Panel devoted a major portion of its study to an analysis of fluid flow models in the fractured vadose zone because (1) understanding flow is a prerequisite for understanding transport, (2) questions regarding the nature of fast pathways are a major concern, and (3) recent studies, especially investigations at the potential site for a nuclear waste repository at Yucca Mountain, have accumulated a significant body of knowledge applicable to fractured vadose zone models. In its consideration of transport in the vadose zone, the Panel focused on the application of environmental tracers (such as tritium and chlorine-36) because they provide integrated responses that are difficult to determine by point measurements of fluid potential or moisture content. The Panel briefly reviewed approaches for modeling transport of conservative solutes, but the scope of the study did not include reactive solutes or water-rock interactions.

The Panel report is composed of three main sections. First, we discuss general considerations applicable to the development and testing of conceptual models. Second, we summarize the current state of knowledge of flow and transport processes in the fractured vadose zone. Third, we present our conclusions and recommendations. Appended to the Panel report are the invited papers based on presentations at the workshop. These papers served as the starting point for workshop discussions, and form much of the background material used in preparation of this report. Although the papers have undergone peer review, their inclusion in this report does not constitute any specific endorsement of their contents, either by the Panel or by the National Research Council.

DEVELOPMENT AND TESTING OF CONCEPTUAL MODELS

Definition of Conceptual Model

Models representing natural systems are often viewed as composed of two components, a conceptual model and a mathematical model. In general terms, a conceptual model is qualitative and expressed by ideas, words, and figures. A mathematical model is quantitative and expressed as mathematical equations. The two are closely related. In essence, the mathematical model results from translating the conceptual model into a well-posed mathematical problem that can be solved.

Various definitions of conceptual models can be found in the scientific and technical literature. These definitions are generally consistent in their fundamental meaning, and differ mainly in scope, detail, and context. The statement of the conceptual model often reflects the key questions and unknowns to be investigated. For example, Anderson and Woessner (1992) give the following definition for the purpose of modeling groundwater flow in aquifers:

> A conceptual model is a pictorial representation of the groundwater flow system, frequently in the form of a block diagram or a cross section.

This relatively brief statement is suitable for modeling groundwater flow because the process is well understood, and the principal concern is the delineation of subsurface structure, such as stratigraphy, and the spatial distribution of hydraulic properties, such as hydraulic conductivity. By contrast, Tsang (1991) gives the following definition in the context of modeling a broad range of coupled physical-chemical processes:

> A site-specific conceptual model consists of three main components: structure, processes, and boundary and "initial" conditions. "Structure" refers to the geometric structure of the system, such as stratigraphy, faults, heterogeneity, fracture density and lengths, and other geometric and geologic characteristics. "Processes" are physical and chemical phenomena such as buoyancy flow, colloidal transport, matrix diffusion, and dissolution and precipitation. "Boundary conditions" are constant or time-dependent conditions imposed on the boundaries of the model domain. "Initial conditions" are the physical and chemical conditions over the model domain at a particular instant of time. This is usually taken at the initial instant of time, though in general it can be any specified point in time.

The more detailed statement reflects the complexity of the modeled processes, some of which may be poorly understood.

In the present study, we define a conceptual model as follows: **A conceptual model is an evolving hypothesis identifying the important features, processes, and events controlling fluid flow and contaminant transport of consequence at a specific field site in the context of a recognized problem.** This definition stresses several ideas. A conceptual model is a hypothesis because it must be tested for internal consistency and for its ability to represent the real system in a meaningful way. The hypothesis evolves (is revised and refined) during testing and as new information is gathered. Although a conceptual model is by necessity a simplification of the real system, the degree of simplification (or conversely, the amount of complexity retained in the model) should be commensurate with the problem being addressed. In order to present a clear and easily understandable definition, we have used the term 'events' rather than the mathematically explicit terms 'initial conditions' and 'boundary conditions' that have been used in earlier definitions. The context in which the model is developed constrains the range of applicability of the model.

The Modeling Process

We refer to the modeling process as an iterative sequence of actions that includes (1) identifying a site-specific problem; (2) conceptualizing important features, processes, and events; (3) implementing a quantitative description; (4) collecting and assimilating field data that are used to calibrate the model and evaluate its predictive capabilities; and (5) developing predictions that are used to resolve the identified problem. This process is illustrated by the flow chart in Figure 1-1.

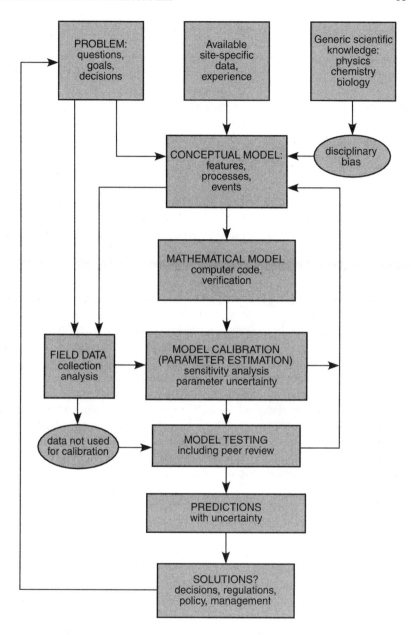

FIGURE 1-1 Flow chart illustrating the elements of the modeling process.

The modeling process usually begins with some initial perception of a field problem, expressed as questions to be answered or decisions to be made (as indicated in the upper left of Figure 1-1). Available site-specific data, related experience, and generic scientific knowledge are then combined to identify the factors (features, processes, and events) that are important for the identified problem. This typically involves preliminary calculations using generic parameters to understand the relative importance of various aspects of the problem. The result of this assessment is the initial conceptual model.

The formulation of this initial conceptual model is arguably the most important step in the modeling process. The conceptual model is the foundation of the mathematical model, and strongly influences the type of computer code to be used and the design and priority of site characterization activities. Philips (this report, Chapter 9) describes a case study where processes that were neglected in the initial conceptual model were later found to be of fundamental importance, leading to a substantial revision of the conceptual model. Such revisions, possibly with additional iterations, may be expected in complex environments where often poorly understood physical, chemical, and biological processes interact. Therefore, the conceptual modeling process should consider a broad range of reasonable alternative hypotheses. It is also important to employ a variety of different types of data. For example, an initial conceptual model that is based on regional water budgets, general hydraulic properties, and sparse environmental tracer data is more likely to include relevant and significant processes than if only general hydraulic properties were considered. The conceptual model is the foundation upon which all aspects of the modeling process are constructed, and this should be clearly understood by all members of a project team, from data gatherers to modelers to managers. In any situation where a particular field problem requires a linkage between site characterization and performance assessment, the development of a conceptual model and the continued testing and assessment of the model should be an explicitly required component of efforts to address the problem.

The next step in the modeling process is the development of a quantitative description of the conceptual model in terms of mathematical equations to be solved—the mathematical model. Except for simple problems, mathematical equations are usually solved by use of a computer code, also called a numerical model. A newly developed computer code must be tested to assure that it correctly solves the equations of the mathematical model. This "verification" process typically involves comparisons of the numerical results with known solutions for related simple configurations. Because the verification process of a complex code usually involves piecemeal testing of the individual transport processes represented by the model, without simultaneous testing of all or most features of a code, it is rarely possible to be assured that a code is fully verified. In addition, even a well-established computer code should be used carefully by knowledgeable analysts to avoid potential problems such as numerical instability.

A mathematical model always includes a number of parameters, such as the distribution of hydraulic properties within the modeled region. At the early stages of the modeling process, the values of some of these parameters are unknown. The process of estimating these parameters is known as model calibration or parameter estimation. The calibration process will typically include sensitivity analyses, designed to elucidate how changes in parameters can influence the simulation results. Calibration procedures based on statistical/stochastic formulations can also provide quantitative measures of uncertainty in the parameter estimates. Such information is extremely valuable for guiding the collection of additional field data to refine the model. As shown in Figure 1-1, the calibration step can involve significant feedback to the conceptual model.

Model calibration is often achieved by adjusting model parameters, either manually or by automated methods, until the model simulated results agree with field data to an acceptable level. As the model is a simplification of the real system, a perfect fit between the simulated results and field data is unlikely to be achieved. A serious lack of fit indicates that the conceptual model should be re-examined. On the other hand, a good fit does not necessarily prove that the conceptual model is adequate to address the issues in question. Another complicating factor may be that a measured parameter may not be the same as a modeled parameter, e.g., it is difficult to determine what the measured moisture tension at a porous tip actually represents for a situation where the moisture originates as film-flow in a fracture. Additional model testing, which creates another feedback loop to the conceptual model element in Figure 1-1, is essential to gain confidence that the model provides meaningful results that can be used for decision-making and problem resolution. Model testing is discussed in further detail in the next section.

It is important to recognize that model predictions require assumptions about future events or scenarios. The importance of this "event" component of the conceptual model is made clear by postaudit studies of groundwater models (e.g., Anderson and Woessner, 1992, p. 288-293). These studies suggest that, when model predictions were compared to actual outcomes, a major reason for inaccurate prediction was that the assumed future scenarios (e.g., future stresses on the system) did not occur. This finding points to the need for developing model predictions for many different possible future scenarios.

Another important aspect of model prediction is uncertainty, which arises from many factors, such as simplification of the real system, limited field data, measurement errors, and multiple conceptualizations or interpretations that cannot be resolved by existing data. Even if the modeling process has undergone several cycles of revision, testing, and gathering of additional field data, prediction uncertainty is reduced but not entirely eliminated. Decisions based on model predictions must take this into account by approaches such as analysis of risks, worst-case scenario, or incorporation of safety margins. If the conceptual model uncertainty is too great, the project goals may not be achievable within a reasonable time or at acceptable cost. Alternatively, it may be necessary to redefine the original problem to be resolved.

Model Testing

The formulation of a conceptual model is inherently subjective in that it relies on limited available site-specific observations and data, as well as experience and insights developed through work on similar sites and/or related problems. A conceptualization is susceptible to biases arising from the disciplinary background and experience of the analyst, and/or by differing perceptions of the problem as influenced by external social and political forces. Although model calibration represents one level of testing by requiring the model to reproduce one set of field data, the model may be applied to field conditions that are significantly different than the conditions under which calibration data were collected. For these reasons, it is important to test the predictive capabilities of a model.

Due to the large variability in the objectives of modeling projects, it is not possible to provide a prescriptive step-by-step procedure for model testing. The amount of effort devoted to model testing will strongly depend on the question to be resolved, the scope of the investigation, available resources, and the consequence of an inadequate or inappropriate model. In this section, we discuss some general issues that should be considered in conceptual model testing. We assume at the outset that the computer code used for simulation is already verified, in other words, that the program logic and numerical algorithm are correctly implemented, and the results are free of computational errors.

A traditional procedure of model testing is to use the calibrated model to simulate a set of field data that was not used during the calibration process. For example, if historical data exist for the site, a portion of the data may be used for calibration and the remaining data used for testing. Alternatively, the model is used to predict the result of a field experiment, and the experiment is then carried out as a check against model predictions. The test is successful if the model simulation agrees with the test data, within reasonable limits, without the need to further adjust the model parameters. A model that passes one or a series of such tests would have demonstrated a certain level of predictive capability. If the model fails the test, then the test data are used to further calibrate the model and additional field data are needed for a new round of testing.

Although the above test procedure is conceptually appealing, it may not always be feasible. In most cases, field data are limited and all the data must be used for calibration, leaving none for testing. Therefore, a broader view of model testing is often necessary. One approach is to evaluate how well the conceptual model represents the real system in terms of features, processes, and events. This involves both examining the model assumptions and evaluating alternative hypotheses. The underlying rationale is that a model that has undergone such evaluations can be used with an increased degree of confidence.

As an example of evaluating a "feature" component of the conceptual model, consider the question of whether a geologic stratum that is represented in the model as a continuous layer underlying the entire model region may in fact be discontinuous and absent at certain locations. The first step in evaluating this alternative hypothesis is to consider its consequence. This can be done by modifying the model according to the alternative hypothesis (discontinuous layer), recalibrating the model, and developing new predictions. If the modified model with changed parameters cannot match the data, this is an indication that the alternative hypothesis is inconsistent with available field data. If recalibration is successful but the new predictions are similar to the old predictions, this is an indication that the alternative hypothesis is of little consequence (within the context of the question to be resolved). In both cases, the alternative hypothesis may not warrant additional consideration. However, if the recalibrated model leads to new predictions that are significantly different from the old predictions, then further investigations are needed. The continuity of the layer in question may be evaluated by geophysical methods to image the subsurface, by test borings at strategic locations, or by using an understanding of the depositional environment to infer the likelihood that the layer may be continuous or discontinuous.

Whether or not a model can be "validated" is a topic that has seen substantial debate in recent years. Certain authors (e.g., Konikow and Bredehoeft, 1992; Oreskes et al., 1994) have presented the opinion that models cannot be validated, because the truth of scientific theory can never be proven. Describing a model as "validated" implies, especially to a nontechnical audience, a level of correctness and certainty that is unattainable. Other authors (e.g., see discussion by Jarvis and Larsson, this report, Chapter 6) argue that a "validated model" can be taken to mean that a model is acceptable for its intended use. Currently, there is a lack of consensus on the definition of "validation."

From an operational point of view, model testing and evaluation can be viewed as activities designed to establish the credibility of a model. In this regard, peer review is an important part of the modeling process. Review by a group of objective, independent, and respected experts can utilize knowledge and opinions beyond those of the study team. Maintaining a free flow of information (e.g., field data, model results) can also add a significant measure of credibility, because the model is open to scrutiny by concerned parties (e.g., regulators, license applicants, government agencies, public citizen groups, and the general scientific community). In the final analysis, whether or not a model is acceptable to concerned parties depends on whether or not these parties have confidence that the model predictions provide meaningful input for decision-making and problem resolution.

FLOW AND TRANSPORT IN THE FRACTURED VADOSE ZONE

Nature of Problem

A major issue in the investigation of the vadose zone is estimating fluid flux and solute travel time from land surface to the water table. Field observations suggest that infiltrating water and solute may not necessarily advance downward as a uniform infiltration front, as would be simulated by a model of infiltration into a homogeneous medium. Instead, fluid and solute may travel at widely different fluxes and velocities through different parts of the vadose zone. Field evidence for uneven water movement includes the detection of environmental tracers indicating that relatively young water has moved to significant depths in arid regions, where infiltration rates are expected to be small. For example, at the Exploratory Studies Facility, Yucca Mountain, Nevada, water with a bomb-pulse chlorine-36 signature (that is, water less than 50 years old) was found several hundred meters below the land surface (Fabryka-Martin et al., 1998; see also discussion by Phillips, this report, Chapter 9), whereas a simple model of infiltration through a homogeneous vadose zone would predict travel times of hundreds to thousands of years to reach such a depth.

In this study, we focus on five major issues that cause difficulties for estimating fluid flux and travel times through the fractured vadose zone. First, flow in unsaturated fractures may occur as either capillary flow or film flow. Recent studies have also reported intermittent flow behaviors that are not considered by classical theory. The field conditions under which these flow mechanisms occur (either simultaneously or one predominating over the other) are poorly understood. Second, preferential flow can occur in the vadose zone as a result of heterogeneities and/or flow instability. The factors controlling preferential flow are difficult to characterize and quantify. Third, as water moves through variably saturated fractures, a portion of the flow is imbibed into the rock matrix. Although the degree of imbibition exerts a strong influence on fluid movement, the nature of fracture-matrix interaction is not well known. Fourth, solute transport in fractured rocks can exhibit complex behaviors that are difficult to interpret. Development of conceptual models may be more difficult for transport than for flow. Fifth, in a number of important cases, the interpretation of environmental tracers has lead to conclusions that seemingly contradict the initial conclusions based on classical hydrodynamic analysis. Resolution of this apparent conflict is a necessary requirement before robust conceptual models can be developed. In the remainder of this section, we present a summary of the current understanding of these five issues.

Capillary Flow, Film Flow, and Intermittent Behaviors

The basic notion of capillarity, as developed in classical theory of unsaturated flow, is that fluid is held under tension in pore space within an unsaturated porous medium. In drier conditions (lower saturation, larger magnitude of negative pressure head), fluid occupies the smaller interstices between solid grains. With increasingly wet conditions (higher saturation, negative pressure head approaching zero), the fluid occupies increasingly larger pores. By drawing an analogy between an unsaturated porous medium and an unsaturated fracture, an understanding of fluid within interstitial space can be applied to fluid within a fracture aperture. A basic assumption is that the fluid is in contact with both sides of the fracture wall. This analogy suggests that in drier conditions, the fluid is held within small apertures, and that with increasingly wetter conditions, fluid will occupy increasingly larger apertures. The term "capillary flow" refers to flow, either in interstitial pore space or within fractures, under these conditions.

Figure 1-2 illustrates the current conceptualization of capillary flow in a fracture subject to increasing levels of saturation. Figure 1-2a illustrates a fracture that is essentially dry, and there is no fluid flow in the fracture plane. In Figure 1-2b, the shaded areas illustrate islands of water around the contact regions. At this saturation, there is still no flow in the fracture plane because the wetted areas are disconnected. In Figure 1-2c, saturation has increased to a point where the wetted areas are connected to form contiguous pathways within the fracture, thus allowing fluid flow along the fracture plane. The rate of fluid flow is controlled by the local fracture aperture of the wetted areas. In Figure 1-2d, the fracture is close to complete saturation (except for a trapped air bubble), and the flow rate along the fracture plane is higher than in Figure 1-2c.

In the case of an air-filled fracture in the vadose zone (Figure 1-3), "film flow" occurs as a thin film of fluid flowing down the fracture wall. Unlike capillary flow, in which the fluid contacts both fracture walls, the film of fluid contacts only one fracture wall, with an air phase between itself and the opposing fracture wall. The film of fluid flows under the force of gravity and is not affected by capillary forces within the fracture. Consequently, film flow can occur in small-aperture as well as large-aperture fractures, flow rate is not directly controlled by the width of the fracture aperture, and the onset of flow does not require a contiguous fluid-filled pathway within the fracture. Dragila and Wheatcraft (this report, Chapter 7) suggest that film flow is likely to occur in fractures larger than approximately 1 mm, because less energy is required to transport fluid that contacts one fracture wall compared with the transport of fluid that contacts both fracture walls.

Film flow can exhibit behaviors that are not expected in capillary flow. In laboratory experiments on a sample block of Bishop Tuff, Tokunaga and Wan (1997) demonstrated that the average fluid film velocity could be about 1,000 times faster compared to the average pore water velocity if the sample is saturated and subjected to a unit hydraulic gradient. Analysis by Dragila and Wheatcraft (this report, Chapter 7) suggests that the free-surface film flow may behave in a chaotic manner and may develop solitary waves, that is, water traveling in "lumps" over a thin film substrate. These solitary waves can carry mass (as

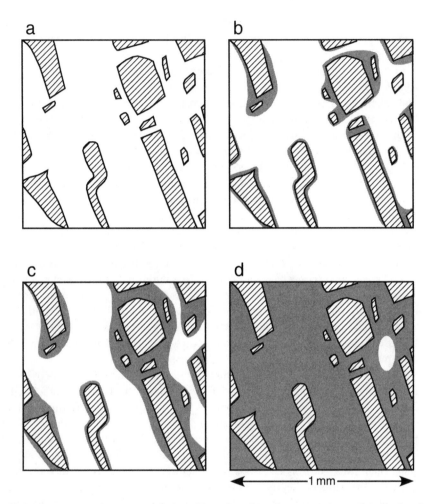

FIGURE 1-2 Conceptualization of capillary flow in a fracture at increasing levels of saturation (plan view of fracture face). Hatched areas denote regions where opposing fracture walls are in contact. Gray areas denote wetted regions. (a) A dry fracture. (b) Wetted areas form islands around the contact regions but are disconnected. (c) Wetted areas are connected to form contiguous pathways in the fracture. (d) Fracture near full saturation with one trapped air bubble. From Peters, R.R., and E.A. Klavetter, 1988. A continuum model for water movement in an unsaturated fractured rock mass. Water Resources Research 24(3): 416-430. Copyright by American Geophysical Union.

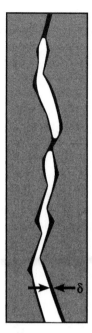

FIGURE 1-3 Film flow in a fracture, except where water contacts both fracture walls at locations of small fracture aperture. Film thickness is denoted by δ. From Tokunaga, T.K., and J. Wan, 1997. Water film flow along fracture surfaces of porous rock. Water Resources Research 33(6): 1287-1295. Copyright by American Geophysical Union.

opposed to just energy), and travel at a speed about twice the average film speed. Solitary waves may also lead to sporadic flow behavior. If a solitary wave grows in amplitude or enters a region of smaller fracture aperture, the film may contact the opposite fracture wall. This produces a situation in which capillary force becomes important, and results in a temporary change in flow rate.

In both capillary flow and film flow situations, a portion of the infiltrating fluid can travel at velocities significantly higher than the remainder of the fluid. Thus, estimation of solute travel time based on the average advance of an infiltration front may not accurately represent the fastest solute travel. In capillary flow, heterogeneity and flow instability can cause large variations in velocity, resulting in "preferential flow" (discussed at greater length below). The combination of film flow in fractures and capillary flow in the matrix may also increase transport rates. Dragila and Wheatcraft (this report, Chapter 7) hypothesize a scenario in which fractures that are not connected with each other can nonetheless enhance infiltration rates. During an infiltration event, a fracture in the vadose zone not exposed to the surface may generate seepage by creating a capillary barrier against the unsaturated porous matrix. If the fracture is inclined towards the vertical, the seepage may generate a free-surface film that will flow at very rapid speeds to the lower terminus of the fracture, where the fluid is again absorbed into the matrix. By this mechanism, the fracture becomes a short circuit to the porous matrix flow. A series of fortuitously placed fractures could substantially increase the transport rate.

At the field scale, the approach for modeling film flow (in a fracture network) is unclear. Current studies of film flow have been conducted primarily at the laboratory scale (in a single fracture). In Chapter 7, Dragila and Wheatcraft model film flow by a set of ordinary differential equations that is a simplified form of the momentum and conservation equations of fluid mechanics. By contrast, Tokunaga and Wan (1997) characterize film flow by measuring how "surface transmissivity" and average film thickness vary with (negative) pressure head. The resultant curves appear analogous to the hydraulic conductivity and retention curves used in traditional models of capillary flow (discussed in next section). Nonetheless, it remains an open issue whether a film flow model can be combined with the capillary flow model at the field scale, or whether the two types of flow require a fundamentally different approach.

There is also a diversity of opinions concerning whether film flow plays a significant role in infiltration. Because fluid films can travel at high velocities, Tokunaga and Wan (1997) state that "film flow can be an important mechanism contributing to fast flow in unsaturated fractures and macropores." However, based on order-of-magnitude calculations for conditions applicable to Yucca Mountain, Pruess (1999) suggests that "rates of film flow will be small." Resolution of this and associated issues relating to film flow will have to await future research.

Recent laboratory experiments using a transparent epoxy replica of a rock fracture (Su et al., 1999) have demonstrated intermittent flow behaviors that are not considered by classical theory. Observations of water invasion into an initially dry fracture replica indicated that flow paths in the fracture consisted of broad, water-filled regions, known as "capillary islands," connected by thin threads of water, known as "rivulets." Even though inflow to the fracture replica was kept constant, intermittent flow persisted in the form of cyclic snapping and re-forming of rivulets.

Figure 1-4 illustrates the intermittent flow cycle observed by Su et al. (1999). The cycle begins just after the rivulet has snapped, and water accumulates in a small capillary island just upstream of the snap point (Figure 1-4b). The capillary island grows in size (Figure 1-4c) and then migrates down the fracture replica, leaving behind a new rivulet (Figure 1-4d). Finally, the capillary island drains towards the bottom outlet (Figure 1-4e), and the rivulet eventually snaps, starting a new cycle.

Su et al. (1999) hypothesized that the type of intermittent flow observed in their experiments evolves from liquid flowing through a sequence of small to

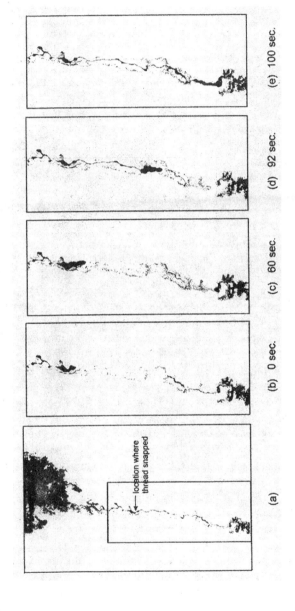

FIGURE 1-4 Liquid distribution during water invasion into initially dry, transparent, epoxy replica of a natural fracture. (a) Full view of 21.5 × 33 cm fracture replica. (b-e) Enlargement of the boxed region in (a). Time indicates seconds after the thread (rivulet) has snapped. From Su, G.W., J.T. Geller, K. Pruess, and F. Wen, 1999. Experimental studies of water seepage and intermittent flow in unsaturated, rough-walled fractures. Water Resources Research 35(4): 1019-1037. Copyright by American Geophysical Union.

large to small apertures. In Chapter 8, Doe analyzes the physics of capillary islands, based on the work by Furmidge (1962) on sliding of liquid drops on solid surfaces. These studies have enhanced the conceptual understanding of intermittent flow in unsaturated fractures. How to incorporate these small-scale mechanisms into field-scale models remains a difficult challenge.

Preferential Flow

While the simplest conceptualization of infiltration in the vadose zone is that of a uniform, downward-advancing wetting front in a homogeneous medium, field observations indicate that infiltration can be non-uniform (e.g., Kung, 1990a, see also discussion by Hendrickx and Flury, this report, Chapter 5). The term "preferential flow" is often used to describe non-uniform (or uneven) water flow characterized by widely different local-scale velocities. The consequence of preferential flow is that a portion of the infiltrating water can be concentrated to move along certain pathways at rates that are significantly faster than the rest of the infiltrating water. With the occurrence of preferential flow, contaminants carried by the infiltrating water can reach a given depth in less time than predicted by calculations assuming a uniform wetting front.

As pointed out by Hendrickx and Flury (this report, Chapter 5), the term "preferential flow," in itself, does not distinguish among the causes of the non-uniform flow. Nonetheless, preferential flow is generally recognized to arise from three factors (Figure 1-5): (a) the presence of macropores (including fractures), (b) the development of flow instability, and (c) "funneling" of flow due to the presence of sloping layers that redirect the downward water movement. More sophisticated geological characterization of any particular site, with a well developed understanding of inhomogeneity, will contribute to the development of models that appropriately describe preferential flow.

Macropores refer to void space whose characteristic dimensions are significantly larger than the characteristic pore size in the rest of the medium. Soil macropores include decayed root channels, earthworm burrows, gopher holes, and drying cracks in fine-textured (clay) soils. In structured soils, where the primary grains are clustered into aggregates, the interaggregate pore space is another macropore example. Rock fractures may be also considered as a kind of macropore. Under certain conditions, the macropores allow water to move rapidly through the subsurface while bypassing the smaller pore spaces. The term "macropore flow" is often used to describe this bypassing process.

Although the physics of macropore flow in near-surface soils and in deep subsurface fractured rock are similar, macropore properties may be dynamically altered to different extents and in different ways. In near-surface soils, processes such as shrink-swell, freeze-thaw, biological activity (leading to earthworm holes and root channels), and physical manipulation (e.g., plowing of an agricultural field) can dynamically alter the preferential flow pathways. Because these pro-

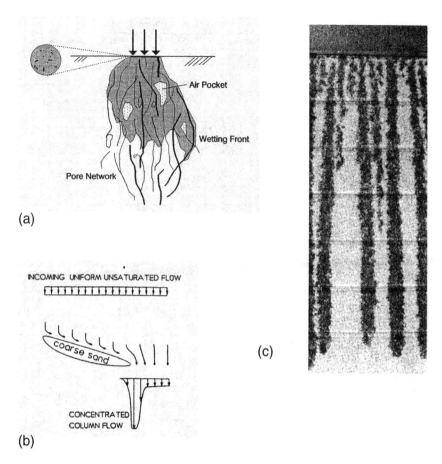

(a)

(b)

(c)

FIGURE 1-5 Different causes of preferential flow. (a) Macropore flow. Thickness of irregular lines indicates macropore size. Modified after Mohanty, B.P., R.S. Bowman, J.M.H. Hendrickx, and M.Th. van Genuchten, 1997. New piecewise-continuous hydraulic functions for modeling preferential flow in an intermittent-flood-irrigated field. Water Resources Research 33(9): 2049-2063. Copyright by American Geophysical Union. (b) Unstable flow in fine-over-coarse-layered sand system. From Glass, R.J., and J-Y. Parlange, 1989. Wetting front instability. 1. Theoretical discussion and dimensional analysis. Water Resources Research 25(6): 1187-1194. Copyright by American Geophysical Union. (c) Funnel flow. Reprint from Geoderma, 46, K-J.S Kung, Preferential flow in a sandy vadose zone. 2. Mechanism and implications. Pp. 59-71, 1990b, with permission from Elsevier Science.

cesses and their dynamics tend to decrease with depth, they are less important in the deeper subsurface, where fractured rocks predominate. Nonetheless, counter to common assumptions, rock fractures can also be dynamically altered. Berkowitz et al. (this report, Chapter 4) indicate that in some situations, unsaturated flow pathways within fractured rock can be altered by chemical dissolution and precipitation.

Preferential flow may also occur in the absence of macropores, in the form of unstable flow in which an initially subhorizontal, downward-moving wetting front breaks into "fingers." This type of flow pattern is sometimes referred to as "fingering." Wetting front instability can occur in a layered soil profile, as the front moves from a fine-textured layer into coarse-textured layer (Hill and Parlange, 1972). Instability and the resultant fingering can also occur in seemingly homogeneous soils, due to water repellency and/or air entrapment, and in fracture planes, as demonstrated by the laboratory experiments of Nicholl et al. (1994).

Another cause of preferential flow is lateral redirection of water through an interbedded soil profile that consists of sloping layers. Where a sloping layer acts as a barrier to flow, the downward infiltrating water will be redirected laterally above that layer. As illustrated in Figure 1.5c, the redirected flow is focused into a narrow column that can move downward at a faster rate than a broad wetting front in a homogeneous medium. The term "funnel flow" is often used for this process (Kung, 1990b). The same mechanism can occur in a subvertical fracture containing a subhorizontal obstacle, as illustrated by the computer simulation of Pruess (1999). In this simulation (Figure 1-6), the wetting front travels down a uniform fracture to a depth of 100 m in 456 days. However, when the fracture contains a subhorizontal obstacle, the redirected flow is funneled into a narrow column that reaches the 100-m depth in significantly less time (203-113 days).

A variety of approaches have been developed for modeling preferential flow. For field application, it is difficult to distinguish among the various mechanisms that cause preferential flow, and these mechanisms often act together to reinforce each other. Many models of preferential flow employ a combination of "fast flow" and "slow flow" components. Figure 1-7 (modified after Altman et al., 1996) summarizes a series of increasingly complex models that have been used to simulate preferential flow. This figure was originally prepared in the context of flow through fractured rocks, but may be viewed in the broader context of preferential flow (that is, fractures may be replaced by macropores, and matrix may be replaced by micropores).

Figure 1-7a illustrates the traditional porous medium model of capillary flow in the vadose zone. The governing equation is Richards' equation, which is derived from a combination of Darcy's law (for unsaturated flow) and the principle of mass conservation. The arrow in the schematic representation indicates fluid flow through the porous medium. The medium is characterized by two functional relationships: (1) the hydraulic conductivity curve, which gives the relation be-

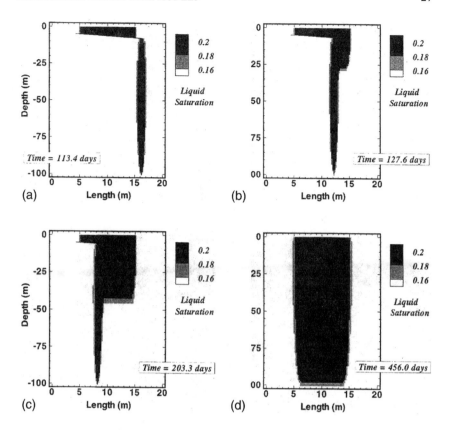

FIGURE 1-6 Result of computer simulation to illustrate funnel flow in a single fracture. The four panels show simulated water saturation at time of breakthrough at –100 m depth for seepage in a subvertical (80°) fracture with a single embedded subhorizontal obstacle of rectangular shape and variable length. The impermeable obstacle starts at the left boundary at a depth of –4 m and slopes downward to the right at an angle of 14°. Different cases were simulated in which the length interval (*L*) blocked by the obstacle was (a) *L* = 16 m, (b) *L* = 12 m, (c) *L* = 8 m, and (d) *L* = 0 m (i.e., no obstacle). Modified from Pruess, K., 1999. A mechanistic model for water seepage through thick unsaturated zones in fractured rocks of low matrix permeability. Water Resources Research 34(4): 1039-1051. Copyright by American Geophysical Union.

tween (unsaturated) hydraulic conductivity and pressure head, and (2) the retention curve, which gives the relation between water content and pressure head. On the right side of Figure 1-7, the hydraulic conductivity curve is schematically drawn as a single curve. However, the relation is known to be hysteretic, which means that the curve during wetting is different than the curve during drying.

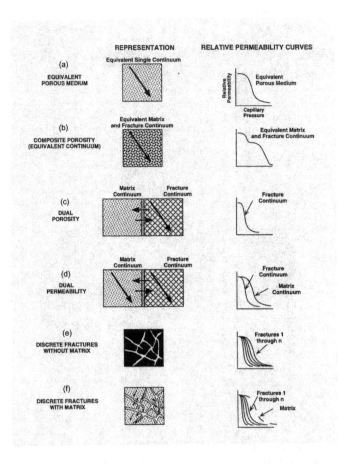

FIGURE 1-7 Alternative conceptual models and their relative permeability characteristic curves (equivalent to hydraulic conductivity function in text) for flow through fractured rocks. From Altman et al. (1996, Figure 2-2).

Figure 1-7b illustrates a simple extension of the traditional porous medium model to account for fracture flow. This is achieved by using a composite hydraulic conductivity function that represents both matrix and fractures. The arrow in the schematic representation indicates fluid flow through the composite (matrix and fracture) system. The underlying assumption is that locally, the pressure head in the fracture is equal to the pressure head in the matrix. At low saturation (greater magnitude of negative pressure head), flow is assumed to occur only in the matrix, and so the conductivity function represents only the matrix. At high saturation (pressure head approaching zero), flow is assumed to occur in both the matrix and the fractures, and the conductivity function represents both components. Near full saturation, the fracture conductivity may be significantly higher than the matrix conductivity. Thus, the composite hydraulic conductivity curve may have a double hump appearance. By using such an unconventional hydraulic conductivity curve, a traditional model of unsaturated flow can be adapted for preferential flow. However, this approach may not always be applicable. Mohanty et al. (1997) successfully used this approach to simulate preferential flow in a flood-irrigated agricultural field. However, Flint et al. (this report, Chapter 2) noted that the assumptions behind this approach might not hold for arid environments such as Yucca Mountain.

In the dual-porosity model (Figure 1-7c), the fracture and the matrix are separately represented by two interacting continua. For saturated conditions, this approach is widely used in modeling flow in fractured porous media. The assumption is that flow takes place only through the fracture network, as indicated by the flow arrow in the fracture continuum. Fluid exchange may occur between fracture and matrix, as indicated by the two smaller arrows between the fracture and matrix continua. However, there is no flow through the matrix blocks. For unsaturated conditions, however, this approach is seldom used because it is generally too restrictive. One exception is modeling of solute transport under steady-state flow where solute exchange may occur between the flowing water (in the fractures) and the immobile water (in the matrix) by diffusion.

The dual-permeability model (Figure 1-7d) is an extension of the dual-porosity model to allow for flow through the matrix blocks, as indicated by the additional flow arrow in the matrix continuum. This type of model is widely used for simulating preferential flow. The model involves two water retention functions, one for the fracture network and one for the matrix, and two hydraulic conductivity functions: $K_f(h_f)$ for the fracture network, and $K_m(h_m)$ for the matrix, where h_f and h_m are the pressure heads in the fracture network and matrix, respectively. The flow between the fracture and matrix, denoted by Γ_w, is often expressed as:

$$\Gamma_w = \alpha_w \, (h_f - h_m), \tag{1.1}$$

where α_w is a transfer coefficient (Gerke and van Genuchten, 1993b). The value of α_w exerts a strong control on flow through the fracture-matrix system, and is discussed in greater detail in the next section on fracture-matrix interaction.

In a discrete fracture model (Figures 1-7e and f), fractures are explicitly represented in the model. This approach was first developed for saturated domains, where the matrix can be assumed to be impermeable (Figure 1-7e). To simulate saturated flow, discrete fracture models require data on the geometry and transmissivity of individual fractures. Such data are almost always far from complete for a given field site. In this regard, discrete fracture models share the data burden of any model that attempts to capture the detailed heterogeneity of the flow system. Field applications of discrete fracture models typically employ

(1) major, flow-controlling features determined by field characterization, (2) stochastically generated fracture networks based on statistics of fracture geometry and transmissivity, and (3) geologic understanding of the fracture origin and growth process. This topic was discussed in detail by the Committee on Fracture Characterization and Fluid Flow (NRC, 1996, Chapter 6).

When applied to the fractured vadose zone, discrete fracture models require the ability to simulate flow in the matrix as well as in the fractures (Figure 1-7f). Because fractures and matrix occur together in the same domain, the flow field can be highly complex, as indicated by the numerous arrows in the schematic representation. Fractures may be represented by two-dimensional elements embedded within a three-dimensional mesh representing the matrix. In addition to geometry data, each fracture in the model must be assigned a hydraulic conductivity curve and a retention curve. Because the data requirements are extremely demanding, unsaturated discrete fracture models have primarily been applied to simple ideal systems for conceptual understanding (e.g., Wang and Narasimhan, 1985), to laboratory experiments (e.g., Kwicklis et al., 1998), or to hypothetical field settings with orthogonal fracture networks (e.g., Therrien and Sudicky, 1996).

Discrete fracture models are also valuable for understanding the effects of fracture geometry. As discussed by Doe (this report, Chapter 8), these geometric effects include (1) diversion, (2) focusing, and (3) discontinuity in the fracture network. Diversion occurs when fractures have a strongly preferred orientation, thus imparting anisotropy to the rock. Fluid flow deviates from the direction of hydraulic-head gradient towards the preferred orientation. Focusing refers to the localization of seepage due to heterogeneities within the flow system. According to Pruess (1999), focusing can be induced by fracture intersections and terminations, as well as by asperity contacts in individual fractures. Discontinuity in the fracture network may result in the absence of a through-going preferential pathway. Thus, fluid may flow rapidly in fractures, but will slowly bleed into the matrix at fracture terminations that do not connect to other fractures.

The models illustrated in Figure 1-7 all assume applicability of Richards' equation, and hence of Darcy's law. This assumption may not be strictly correct for the fracture system. Single or dual continuum models (Figure 1-7b, c, and d) assume that the fracture network can be replaced by a porous medium. Analysis by Kwicklis and Healy (1993) suggests that a continuum representation of a fracture network may be more suitable for some ranges in pressure head than for other ranges. In discrete fracture models, an individual fracture is typically treated essentially as a thin layer of porous medium. However, preferential flow within a fracture (e.g., Figure 1-6) can focus transport to a narrow region. Unless each fracture is represented by a fine mesh in the model, such mechanisms will not be explicitly simulated. Furthermore, at the small-scale level, intermittent flow behaviors observed in recent visualization experiments (e.g., Su et al., 1999) are not characterized by Darcy's law.

For practical applications, however, the real issue may not necessarily be the validity of Darcy's law as such, but whether Darcy's law, even if formally invalid, can still provide an adequate and useful description of the preferential flow process. Alternative descriptions of the flow regime in fractures include Manning's equation for turbulent overland flow (Chen and Wagenet, 1992), kinematic wave theory (e.g., Germann and Beven, 1985), and simple gravity-flow models (e.g., Jarvis et al., 1991, see also Jarvis and Larsson, this report, Chapter 6). These alternatives, however, also involve simplifying approximations. It is unclear whether or not these alternatives would provide substantive enhancements for practical applications.

Fracture-Matrix Interaction

The extent to which fracture-matrix interaction controls unsaturated flow depends on a number of factors. Upon the initiation of fracture flow (e.g., after a precipitation event), water in the fracture may be imbibed into the rock matrix (Figure 1-8). Factors that may create "strong" fracture-matrix interaction include:

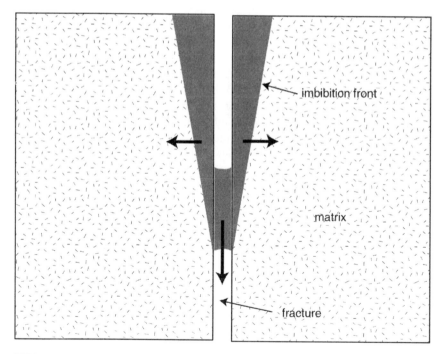

FIGURE 1-8 Imbibition of fluid from fracture to matrix.

high matrix suction, large contact area between water and fracture wall, and absence of fracture coatings that impede matrix imbibition. Under these conditions, water in the fracture will be quickly absorbed into the matrix, and fracture flow cannot be sustained. By contrast, factors that create "weak" fracture-matrix interaction include: low matrix suction, small contact area between water and fracture wall, and presence of fracture coatings. Under these conditions, water in the fracture is not readily absorbed into the matrix, and fracture flow may occur to significant depths.

Although fracture-matrix interaction can, in principle, be analyzed by a detailed simulation of water flow in a fracture and propagation of the imbibition front in the matrix, current models of flow in the fractured vadose zone generally treat fracture-matrix interaction in a simplified manner. Whether or not the simplifications are justified generally depends on site-specific conditions. For example, the equivalent continuum model using a composite hydraulic conductivity function (Figure 1-7b) assumes that locally, the pressure head in the fracture is equal to the pressure head in the matrix. Under this simplification, fracture flow will not occur until the matrix is fully saturated. Although this simplification may be appropriate for certain field conditions (e.g., Mohanty et al., 1997), there are other situations where it cannot be applied (e.g., it may not be suitable for drier environments; see Flint et al., this report, Chapter 2).

In dual porosity and dual permeability models (Figure 1-7c and d), fracture-matrix interaction is represented by equation (1.1), where the transfer parameter α_w describes the extent of fracture-matrix flow. Many investigators, starting with Warren and Root (1963), have studied the nature of α_w. Typically, α_w depends on the matrix hydraulic conductivity and average fracture spacing (or equivalently, the average size of the matrix blocks). It is commonly assumed that fracture-matrix flow can occur over the entire fracture surface. However, recent findings from experiments at Yucca Mountain suggest that the above approach for calculating α_w significantly overestimates the degree of fracture-matrix interaction (see discussion by Flint et al., this report, Chapter 2). A relatively large value for α_w implies little to no fracture flow. By contrast, evidence indicating significant fracture flow at Yucca Mountain includes: detection of an environmental tracer (bomb-pulse chlorine-36) at depth, an estimated infiltration rate that is greater than the saturated matrix hydraulic conductivity of welded tuff units, and geochemical nonequilibrium between perched water and pore water in the rock matrix. These findings suggest that fracture-matrix interaction is relatively weak at Yucca Mountain.

To avoid overestimating the degree of fracture-matrix interaction at Yucca Mountain, Ho (1997) applied a reduction factor to decrease the fracture-matrix interface area. This was based on an assumption that if preferential flow occurs in the fracture plane, then only a fraction of the fracture plane is wetted, and the connection area between fracture and matrix is reduced. In addition, Liu et al. (1998) proposed the "active fracture model," which hypothesized that only a portion of the fractures in a connected network contribute to water flow, while other fractures are simply bypassed. This would further reduce the connection area between fractures and matrix. These factors led Ho (1997) to reduce the fracture-matrix interaction by four orders of magnitude in order to better model infiltration at Yucca Mountain and simulate the fast transport of bomb-pulse nuclides.

Fracture-matrix interaction can also be reduced by the presence of fracture surface coatings that impede matrix imbibition (Thoma et al., 1992). To account for fracture coatings, Gerke and van Genuchten (1993b) characterized the fracture-matrix interface by a hydraulic conductivity function, $K_a(\bar{h})$, where $\bar{h}$ is defined as some type of average (e.g., arithmetic, geometric) of pressure heads in the fracture and in the matrix. This approach can also be applied to structured soils. Soil aggregate can have a higher local bulk density (and hence lower conductivity) near its surface than in the aggregate center, due to deposition of organic matter, fine-texture mineral particles, or various oxides and hydroxides on the aggregate exteriors or macropore walls. Analysis by Gerke and van Genuchten (1993a) suggests that when K_a is roughly equal to K_m (the conductivity of the soil matrix), pressure head in the fractures should be nearly in equilibrium with pressure head in the matrix (except for very large blocks), and this will not result in significant fracture flow. To allow fracture flow over a significant depth, K_a must be smaller than K_m by several orders of magnitude, or the matrix block sizes must be relatively large. These findings emphasize the importance of fracture coatings for controlling fracture flow.

For field-scale models, the parameters describing fracture-matrix interactions will probably have to be determined by model calibration or inversion (see Bodvarsson et al., this report, Chapter 11), as there are no methods to determine these parameters by direct measurements. Although calibration may provide a good match between simulated results and existing field data, use of the model for different stresses or climate conditions may require extrapolation to conditions beyond those used in the calibration. For example, as flux increases, the fracture-matrix interaction factor may change in two ways: by an increase in the contact area between the fracture and the matrix, or by an increase in the number of fractures actively carrying water. How to accommodate such changes remains a challenging topic for future research on fracture-matrix interaction.

Solute Transport

Modeling solute transport in the fractured vadose zone adds several layers of complexity compared with modeling fluid flow. The following discussion is limited to modeling of nonreactive solute of sufficiently low concentration that density effects can be neglected. In the traditional approach to modeling solute transport, the fluid flow problem is first solved to determine the distribution of fluid velocity. The transport of solute by this velocity field is known as advection, which is one of two transport mechanisms recognized in classical transport theory. The second transport mechanism is hydrodynamic dispersion, which describes the spreading and mixing of solute due to (1) microscopic and local-scale variations in velocity that are not explicitly described by the above velocity field, and (2) molecular diffusion. In the classical approach, hydrodynamic dispersion is represented as a Fickian (diffusion-like) process. The resulting equation governing solute transport is known as the advection-dispersion equation.

For modeling solute transport in saturated fractured rocks, a simple extension to the classical model is applied to incorporate the effects of solute diffusion between mobile and immobile fluids. In a number of tracer experiments in saturated fractured rocks, the breakthrough curve (plot of concentration versus time) exhibits a long tail, or skewness towards later times (e.g., Novakowski et al., 1995). This feature is commonly interpreted to result from the solute moving through the fracture network while diffusing into the immobile fluid within the matrix. The conceptual flow model is that of the dual-porosity model illustrated by Figure 1-7c. The long tail in the breakthrough curve is explained by the later diffusion of solute from the matrix back into the fractures. This "matrix diffusion" can be considered as one example of solute exchange between mobile and immobile fluids. In addition to the matrix, regions of immobile fluid may exist within an individual fracture (if flow through the fracture is channeled) and in "dead-end" fractures of a network. Solute exchange between mobile and immobile fluid is commonly represented by a diffusion equation (e.g., Tang et al., 1981). Models of this type typically assume steady-state flow.

For the fractured vadose zone, solute transport models typically assume that fluid flow occurs in the matrix blocks and in the fractures. The conceptual flow model is that of the dual permeability model illustrated by Figure 1-7d. Gerke and van Genuchten (1993a) use two advection-dispersion equations, coupled by a solute transfer term, to describe solute transport in the fracture and in the matrix. Solute transfer between the two domains can occur both by fluid flow (advection) and by diffusion. The latter is often approximated as a first-order process; that is, the diffusive flux is proportional to the difference between solute concentration in the fracture and in the matrix. For certain field settings, the advective mass transfer can act in a direction opposite to the diffusive mass transfer. For example, Gerke and van Genuchten (1993a) simulated the infiltration of solute-free water into a fractured vadose zone that contained initial solute in the matrix blocks. As the solute-free water infiltrated down the fractures, a portion was imbibed into the matrix by capillary forces. At the same time, the solute in the matrix tended to diffuse back into the fractures because of the concentration difference between the two domains. The result was a highly complex response in the fracture-matrix system.

Further refinements of solute transport modeling have led to the development of multiregion models (e.g., Gwo et al., 1995; Hutson and Wagenet, 1995). In these models, the fractured rock or macroporous soil is represented by more than two domains (or regions). These domains may be delineated in a somewhat arbitrary manner, or may be associated with geological features such as different types of fractures and matrix blocks. For example, Gwo et al. (1995) used a three-region model, composed of macropores, mesopores, and micropores, to simulate flow and transport through a laboratory column of soil from a forested watershed. In Chapter 3, Jardine et al. conceptualize the fractured weathered shale at the Oak Ridge National Laboratory to be composed of primary fractures, secondary fractures, and soil matrix. They also describe various experiments to examine the interaction and mass transfer between the different domains. In general, multiregion models are more flexible than dual- or single-domain models. However, this flexibility comes at the cost of additional model parameters. Calibration of multiregion models requires a significant amount of field data.

Solute transport can also be simulated in a discrete fracture model as shown in Figure 1-7f. Substantial insight can be gained by using such a model for examining the complex interactions between solutes in fractures and in the matrix. For example, Therrien and Sudicky (1996) simulated three-dimensional transport from a hypothetical waste facility in a fractured, low-permeability stratum overlying an aquifer. Their results show that where vertical fractures are sparse, the contaminant plume can become discontinuous, making it difficult to interpret the data from a field-monitoring program. Another value of discrete fracture models is for assessing the adequacy of continuum models such as dual-permeability or multidomain models. For a realistic field application, however, a discrete-fracture model may be problematic due to the large data requirement.

Use of Environmental Tracers

As noted by Phillips (this report, Chapter 9), environmental tracers have not been widely applied in hydrologic investigations of fractured rocks, even though they offer a powerful and direct method for assessing solute transport effects. Often, the approach to subsurface investigation is first to characterize fluid flow processes, and then to evaluate the additional processes that affect solute transport. While this approach can lead to a detailed understanding of flow processes, key solute transport mechanisms can be overlooked. In some cases, interpretation of environmental tracers has lead to drastic revisions of a conceptual model based initially solely on hydrodynamic analysis. For example, Phillips (this report, Chapter 9) points out that measurements of tritium in the fractured chalk of southern England (see Foster, 1975, and Foster and Smith-Carrington, 1980) revealed that the process of matrix diffusion was active, and this in turn explained an apparent contradiction between the rapid hydraulic response of the aquifer to recharge versus the highly attenuated movement of dissolved pollutants.

Environmental tracers consist of solutes that have a widespread or global occurrence from both natural and anthropogenic sources in precipitation, and have been entering the subsurface over large spatial and typically long temporal scales (tens to millions of years). The large space and time scales associated with these tracers result in a signal that is naturally integrated. When samples for environmental tracer analyses are collected from a point in the subsurface, the results represent an integration of upstream transport mechanisms. This is in contrast to most hydraulic measurements where point sampling yields a point measurement.

While numerous environmental tracers have been used in subsurface hydrology, the existence of both air and water phases in the vadose zone is likely to complicate the use of some tracers, especially those that utilize dissolved gases. Environmental tracers that are likely to be useful for directly investigating fluid flow in fractured vadose zones include tritium, stable isotopes of oxygen and hydrogen in the water molecule, halides, and chlorine-36; see Phillips (this report, Chapter 9) for a brief discussion of these tracers. Other tracers that may be useful for indirectly examining fluid flow include carbon-14, uranium-series nuclides, strontium isotopes, boron-11, silicon-32, and iodine-129; see Cook and Herczeg (2000) for a discussion of these tracers. In addition, dissolved gas tracers such as helium-3, chlorofluorocarbons (CFCs, synthetic gases that were released into the atmosphere beginning in the early 1940s), and krypton-85 can provide a measure of groundwater age in the saturated zone. These age estimates may also be useful for indirectly characterizing fractured vadose zones.

Characterizing the fractured vadose zone can be accomplished by measuring the distribution of environmental tracers in the subsurface. In the absence of preferential flow, or in stratigraphic intervals of fractured rocks where matrix flow dominates, the vertical profile of tritium in the vadose zone can be used to estimate infiltration rates by identifying the depths of tritium peaks corresponding to atmospheric testing of nuclear weapons. However, if the majority of recharge at a site occurs via fracture flow, a detailed investigation of matrix processes in the vadose zone may be misleading. Furthermore, if the infiltration is localized by a few high porosity features or faults or other heterogeneities, an investigation at a random point in the map may miss the dominant vertical flow paths. In these cases, the vadose zone can be evaluated by sampling at the water table, that is, where the infiltrating fluid and solute enter into the underlying saturated zone. While the details of flow and transport processes will tend to be less well resolved by this approach, the integration (averaging) of processes may be enhanced.

The following three examples illustrate the importance of sampling from both the vadose and the saturated zones. In the first example, Wood and Sanford (1995) found significant differences in the chloride concentrations in pore waters in the unsaturated zone, compared with mean chloride concentrations in underlying groundwater in Texas. These differences were interpreted to mean that the majority of recharge occurred along preferential (macropore) pathways, which were not detected by sampling in the vadose zone. In the second example, Solomon et al. (1995) obtained groundwater ages as a function of depth below the water table, and then calculated vertical fluid velocities and recharge rates through the vadose zone on Cape Cod, Massachusetts. Groundwater ages were zero at the water table because the dating method (tritium/helium-3) utilized a dissolved gas, which exchanged into pore air in the vadose zone. Solomon et al. (1995) also delineated the depth of the tritium bomb peak that occurred in 1963. The difference between the total travel time of water as delineated by the tritium bomb peak (30 years), and the tritium/helium-3 age of the water since it was at the water table (16 years), provided a measure of the travel time in the vadose zone (14 years). In the third example, Busenberg et al. (1993) found that concentrations of CFCs dissolved in groundwater near the water table were much larger than expected based on measurements of the CFC content of pore air immediately above the water table. They concluded that preferential flow was not equilibrating with the vadose zone atmosphere and was responsible for significant recharge to the Snake River Plain aquifer.

A mass balance of environmental tracers within the vadose zone can be useful for evaluating regional-scale solute and fluid fluxes, without necessarily understanding the details of fracture flow, fracture-matrix interaction, etc. (e.g., Cook et al., 1994). This approach assumes that a mass balance can be adequately formulated using subsurface measurements (i.e., that spatial variations in the tracer distribution can be adequately sampled), and the tracer input is sufficiently well known. It can be argued that over long time scales, diffusion processes tend to reduce small-scale spatial variation in tracer concentrations, making a mass balance feasible. However, this approach has not been extensively tested in fractured vadose zones.

Environmental tracers may also be useful for evaluating and testing specific processes that are included in a conceptual model. For example, Desaulniers et al. (1981) evaluated the influence of fractures on solute transport in a clay-rich till in Southern Ontario. They found that profiles of oxygen isotopes and chloride were best explained by molecular diffusion, with minimal advective transport in fractures. Because of their usefulness, environmental tracers should be considered a primary method for investigating fractured vadose zones and for formulating and testing conceptual models, and they should be included in the field investigation strategy from the very beginning of site characterization.

CONCLUSIONS AND RECOMMENDATIONS

This report discusses the process through which conceptual models of flow and transport in the fractured vadose zone are developed, tested, refined, and reviewed. **A conceptual model is defined as an evolving hypothesis identifying the important features, processes, and events controlling fluid flow and contaminant transport of consequence at a specific field site in the context of a recognized problem.** The conclusions presented below are grouped according to the two major topics addressed in this report: (1) general considerations during the development and testing of conceptual models, and (2) flow and transport in the fractured vadose zone. These conclusions are followed by the Panel's suggestions for research activities that will contribute to the conceptual modeling process.

Conclusions on Development and Testing of Conceptual Models

1. Development of the conceptual model is the most important part of the modeling process. The conceptual model is the foundation of the quantitative, mathematical representation of the field site (i.e., the mathematical model), which in turn is the basis for the computer code used for simulation. Given a sufficiently robust conceptual model, different mathematical formulations will probably produce similar results. By contrast, an inappropriate conceptual model can easily lead to predictions that are orders of magnitude in error.

2. The context in which a conceptual model is developed constrains the range of its applicability. A conceptual model is by necessity a simplification of the real system, but the degree of simplification must be commensurate with the problem being addressed. Thus, a conceptual model developed for addressing one type of problem may not be adequate for another type of problem. For example, a conceptual model developed for estimating recharge flux may not be adequate for estimating contaminant travel time from land surface to the water table.

3. It is important to recognize that model predictions require assumptions about future events or scenarios, and are subject to uncertainty. Quantitative assessment of prediction uncertainty should be an essential part of model prediction. A suite of predictions for a range of different assumptions and future scenarios is more useful than a single prediction. Uncertainty in model predictions can be partially quantified by sensitivity analysis (how uncertainties in estimated model parameters affect model predictions), by using a statistical- or stochastic-based calibration procedure, or by formulating the mathematical model in a probabilistic framework. However, it is difficult to quantitatively assess the possibility that the conceptual model might not adequately represent the major features and processes in the real system.

4. Testing and refinement of the conceptual model are critical parts of the modeling process. The initial conceptual model is developed based upon limited field data and is susceptible to biases reflected by the disciplinary background and experience of the analyst. Therefore, site investigation should not be designed solely to support the initial conceptual model. Reasonable alternative conceptualizations and hypotheses should be developed and evaluated. In some cases, the early part of a study might involve multiple conceptual models until alternatives are eliminated by field results.

5. Although model calibration does provide a certain level of model testing, a good fit to the calibration data does not necessarily prove that the

model is adequate to address the issues in question. The significance of a good fit to calibration data generally depends on the nature of the data. A model that matches different types of calibration data (e.g., heads and fluxes) collected under different field conditions (e.g., at different water contents) is likely to be more robust than a model that matches a limited range of calibration data. However, if the model cannot be calibrated to match the calibration data, this is an indication that the conceptualization should be re-examined.

6. Checking model simulation results against field data (that were not used for calibration) is one, but not the exclusive, approach to model testing. In some cases, all field data are needed for calibration, and none are left for further testing. However, this does not mean that the model cannot be used for prediction. A broader view of model testing is to develop greater confidence that the model provides a good representation of the real system. This can be achieved by strengthening the justifications for model assumptions, and by evaluating alternative hypotheses.

7. From an operational perspective, the goal of model testing is to establish the credibility of the model. A credible model is essential if it is to gain acceptance by parties involved in decision-making or problem resolution. In addition to testing and evaluation, the credibility of a model can be enhanced by peer review undertaken by an independent panel of experts, and by maintaining an open flow of information so that the model is available for scrutiny by concerned parties.

Conclusions on Flow and Transport in the Fractured Vadose Zone

1. There exists a body of field evidence indicating that infiltration through fractured rocks and structured soils does not always occur as a wetting front advancing at a uniform rate. Large variations in fluid velocity (i.e., preferential flow) may be caused by (a) the presence of macropores and fractures, (b) flow instability, or (c) funneling effects. Thus, model simulation based upon a uniform wetting front advancing down a homogeneous medium may provide erroneous estimates of flux and travel times through the vadose zone. Sophisticated characterization of geological inhomogeneity within the vadose zone increases the likelihood that non-uniform flow can be appropriately modeled.

2. The current state of knowledge is not adequate to determine which processes are likely to control unsaturated flow and transport at a given field site. Laboratory and theoretical analyses demonstrate that film flow in fractures can transport fluid and solute at rates substantially higher than transport by capillary flow. However, at the field scale, the significance of film flow and the modeling approach are topics of controversy. The field environments in which film flow plays a significant role during infiltration are poorly understood. In addition, it is unclear whether film flow can be incorporated into traditional models of capillary flow by defining effective curves for hydraulic conductivity and retention, or whether it requires a fundamentally different set of governing equations to describe the dynamics of a water film.

3. Although not identical, structured soils and fractured rocks exhibit many similarities in flow and transport processes. Macropores and aggregates in structured soils are respectively analogous to fractures and matrix blocks in rock. However, soil studies are typically conducted in the shallow subsurface, where macropores may be dynamically altered to a greater degree than rock fractures at greater depths. Nonetheless, knowledge gained from study of one medium may be useful for the other. Communication between workers in the soil science field and in the fractured rock field will be of benefit to both groups.

4. Models of varying complexity have been developed for preferential flow, but their adequacy for field-scale application requires further testing. The approach in many current models is to avoid explicitly simulating the mechanisms that cause preferential flow. Instead, the model is implemented to simulate fast and slow flow, by use of a composite hydraulic conductivity curve or by dual-permeability domains. Such approaches have been successfully applied to laboratory and small-scale experiments. However, further testing is needed to examine whether these models are adequate for field-scale application over a broad range of field conditions. This issue is of particular concern in the fractured vadose zone because of the inherently nonlinear nature of processes involved. As flow conditions change, different flow and transport mechanisms, not represented in the model, may become important, leading to large errors in predictions.

5. The interaction between fracture and matrix exerts a strong control on fluid and solute movement. However, the strength of this interaction in the field is not well known. The simplified representation of this interaction in current models also requires further evaluation. Factors controlling fracture-matrix interaction include the density of water-transmitting fractures, the amount of wetted area on the fracture surface, the hydraulic conductivity of the matrix, and hydraulic conductivity at the fracture-matrix interface. Current models lump these factors into a transfer coefficient that is determined by model calibration rather than by direct measurements. Whether or not this approach can adequately simulate flow under a range of field conditions requires further evaluation.

6. Solute transport in the fractured vadose zone can exhibit complex behavior due to the large variations in fluid velocity, and the interplay of advective and diffusive transport between fractures and matrix. Better understanding of such systems is required in order to effectively analyze complex responses. Solute transport models are more complex than flow models, and can involve multiple regions to represent the diversity of macropore and micropore sizes. To apply these models, greater guidance is needed on how to delineate different pore regions, and how to determine the parameters that characterize solute exchange between pore regions.

7. Environmental tracers should be included in field investigation strategies from the very beginning of a site characterization program. In a number of studies, geochemistry and environmental tracer data have led to substantial revisions of the conceptual models initially developed based upon hydrodynamic analysis. These experiences emphasize the need for better integration of geochemistry and environmental tracers early in the model development process.

Recommended Research

Flow and transport in the fractured vadose zone have been and will continue to be an active area of research in both the soil science and subsurface hydrology disciplines. The research recommended in this report is not meant to be inclusive. Instead, the list below reflects topics that deserve greater attention so that conceptual models of flow and transport can be improved, and to address the issues identified in this report.

1. Fundamental research to understand flow and transport processes in unsaturated fractures should continue. Better understanding of fundamental processes will improve the model representation of the real system. Particular emphasis should be placed on understanding mechanisms that cause non-uniform (preferential) flow, film flow, and intermittent behavior.

2. Research is needed to understand the spatial variability in vadose zone properties, and to develop upscaling methods. Spatial variability is a key cause of model uncertainty, because the subsurface cannot be exhaustively sampled. Furthermore, vadose zone properties are typically determined by small-scale laboratory measurements. To use these small-scale measurements, upscaling methods are needed to derive field-scale flow and transport properties needed in models. Such upscaling methods should be based on a thorough understanding of small-scale processes, together with an understanding of how these interact and contribute to large-scale phenomena.

3. There is a need for comprehensive field experiments in several fractured vadose zone geologic environments. These experiments should be designed to understand the controlling processes (capillary flow, film flow, and intermittent behavior) for a broad range of field conditions, to evaluate methods of parameter upscaling, and to test alternative conceptual models.

4. Current models should be evaluated for their adequacy for simulating flow and transport in the presence of fingering, flow instability, and funneling. One approach is to construct a model with detailed representation of small-scale heterogeneities based on high-resolution field or synthetic data, so that the processes causing preferential flow are explicitly simulated. The model results can then be compared to results from simpler models that do not explicitly simulate preferential flow. Of particular importance is the evaluation of transfer coefficients to represent fluid and solute exchange between fracture and matrix.

5. There is a need to develop quantitative assessment of prediction uncertainty for models of flow and transport in the fractured vadose zone. Meaningful quantification of uncertainty should be considered an integral part of any modeling endeavor, as it establishes confidence bands on predictions given the current state of knowledge about the system. If prediction uncertainties are realistically quantified, postaudit studies can be carried out in a systematic hypothesis-testing framework, which can provide a great deal of insight about the predictive capabilities of the model.

6. Research should be undertaken to develop improved techniques for geochemical sampling from the fractured vadose zone. Current sampling technology is very limited, especially for sampling at depth. Destructive core sampling followed by fluid extraction is a viable approach for sampling matrix water in the unsaturated zone. Sampling fluids directly from fractures remains problematic. Improved sampling techniques will facilitate the use of environmental tracers and geochemical data for conceptual model building.

REFERENCES

Altman, S. J., B. W. Arnold, R. W. Barnard, G. E. Barr, C. K. Ho, S. A. McKenna, and R. R. Eaton, 1996. Flow Calculations for Yucca Mountain Groundwater Travel Time (GWTT-95). Report SAND96-0819. Albuquerque, N. Mex.: Sandia National Laboratories. 170 pp.

Anderson, M. P., and W. W. Woessner, 1992. Applied Groundwater Modeling. New York: Academic Press. 381 pp.

Busenberg, E., E. P. Weeks, L. N. Plummer, and R. C. Bartholemay, 1993. Age Dating Ground Water by Use of Chlorofluorocarbons (CCl_3F and CCl_2F_2), and Distribution of Chlorofluorocarbons in the Unsaturated Zone, Snake River Plain Aquifer, Idaho National Engineering Laboratory, Idaho. U.S. Geological Survey Water-Resources Investigations Report 93-4054. Reston, Va.: U.S. Geological Survey. 47 pp.

Chen, C., and R. J. Wagenet, 1992. Simulation of water and chemicals in macropore soils. I. Representation of the equivalent macropore influence and its effect on soil-water flow. Journal of Hydrology 130: 105-126.

Cook, P., and A. L. Herczeg, 2000. Environmental Tracers in Subsurface Hydrology. Boston: Kluwer Academic Publishers. 529 pp.

Cook, P. G., Jolly, I. D., Leaney, F. W., Walker, G. R., Allan, G. L., Fifield, L. K., and Allison, G. B., 1994. Unsaturated zone tritium and chlorine-36 profiles from southern Australia: their use as tracers of soil water movement. Water Resources Research 30(6): 1709-1719.

Desaulniers, D.E., J.A. Cherry, and P. Fritz, 1981. Origin, age and movement of pore water in argillaceous quaternary deposits at four sites in southwestern Ontario. Journal of Hydrology 50: 231-257.

Fabryka-Martin, J. T., A. V. Wolfsberg, S. S. Levy, J. L. Roach, S. T. Winters, L. E. Wolfsberg, D. Elmore, and P. Sharma, 1998. Distribution of Fast Hydrologic Paths in the Unsaturated Zone at Yucca Mountain. Paper presented at 8th Annual International High-Level Radioactive Waste Management Conference, Las Vegas, Nev. American Nuclear Society, La Grange Park, Ill. p. 264-268.

Foster, S. S. D., 1975. The chalk groundwater tritium anomaly, a possible explanation. Journal of Hydrology 25: 159-165.

Foster, S. S. D., and A. Smith-Carrington, 1980. The interpretation of tritium in the chalk unsaturated zone. Journal of Hydrology 46: 343-364.

Furmidge, C. G. L, 1962. Studies at phase interfaces, 1. The sliding of liquid drops on solid surfaces and a theory for spray retention. Journal of Colloidal Science 17: 309-324

Gerke, H. H., and M. Th. van Genuchten, 1993a. A dual-porosity model for simulating the preferential movement of water and solutes in structured porous media. Water Resources Research 29(2): 305-319.

Gerke, H. H., and M. Th. van Genuchten, 1993b. Evaluation of a first-order water transfer term for variably saturated dual-porosity flow models. Water Resources Research 29(4): 1225-1238.

Germann, P., and K. J. Beven, 1985. Kinematic wave approximation to infiltration into soils with sorbing macropores. Water Resources Research 21(7): 990-996.

Glass, R. J., and J-Y. Parlange, 1989. Wetting front instability. 1. Theoretical discussion and dimensional analysis. Water Resources Research 25(6): 1187-1194.

Gwo, J. P., P. M. Jardine, G. V. Wilson, and G. T. Yeh, 1995. A multiple-pore-region concept to modeling mass transfer in subsurface media. Journal of Hydrology 164: 217-237.

Hill, D. E., and J-Y. Parlange, 1972. Wetting front instability in layered soils. Proceedings Soil Science Society of America 36: 697-702.

Ho, C. K., 1997. Models of Fracture-Matrix Interactions During Multiphase Heat and Mass Flow in Unsaturated Fractured Porous Media. Paper presented at 6th Symposium on Multiphase Transport in Porous Media, ASME International Mechanical Engineering Congress and Exposition, American Society of Mechanical Engineering, Dallas, Texas.

Hutson, J. L., and R. J. Wagenet, 1995. A multiregion model describing water flow and solute transport in heterogeneous soils. Soil Society of America Journal 59: 743-751.

Jarvis, N. J., P-E. Jansson, P. E. Dik, and I. Messing, 1991. Modelling water and solute transport in macroporous soils. I. Model description and sensitivity analysis. Journal of Soil Science 42: 59-70.

Konikow, L. F., and J. D. Bredehoeft, 1992. Ground-water models cannot be validated. Advances in Water Resources 15(1): 75-83.

Kung, K-J. S., 1990a. Preferential flow in a sandy vadose zone. 1. Field observations. Geoderma 46: 51-58.

Kung, K-J. S., 1990b. Preferential flow in a sandy vadose zone. 2. Mechanism and implications. Geoderma 46: 59-71.

Kwicklis, E. M., and R. W. Healy, 1993. Numerical investigation of steady liquid water flow in a variably saturated fracture network. Water Resources Research 29(12): 4091-4102.

Kwicklis, E. M., F. Thamir, R. W. Healy, and D. Hampson, 1998. Numerical Simulation of Air- and Water-Flow Experiments in a Block of Variably Saturated, Fractured Tuff from Yucca Mountain, Nevada. U.S. Geological Survey Water-Resources Investigations Report 97-4274. Denver, Colo.: U.S. Geological Survey. 64 pp.

Liu, H. H., C. Doughty, and G. S. Bodvarsson, 1998. An active fracture model for unsaturated flow and transport in fractured rocks. Water Resources Research 34(10): 2633-2646.

Mohanty, B. P., R. S. Bowman, J. M. H. Hendrickx, and M. Th. van Genuchten, 1997. New piecewise-continuous hydraulic functions for modeling preferential flow in an intermittent-flood-irrigated field. Water Resources Research 33(9): 2049-2063.

National Research Council (NRC), 1996. Rock Fractures and Fluid Flow: Contemporary Understanding and Applications. Washington, D.C.: National Academy Press. 551 pp.

Nicholl, M. J., R. J. Glass, and S. W. Wheatcraft, 1994. Gravity-driven infiltration instability in initially dry nonhorizontal fractures. Water Resources Research 30(9): 2533-2546.

Novakowski, K. S., P. A. Lapcevic, J. Voralek, and G. Bickerton, 1995. Preliminary interpretation of tracer experiments conducted in a discrete rock fracture under conditions of natural flow. Geophysical Research Letters 22(11): 1417-1420.

Oreskes, N., K. Shrader-Frechette, and K. Belitz, 1994. Verification, validation, and confirmation of numerical models in the earth sciences. Science 263: 641-646.

Peters, R. R., and E. A. Klavetter, 1988. A continuum model for water movement in an unsaturated fractured rock mass. Water Resources Research 24(3): 416-430.

Pruess, K., 1999. A mechanistic model for water seepage through thick unsaturated zones in fractured rocks of low matrix permeability. Water Resources Research 34(4): 1039-1051.

Solomon, D. K., R. J. Poreda, P. G. Cook, and A. Hunt, 1995. Site characterization using ^{3}He/^{3}He ground water ages, Cape Cod, Mass. Ground Water 33(6): 988-996.

Su, G. W., J. T. Geller, K. Pruess, and F. Wen, 1999. Experimental studies of water seepage and intermittent flow in unsaturated, rough-walled fractures. Water Resources Research 35(4): 1019-1037.

Tang, D. H., E. O. Frind, and E. A. Sudicky, 1981. Contaminant transport in fractured porous media: Analytical solution for a single fracture. Water Resources Research 17(3): 555-564.

Therrien, R., and E. A. Sudicky, 1996. Three-dimensional analysis of variably-saturated flow and solute transport in discretely-fractured porous media. Journal of Contaminant Hydrology 23: 1-44.

Thoma, S. G., D. P. Gallegos, and D. M. Smith, 1992. Impact of fracture coating on fracture/matrix flow interactions in unsaturated porous media. Water Resources Research 28(5): 1357-1367.

Tokunaga, T. K., and J. Wan, 1997. Water film flow along fracture surfaces of porous rock. Water Resources Research 33(6): 1287-1295.

Tsang, C. F., 1991. The modeling process and model validation. Ground Water 29(6): 825-831.

Wang, J. S. Y., and T. N. Narasimhan, 1985. Hydrologic mechanisms governing fluid flow in a partially saturated, fractured, porous medium. Water Resources Research 21(12): 1861-1874.

Warren, J. E., and P. J. Root, 1963. The behavior of naturally fractured reservoirs. Society of Petroleum Engineers Journal 3(5): 245-255.

Wood, W. W., and W. E. Sanford, 1995. Chemical and isotopic methods for quantifying groundwater recharge in a regional, semiarid environment. Ground Water 33(3): 458-468.

Technical Papers

2

Development of the Conceptual Model of Unsaturated Zone Hydrology at Yucca Mountain, Nevada

Alan L. Flint,[1] Lorraine E. Flint,[1] Gudmundur S. Bodvarsson,[2] Edward M. Kwicklis,[3] and June Fabryka-Martin[3]

ABSTRACT

Yucca Mountain is an arid site proposed for consideration as the nation's first underground high-level radioactive waste repository. Low rainfall and a thick unsaturated zone are important physical attributes of the site because the quantity of water likely to reach the waste and the paths and rates of movement of the water to the saturated zone under likely future climates will be major factors in estimating the concentrations and times of arrival of radionuclides at the surrounding accessible environment. The framework for understanding the hydrologic processes that occur at this site and that control how quickly water will penetrate through the unsaturated zone to the water table has evolved during the past 15 years. Early conceptual models assumed that very small volumes of water infiltrated into the bedrock, that much of the infiltrated water flowed laterally within the upper nonwelded units because of capillary barrier effects, and that the remaining water flowed down faults with a small amount flowing through the matrix of the lower welded, fractured rocks. When evidence accumulated indicating that infiltration rates were higher than initially estimated, and that mechanisms supporting lateral diversion did not apply at these higher fluxes, the flux calculated in the lower welded unit exceeded the conductivity of the matrix. This required water to flow vertically in the high-permeability fractures of the potential repository host rock.

[1] U.S. Geological Survey, Sacramento, California
[2] Lawrence Berkeley National Laboratory, Berkeley, California
[3] Los Alamos National Laboratory, Los Alamos, New Mexico

The development of numerical modeling methods evolved concurrently with the conceptual model in order to account for the observations made at the site, particularly fracture flow deep in the unsaturated zone. This paper presents the history of the evolution of conceptual models of hydrology and numerical models of unsaturated zone flow at Yucca Mountain, Nevada.

INTRODUCTION

On-land geologic disposal of high-level nuclear waste has been an issue in the United States for nearly half a century. In 1958, the National Academy of Sciences recommended considering geologic disposal of high-level nuclear waste (HLW). In 1959, concerns about the thermal effects of nuclear-waste disposal were added to the recommendation. In the early 1970s, Winograd (1972, 1974) proposed storing nuclear waste in the unsaturated zone, although it was not until the early 1980s that such a design was seriously considered. In 1976, the Director of the U.S. Geological Survey (USGS) suggested to the U.S. Energy Research and Development Administration [ERDA, the predecessor to the U.S. Department of Energy (DOE)] that a nuclear test site in Nevada [Nevada Test Site (NTS)] be examined for potential sites for HLW disposal. The major attributes of the NTS as a potential site for disposal are that it is in a remote location, it is a large contiguous block of land under federal ownership, there is much information on the unsaturated zone based on studies of the underground nuclear testing and the associated presence of radionuclides in the subsurface, rainfall is low, and there is a thick unsaturated zone with a variety of rock types (Winograd, 1971). Initially, however, use of the saturated zone as a nuclear waste repository was the prevailing choice (Roseboom, 1983; Hanks et al., 1999). By 1978, the first boreholes were being drilled on Yucca Mountain to explore the character of the saturated zone for disposal of nuclear waste. The high fracture transmissivity and elevated groundwater temperature of the saturated zone below Yucca Mountain made this zone undesirable as a repository site. In 1982, USGS scientists suggested to DOE that the unsaturated zone at Yucca Mountain be considered instead. Because Nuclear Regulatory Commission (NRC) draft regulations 10 CFR 60 "Disposal of High Level Waste in Geologic Repositories," published in 1981, covered only repositories in the saturated zone, the USGS also suggested to the NRC that the regulations be modified to include the unsaturated zone, and Roseboom (1983) pointed out how such a repository would differ from one in the saturated zone. After extensive public comment and review, the final version of 10 CFR 60 that included the unsaturated zone was released in 1985.

The purpose of this paper is to describe and trace the evolution of the conceptual model of groundwater flow in the unsaturated zone at Yucca Mountain. For this discussion, a conceptual model is simply a relevant set of concepts that describe, in a qualitative way, the behavior of a natural system. Numerical models of the same system, which also will be discussed in this paper, are based on the same set of concepts but describe the behavior of the system in a quantitative manner. It is important to note that the history of the characterization of Yucca Mountain or, in particular, the evolution of a conceptual model of groundwater flow at the Yucca Mountain site, cannot be accurately reconstructed solely on the basis of citable literature. To fully understand this history requires reference to unpublished or draft reports, memoranda, and rough notes. In addition, many of the concepts for the model were developed during discussions between DOE and such entities as the Nuclear Waste Technical Review Board (NWTRB), the Advisory Committee on Nuclear Waste (ACNW), or NRC technical interchanges. In many cases, ideas were developed and worked out during informal get-togethers and, as such, many important ideas and information used in the development are not readily available or directly citable.

YUCCA MOUNTAIN SITE DESCRIPTION

Yucca Mountain is located in southern Nevada about 145 km northwest of Las Vegas (Figure 2-1). The study area covers approximately 45 km², of which approximately 5 km² covers the potential repository site. Beneath the crest of Yucca Mountain, the water table ranges from approximately 350-750 m below land surface, with an average of 500 m. The potential repository host rock is the Topopah Spring Tuff of the Paintbrush Group, a densely welded and fractured tuff located in the unsaturated zone at an average depth of 300 m below land surface (Hanks et al., 1999).

Climate and Precipitation

An understanding of the response of the hydrologic system to current climatic conditions is a prerequisite for predicting the response of the system to potential future climatic conditions (Botkin et al., 1991). The climate in the Yucca Mountain area is arid to semiarid. Weather patterns vary seasonally. Summer precipitation comes primarily from the south and southeast. Winter winds bring moisture from the west, and hence the climate is subject to a regional rain shadow east of the Sierra Nevada and has been for the entire geologic history of Yucca Mountain, more than 13 million years. Topographic effects cause substantial variability in average annual precipitation in the Yucca Mountain area. Precipitation averages from less than 130 mm for lower elevation locations in the south to more than 280 mm for higher elevation locations in the north, with an estimate of 170 mm directly over the potential repository location (Hevesi and Flint, 1996).

Regional Hydrogeology

Yucca Mountain is located within the Basin and Range physiographic province (Grayson, 1993). The linear mountains and valleys of this area that have a

50

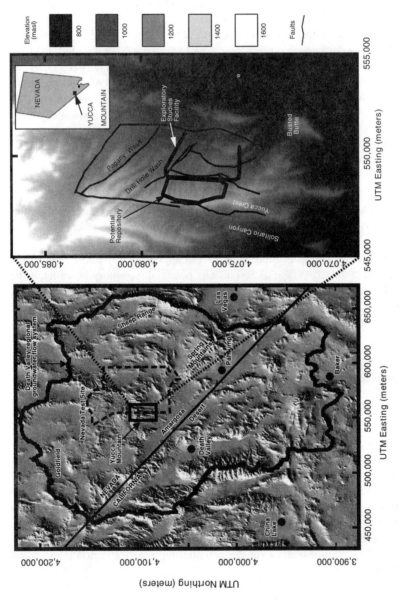

FIGURE 2-1 Yucca Mountain regional (on left) and site-scale study areas (expanded on right). Major block bounding faults (as represented by the project lithostratigraphic model), the Exploratory Studies Facility, and the potential repository boundary are marked.

distinct north to northwest trend define the Basin and Range physiography. Within the Basin and Range physiographic province, there are several topographic regions. Yucca Mountain is in the Death Valley region, which has the largest and most prominent desert basin in the Basin and Range physiographic province. The Death Valley region is primarily in the northern Mojave Desert; the region extends northward into the Great Basin Desert and lies in the rain shadow of the Sierra Nevada. Death Valley itself is the ground-water discharge area for a large part of the Death Valley region. The Death Valley region is composed largely of closed topographic basins that apparently coincide with several closed shallow groundwater flow systems (Winograd and Thordarson, 1975). Recharge in these systems is sparse, and is derived mostly from the higher altitudes and comes as infiltration of precipitation or the infiltration of ephemeral runoff. Discharge occurs primarily by spring flow and by evaporation and transpiration of shallow ground water from playas. The deepest part of the saturated flow system consists of extensive Paleozoic carbonate aquifers that connect the closed shallow groundwater systems at depth. Discharge from the system occurs in several intermediate areas that are geomorphically, stratigraphically, and structurally controlled; but ultimately, most groundwater flow discharges to Death Valley. The predominant direction of drainage for surface-water and groundwater flow in the Death Valley region is generally from north to south because of a decrease in the average altitude from north to south in the southern Basin and Range area.

Site Geology

Yucca Mountain consists of a 1-3-km-thick sequence of ash flow and ash fall tuffs erupted from Timber Mountain, a source caldera complex located directly to the north. The unsaturated zone at Yucca Mountain is about 500-750 m thick (Snyder and Carr, 1982; Buesch et al., 1996), characterized by pyroclastic flows that consist of separate formations. From youngest to oldest, the formations are the Rainier Mesa Tuff (11.6 million years) of the Timber Mountain Group; the Tiva Canyon, Yucca Mountain, Pah Canyon, and Topopah Spring Tuffs of the Paintbrush Group (12.7 million years); the Calico Hills Formation (12.9 million years); and the Prow Pass, Bullfrog, and Tram Tuffs of the Crater Flat Group (13.5 million years) (Carr et al., 1986; Sawyer et al., 1994) (Figure 2-2). Interstratified with these formations are bedded tuffs that consist primarily of fallout tephra deposits and small amounts of pyroclastic flow deposits and reworked material (Moyer and Geslin, 1995; Buesch et al., 1996). The bottom and top of the Tiva Canyon and Topopah Spring Tuffs contain vitric, nonwelded to densely welded tuff; the interiors of the tuffs are thick, crystallized, and moderately to densely welded and fractured. Most of the lithostratigraphic units in the Tiva Canyon and Topopah Spring Tuffs are laterally continuous and stratiform (Scott and Bonk, 1984). The Yucca Mountain and Pah Canyon Tuffs are relatively thick to the north of the potential repository location near Yucca Wash and contain

	Currently used nomenclature	Lithology	Scott and others (1983)	Montazer and Wilson (1984)	U.S. DOE (1984)
Paintbrush Group	Tiva Canyon Tuff (TCw)	welded	Tiva Canyon Member	Tiva Canyon welded unit (TS)	TCw
	Yucca Mountain Tuff, Pah Canyon Tuff and bedded tuffs (PTn)	nonwelded	Nonwelded, argillic and vitric tuffs	Paintbrush nonwelded unit (P)	PTn
	Topopah Spring Tuff (TSw)	welded	Topopah Spring Member	Topopah Spring welded unit (TS)	TSw1 TSw2 TSw3
	Calico Hills Formation (CHn) vitric (CHv) zeolitized (CHz)	nonwelded	Nonwelded, vitric and zeolitized tuffs	Calico Hills nonwelded unit (CH)	CHn
Crater Flat Group	Prow Pass Tuff (PPW)	welded/ nonwelded	Nonwelded, vitric and zeolitized tuffs	Crater Flat unit (CF)	PPw CFUn
	Bullfrog Tuff (BFW)	welded/ nonwelded	Nonwelded, vitric and zeolitized tuffs	Crater Flat unit (CF)	BFw CFUn
	Tram Tuff (TRW)	welded	Not included	Crater Flat unit (CF)	TRw

FIGURE 2-2 Hydrogeologic units and lithostratigraphy currently used at Yucca Mountain, and as used in earlier publications.

both nonwelded and welded intervals. The welded intervals of these units, however, thin southward starting near Drill Hole Wash, and, therefore, only thin welded intervals occur in the center of the potential repository location. These welded intervals are absent altogether from the southern half of the repository location (Moyer et al., 1996). The nonwelded tuffs of the Paintbrush Group, including the nonwelded intervals of the Yucca Mountain and Pah Canyon Tuffs, the interstratified bedded tuffs, the nonwelded base of the Tiva Canyon Tuff, and the nonwelded top of the Topopah Spring Tuff collectively are commonly referred to as the Paintbrush nonwelded hydrologic unit (PTn). The Calico Hills Formation (CHn) is composed of nonwelded pyroclastic flow and fallout deposits (Moyer and Geslin, 1995). Tuffaceous rocks have been zeolitized (CHz) at the north end of Yucca Mountain, yet parts of the formation remain largely vitric

(CHv) towards the south end of the mountain. The Prow Pass Tuff is a compound cooling unit and consists of nonwelded to partially welded tuff at the top and bottom with intervals of welded tuff. The vitric parts of this unit are typically zeolitized in the north, but only in the southwestern part of Yucca Mountain does a significant part of this unit remain vitric, with partially to moderately welded, crystallized tuff in the interior of the unit.

Site Geomorphology

The hydrology of Yucca Mountain has largely been influenced by interrelationships between tectonic and geomorphic processes. Faults and fault scarps, and erosional processes on the eastern sloping ridge, have defined the topography of the mountain, and have created a series of washes (Figure 2-1) that are downcut to varying degrees into different bedrock layers. The topography generally is controlled by high-angle faults that tilt the resistant volcanic strata eastward. Locally, slopes are steep on the west-facing escarpments of the Solitario Canyon Fault and in some of the valleys that cut into the more gentle eastward-facing dip slopes. Narrow valleys and ravines have been cut into the bedrock. Floors of wider valleys consist of alluvial deposits that have formed terraces into which intermittent streams have cut channels. Locally, small sandy fans flank the lower slopes and spread out on the valley floors. East of the crest of Yucca Mountain, drainage is into Fortymile Wash; west of the crest of the mountain, streams flow southwestward down fault-controlled canyons and discharge in Crater Flat. The study site area can be divided into two parts north and south of Drill Hole Wash. The washes in the southern area trend eastward, are relatively short (less than 2 km), and are defined by erosional channels that produce gently sloping sideslopes. The washes north of Drill Hole Wash are controlled by faults, are northwest trending, and are approximately 3-4 km long with steep sideslopes.

Alluvial deposits in the valley floors and washes include fluvial sediments and debris-flow deposits. Soil development and thickness of the alluvial deposits are variable, and the soils are gravelly in texture. The deposits range from 100 m thick in the valleys to less than 30 m thick in the mouths of the washes. Midway up the washes, most alluvial fill is less than 15 m deep in the center of the wash. Many of these deposits have developed cemented calcium carbonate layers (Flint and Flint, 1995).

DEVELOPMENT OF INITIAL CONCEPTUAL AND NUMERICAL MODELS (1983-1990)

A conceptual model describes the physical processes that are part of an environment, how they relate to each other, and which processes dominate the system. It describes the physical framework within which the processes can be understood and numerical relations can be developed.

Conceptual Model Issues at Yucca Mountain

Many current and historical issues are relevant to the discussion of the conceptual model of unsaturated zone hydrology at Yucca Mountain. We discuss the most significant issues and how the associated components of the conceptual model developed or were modified as the conceptual model changed. A simplified schematic that highlights these issues is presented in Figure 2-3.

In general, the major components of the conceptual model include the following processes and features: (1) surface infiltration rates and their distributions in space and time, (2) lateral flow in the nonwelded PTn, (3) lateral flow at the vitric-zeolitic interface in the matrix of the deep nonwelded tuffs (CHv and CHz), (4) the role of faults as conduits or barriers to flow, (5) the occurrence and stability of perched water, (6) the distribution and significance of fast pathways, and (7) the flux between fractures and matrix in unsaturated rock. Most conceptual models for Yucca Mountain include these components, but advances in our scientific understanding of these processes and features have greatly influenced the way the conceptual and numerical models have developed over the years.

Initial Data Collection

Initial data collection at Yucca Mountain consisted of mapping the bedrock surface and drilling boreholes to describe the geology and water table depths at the site. By 1986, more than 100 boreholes had been drilled at or near Yucca

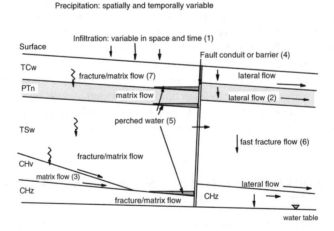

FIGURE 2-3 Generalized conceptual model of the hydrology for Yucca Mountain, Nevada. Arrows denote direction of flow; numbers denote the major components described in the text; and abbreviations for lithostratigraphy are described in Figure 2-2.

Mountain. Many of the early holes drilled during the late 1970s extended to the water table, yielding important data such as potentiometric surfaces, groundwater chemistry, detailed lithostratigraphy, and matrix properties (Anderson, 1981; Rush et al., 1983; Weeks and Wilson, 1984; Whitfield et al., 1984; Flint and Flint, 1990). These were followed in 1984 by an extensive series of shallow boreholes that were drilled to investigate shallow infiltration processes (Hammermeister et al., 1985). Studies of the surface geology at Yucca Mountain had already been ongoing for more than 10 years (Byers et al., 1976; Scott et al., 1983; Scott and Bonk, 1984) prior to the time that investigations of infiltration and percolation processes began in earnest.

A stop-work order for most site characterization activities was issued by the DOE in early 1986 because of concerns related to the quality assurance of data collection; however, selected surface and laboratory investigations (those considered to be collecting irretrievable data) were allowed to continue in order to characterize the geology, faults, and matrix and fracture properties of Yucca Mountain (Klavetter and Peters, 1987; Istok et al., 1994; Flint et al., 1996b). After the stop-work order was lifted in late 1991, a series of shallow, cored neutron boreholes were drilled to study infiltration processes (Flint and Flint, 1995). On completion of the neutron boreholes, deep boreholes were drilled for long-term monitoring and geotechnical boreholes were drilled along the surface projection of the underground Exploratory Studies Facility (ESF) prior to its construction (Rousseau et al., 1998) to provide design information for the construction of the ESF. The results of core analysis and borehole instrumentation and geophysics, which measured subsurface conditions, have aided in the development of the conceptual model used to help understand infiltration and percolation rates and processes at Yucca Mountain. The measurements and analyses have provided detailed data sets needed for the development and testing of the site-scale numerical flow model.

Early Conceptual Models of Hydrology at Yucca Mountain

The earliest detailed conceptual model of the unsaturated zone at Yucca Mountain was published by Scott et al. (1983) (Figure 2-4). Their conceptual model of hydrology at Yucca Mountain is a component of a larger geologic/hydrologic framework model that is presented in a very straightforward manner. First, they identified the problem and stated that groundwater was one of the most critical parameters for nuclear waste isolation. Second, they described the stratigraphic, structural, and hydrologic framework that is the basis of their geologic/hydrologic framework model by presenting the detailed geologic setting. Third, they identified the relevant hydrologic processes needed to describe the hydrology for their geologic/hydrologic framework model. Finally, using these processes and applying them to their model, they described the hydrologic consequences of groundwater flow.

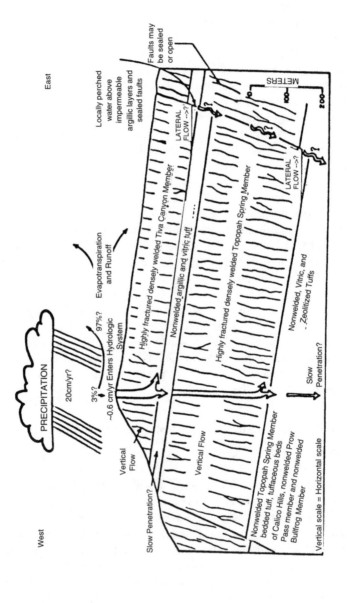

FIGURE 2-4 Conceptual model of the hydrology of Yucca Mountain. From Scott et al. (1983, Figure 18). Nomenclature is described in Figure 2-2.

Although Scott et al. (1983) acknowledged that there were uncertainties in their estimates, their conceptual model of the hydrologic system was as follows: approximately 200 mm/yr precipitation falls on Yucca Mountain, of which about 3 percent enters the hydrologic system (6 mm/yr net infiltration). This infiltrating water moves vertically through fractures in the welded Tiva Canyon Tuff (TCw; in 1983 the accepted nomenclature was Tiva Canyon Member), vertically as matrix flow in the nonwelded PTn (although they believed some potential existed for lateral flow at the TCw-PTn interface), vertically through fractures in the welded Topopah Spring Tuff (TSw) and through the matrix in the lower vitric nonwelded tuffs, laterally at the vitric-zeolitic interface of the nonwelded tuffs, and vertically in lower zeolitic nonwelded tuffs. They also believed that lateral flow occurred along the tilted strata and that the potential existed for perched water to accumulate on the upgradient side of faults. Although they believed that faults could be either sealed or open to flow, they cautioned that faults should be assumed to be open in the absence of evidence to the contrary. This first and simple conceptual model is perhaps closer to the current (2000) conceptual model for Yucca Mountain than any of those formulated in the intervening years.

A somewhat more simplified conceptual model was proposed by Roseboom (1983) as a generic case for considering the unsaturated zone in the arid southwest for HLW disposal. As explained in the Introduction, the NRC's existing draft regulations assumed that all repositories would be in the saturated zone and the regulations would have to be changed to accommodate unsaturated sites. Thus, Roseboom presented the first diagrammatic representations of a repository in unsaturated rock (Figure 2-5) and pointed out fundamental differences between repositories in the saturated and the unsaturated zone. Although the model was generic for the most part, specific discussions in Roseboom's paper of Yucca Mountain presented a conceptual model similar to that of Scott et al. (1983). Assuming that precipitation is 127 mm/yr and that 3 percent of this precipitation becomes net infiltration, Roseboom estimated a recharge rate of 3.8 mm/yr to the local water table; he said that the flow path was downward, with no lateral component, and that flow was rapid through the fractured zones. The amount of flux through the repository, therefore, would be directly related to the average precipitation. Roseboom also believed that even higher infiltration rates for short-term climate change could be easily drained from the stratigraphic horizon in which the repository is located because of the high fracture permeability. Rapid drainage of water through fractures would minimize contact time between water and the waste canisters and therefore repositories in the unsaturated zone would be preferable to repositories in the saturated zone if the host rock had high fracture permeability.

In 1984, Montazer and Wilson (1984) presented a detailed conceptual model of Yucca Mountain (Figure 2-6) that was based on soil physics; this model was similar to the model of Scott et al. (1983) but with several exceptions. Montazer and Wilson estimated that the average annual precipitation at Yucca Mountain is

58

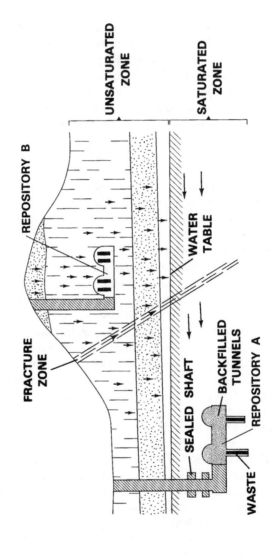

FIGURE 2-5 Conceptual model of the hydrology of the Yucca Mountain. From Roseboom (1983, Figure 1).

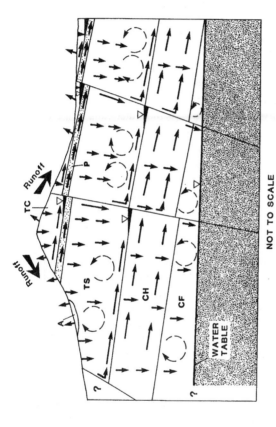

NOT TO SCALE

EXPLANATION

A   ALLUVIUM

TC   TIVA CANYON WELDED UNIT

P   PAINTBRUSH NONWELDED UNIT

TS   TOPOPAH SPRING WELDED UNIT

CH   CALICO HILLS NONWELDED UNIT

CF   CRATER FLAT UNIT

DIRECTION OF LIQUID FLOW

DIRECTION OF VAPOR MOVEMENT

PERCHED WATER

FIGURE 2-6 Conceptual model of the hydrology at Yucca Mountain, Nevada. From Montazer and Wilson (1984, Figure 14). Nomenclature is described in Figure 2-2.

approximately 150 mm. Using the same 3 percent rate for net infiltration as Scott et al., they calculated a net infiltration of 4.5 mm/yr but suggested that recharge ranged from 0.0-4.5 mm/yr. They believed that water probably infiltrated either directly into fractures within bedrock exposures or as surface runoff seeping into alluvium beneath the channels of washes. As water saturated the walls of the fractures, it could continue flowing to greater depths without increasing the saturation of interior matrix blocks.

Despite their belief that complete matrix saturation was not necessary to induce and propagate fracture flow, they introduced a relation between effective (bulk) permeability and matrix potential for a combined fracture-matrix medium (composite porosity, see Figure 2-10, which is discussed later) that required high matrix saturations before fractures would conduct water. Figure 2-7 shows their analysis of flow through a single fracture (curve 2) transecting a porous matrix (curve 1) with much lower effective permeabilities. At very low levels of saturation, flow due to gravity or "drainage" is limited to the matrix and follows curve 1 to lower levels of permeability and saturation. As the water content increases to higher levels of partial saturation, this curve crosses the curve for the fracture. At higher saturation levels, fracture flow dominates the drainage. Scott et al. (1983) suggest that with rapid wetting from periods of intense precipitation, less complete saturation of the matrix might occur. Curves 1a and 1b illustrate the "hysteresis effect" of more rapid wetting, which effectively moves curve 1 to lower values because air becomes trapped in the matrix. This would reduce the permeability to lower values as the wetting rate increases. Thus with more rapid wetting, the fractures become more dominant in draining the rock mass. The combined uppermost parts of a fracture curve and a matrix curve can be combined to represent a single medium with composite porosity, and is termed an "equivalent continuum."

Montazer and Wilson (1984) maintained that most of the infiltrating water (4 mm/yr) was diverted laterally within the Paintbrush nonwelded unit (PTn) and that the remaining water (0.5 mm/yr) moved downward through the matrix of the Topopah Spring welded unit (TSw). They cited work by Weeks and Wilson (1984), who estimated a downward flux of between 0.003 and 0.2 mm/yr in the matrix of the TSw, as support for the matrix flow component of their conceptual model. Montazer and Wilson suggested that lateral flow would lead to perched water on the upgradient side of faults, but that the water would ultimately move into the fault and recharge the saturated zone. Noting the sharp contrast in pore size distributions of the TSw compared with those of the underlying nonwelded and bedded vitric tuffs, they believed that the vitric tuffs with their much larger pores would act as a capillary barrier to vertical flow, leading to lateral flow in the TSw, whereas zeolitic tuff, which has much smaller pores, would not divert flow. This perspective differs from the one held by Scott et al. (1983), who proposed that the zeolitic boundary would be a barrier because of the low conductivity of this unit. The most significant differences between the conceptual model by

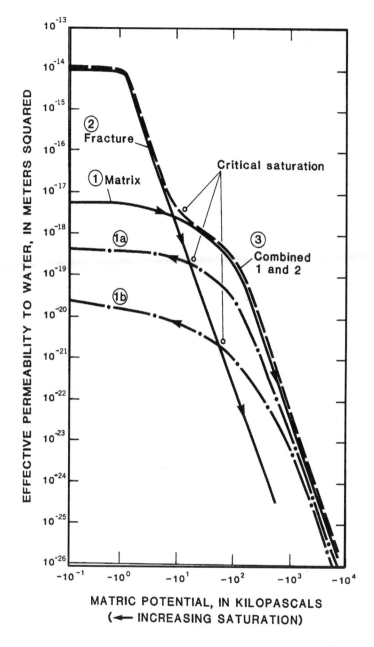

FIGURE 2-7 Hypothetical relationship between effective permeability and matric potential for a double-porosity medium. Curves 1a and 1b are wetting curves for the matrix; and downward arrows show drainage curves. From Montazer and Wilson (1984, Figure 6).

Montazer and Wilson compared with the earlier model by Scott et al. were a lack of fracture flow in the TSw and the prevalence of lateral flow above and within the PTn. The views of Montazer and Wilson greatly influenced the direction of subsequent hydrologic research at Yucca Mountain for several years.

In a general review of the hydrology in and around Yucca Mountain, Waddell et al. (1984) suggested that recharge to the water table probably was less than 5 mm/yr, although they also cited the work of Sass and Lachenbruch (1982), who estimated 8 mm/yr flux through the TSw on the basis of an analysis of temperature profiles (analyses that would be redone in the 1990s). Similar to Montazer and Wilson (1984), they believed that infiltration occurred predominantly in the washes or by direct entry into fractures exposed at the surface.

Early Numerical Process Models

Four major concepts would strongly influence further development of the conceptual model of Yucca Mountain and would control the thinking of most Yucca Mountain hydrologists and modelers for the next 10 years. These concepts were (1) that the matrix must be fully saturated before fracture flow could be initiated or sustained (Wang and Narasimhan, 1985), (2) that overall, flux is low (Scott et al., 1983; Montazer and Wilson, 1984), (3) that only matrix flow (i.e., fluxes less than 0.5 mm/yr) is assumed to occur in the TSw (Montazer and Wilson, 1984), and (4) that lateral diversion of most of the net infiltration occurs within or above the PTn.

Wang and Narasimhan (1985) presented a conceptual and a numerical theory of fracture flow and fracture-matrix interaction in fractured rock that showed that the matrix must be fully saturated before fracture flow could be initiated or sustained. They suggested that there is virtually no downward flow in fractures beneath the PTn and that flow between adjacent matrix blocks would occur across fractures only at asperities (Figure 2-8). They proposed that fracture flow occurs only when the water potential of the fracture is in equilibrium with the water potential of the matrix, which, in the unsaturated zone, occurs only during periods of surface flooding when there is a transition between saturated and unsaturated fracture conditions. They numerically modeled flow through a matrix system with discrete fractures and suggested that, for the most part, unsaturated fractured systems could be simulated numerically without taking fractures explicitly into account.

Reports by Sinnock et al. (1984, 1987), Klavetter and Peters (1986), and Peters and Klavetter (1988) continued to support the assumption that net infiltration was low (4 mm/yr), that lateral movement occurred at the base of the PTn, and that although no perched water was expected in the PTn, most of the water flowed downward through the faults, with less than 0.5 mm/yr of flux flowing through the TSw. Klavetter and Peters (1986) and Peters and Klavetter (1988) agreed, on the basis of the work of Wang and Narasimhan (1985), that the matrix

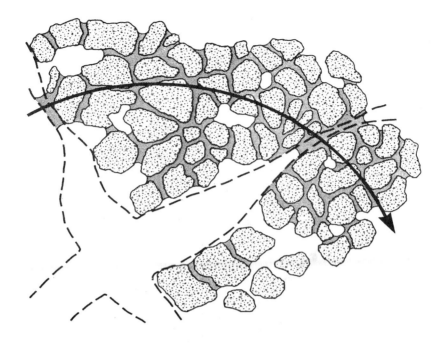

FIGURE 2-8 Conceptual model of partially saturated, fractured, porous medium showing schematically the flow lines moving around the dry portions of the fractures. From Wang, J. S. Y., and T. N. Narasimhan, 1985. Hydrologic mechanisms governing fluid flow in a partially saturated, fractured, porous medium. Water Resources Research 21(12): 1861-1874. Copyright by American Geophysical Union.

must be saturated in order for fracture flow to occur. In the results of the first Total System Performance Assessment (TSPA) for the Yucca Mountain site, Sinnock et al. (1984) argued that distributed fracture flow would not occur if fluxes were less than 0.5 mm/yr and would result in a relatively dry TSw (less than –2 MPa) despite the fact that Weeks and Wilson (1984) had estimated water potentials to be between –0.06 and –0.26 MPa on the basis of experimental data. Sinnock et al. (1984, 1987), Klavetter and Peters (1986), and Peters and Klavetter (1988) based their numerical analyses on the conceptual model of the U.S. Department of Energy (1984) (Figure 2-9), which essentially was a simplified version of the Montazer and Wilson (1984) model.

The first numerical two-dimensional flow model of the unsaturated zone at Yucca Mountain was presented by Rulon et al. (1986) and supported Montazer and Wilson's (1984) concept of lateral flow in the PTn at low fluxes with as much

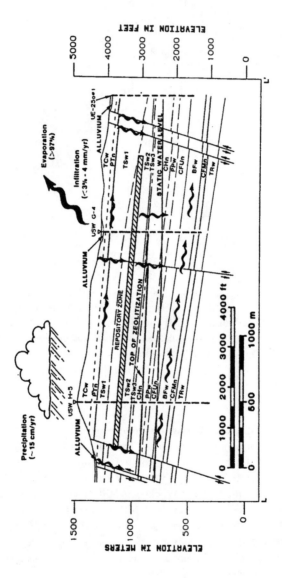

FIGURE 2-9 Conceptual model of the hydrology of Yucca Mountain, Nevada. From U.S. Department of Energy (1984). Nomenclature is described in Figure 2-2.

as 50 percent of infiltration being laterally diverted for fluxes of less than 1 mm/ yr. However, at a higher flux of 4.5 mm/yr, the model predicted only about 1 mm/ yr of lateral flow; this prediction was based partly on the assumed high permeability of the faults. Using the combined fracture-matrix characteristic curve (see Figure 2-7) from Montazer and Wilson (1984), which essentially was an equivalent continuum model, Rulon et al. (1986) defined a critical saturation above which fracture flow would dominate over matrix flow. [The shape of this curve was demonstrated experimentally by Peters et al. (1984).] The value for critical saturation varied depending on the assumed matrix properties. For their two sets of fracture-matrix simulations, the critical saturation was either 0.3 or near 1; the latter represented the concept of Wang and Narasimhan (1985).

Rockhold et al. (1990) produced the first three-dimensional numerical model, but only for a small areal extent around the potential repository location. They modeled fracture flow in the TCw but matrix flow in the TSw and the PTn, where the PTn diverted most of the vertical flow laterally.

Early TSPA models for Yucca Mountain investigated only one-dimensional columns (Barnard et al., 1992). Their analysis suggested there was less than a 1 percent chance that flux through the TSw was more than 3 mm/yr and an 80 percent chance that the flux was less than 1 mm/yr. During this same time frame, in an effort to evaluate the possibility of significant fracture flow in the TSw, Gauthier et al. (1992) developed a "weeps" model that represented only fractures; their model rapidly drained the fractured TSw and accommodated higher infiltration fluxes without causing significant lateral diversion in the PTn.

Altman et al. (1996) presented a general view of the conceptual models of flow between matrix and fractures and the resultant relative permeability curves (assuming Richards' equation-type flow) that are typically required by numerical models of flow through fractured rock. Figure 2-10 demonstrates the evolution of numerical modeling from the early composite porosity used by Rulon et al. (1986), but first presented by Montazer and Wilson (1984), to the more sophisticated dual-permeability models used currently; the figure gives common terminology for further discussion.

DEVELOPMENT OF CURRENT CONCEPTUAL AND NUMERICAL MODELS (1991-2000)

Three-Dimensional Site-Scale Numerical Model

The basis for the current site-scale model of the unsaturated zone at Yucca Mountain was established during meetings in 1991 between the USGS and Lawrence Berkeley National Laboratory (LBNL); these meetings resulted in the first three-dimensional numerical site-scale model (Wittwer et al., 1992). Although the model was designed to allow for the spatial distribution of infiltration and for discrete faults (Figure 2-11), it also allowed for the use of the equivalent

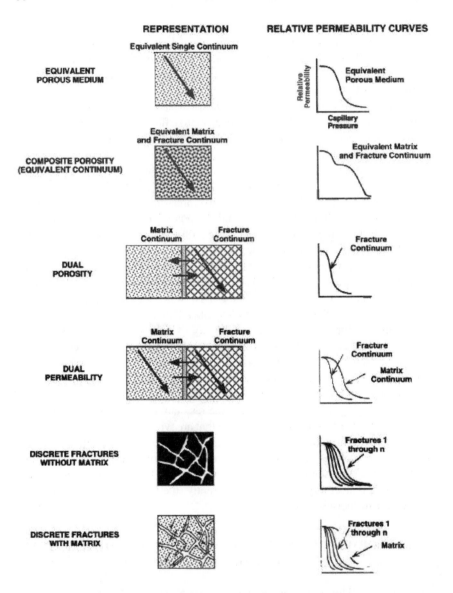

FIGURE 2-10 Alternative conceptual models and their corresponding characteristic curves for relative permeability for flow through fractured rock. From Altman et al. (1996, Figure 2.2).

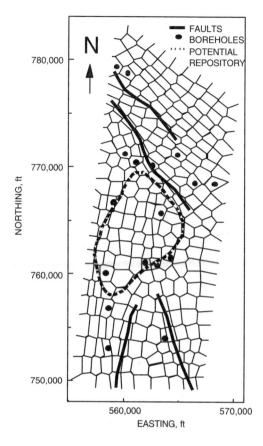

FIGURE 2-11 Three-dimensional site-scale model grid with early version of potential repository boundary. From Flint and Flint (1994, Figure 2). Copyright 1994 by the American Nuclear Society, La Grange Park, Illinois.

continuum concepts of Montazer and Wilson (1984). Initially, simulations using the three-dimensional model used a uniform (i.e., not spatially distributed) surface infiltration rate of 0.1 mm/yr. This model was used to explore the role of faults and the redistribution of moisture at depth (Wittwer et al., 1995).

Development of Conceptual Model of Spatially Distributed Infiltration

Hevesi et al. (1992) used a geostatistical approach to construct the first detailed map of precipitation for the Yucca Mountain site; they estimated an average annual precipitation of 170 mm/yr. This map was used with the Maxey-

Eakin method (Maxey and Eakin, 1950), which postulates a simple correlation between precipitation and recharge on a regional scale, to estimate the spatial distribution of recharge for the Yucca Mountain area and the Death Valley region. Recognizing that there is a large variability in the spatial and temporal distribution of precipitation in the study area, it was noted that spatial and temporal distributions of infiltration could be the most important factors influencing the distribution and rate of percolation in the unsaturated zone (U.S. Department of Energy, 1992).

With the ongoing development of a three-dimensional numerical site-scale model, it became clear that quantitative estimates of the spatial distribution of infiltration would be needed to define the upper boundary condition for the flow model. The flow model of Wittwer et al. (1992) had incorporated three topographic zones that served as the initial estimated zones of infiltration. The topographic zones were chosen on the basis of field observations and preliminary analyses of neutron borehole logs that indicated a qualitative correlation between topographic zones and depth of alluvium. Distinctly different changes in subsurface water content with time and depth of water penetration following precipitation characterized the different infiltration zones (Flint et al., 1993; Flint and Flint, 1994; Flint and Flint, 1995).

The initial approach taken to spatially distribute surface infiltration on a site scale is explained in Flint and Flint (1994). For that study, they relied on maps of surface bedrock geology and alluvium (Scott and Bonk, 1984) to distribute material properties and applied the Darcy flux approach to neutron log water-content data for shallow boreholes and core property data. According to this approach, the depth at which the water content does not change seasonally is assumed to represent steady-state moisture conditions, and the effective hydraulic conductivity at that moisture content was assumed to be equal to flux. A site-scale map was then prepared of the bedrock units at the steady-state depths. These units were classified into five representative categories for which water content and hydraulic properties were used to calculate a flux (Plate 1), which then was distributed among the site-scale grid blocks. Fluxes ranged from 0.02 mm/yr for areas exposed with TCw to 13.4 mm/yr for areas exposed with PTn, with a site-scale mean of 1.4 mm/yr. Fluxes were highest in the washes and along the exposure of the permeable PTn in the Solitario Canyon escarpment.

Hudson and Flint (1995) refined this approach using estimates from precipitation maps (Hevesi and Flint, 1996), soil thickness, topographic position, and calculations of flux in boreholes to develop correlation matrices and to statistically distribute infiltration for a larger area than the area of the site-scale model (Figure 2-12). Infiltration rates ranged from 0-45 mm/yr, with a mean of 11.6 mm/yr in the study area. They suggest that most of the infiltration probably enters the system on ridgetops and sideslopes where soil cover is thin.

Estimates of the spatial distribution of infiltration indicated that some areas of the site can have high rates of infiltration while other areas can have little to no

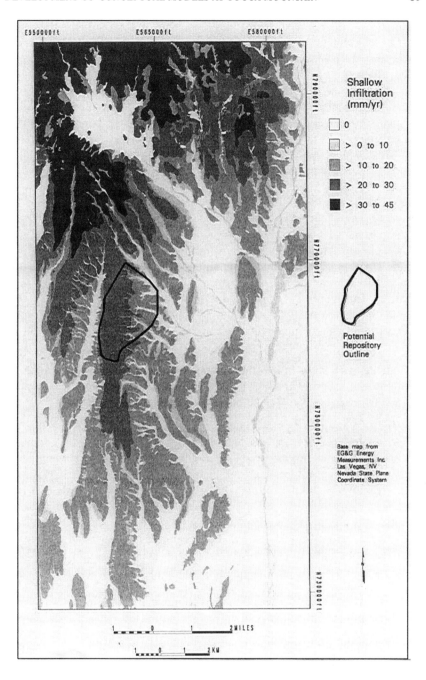

FIGURE 2-12 Infiltration distributed statistically. From Hudson and Flint (1995, Figure 8).

infiltration. Initial efforts to use the infiltration maps of Flint and Flint (1994) and Hudson and Flint (1995) in numerical models resulted in a saturated matrix and sublayers, which was inconsistent with measured saturation data (Wittwer et al., 1995; Ho et al., 1995) because of invalid coupled matrix-fracture assumptions and an equivalent continuum approach. In many areas of the models, fluxes exceeded the saturated conductivity of the matrix, which indicated the necessity to modify the models to allow for fracture flow in units with a low-permeability matrix.

The final iteration of approaches to estimate infiltration was done by incorporating a detailed conceptual model of infiltration (Flint et al., 1993; Flint and Flint, 1995) into a numerical model (Flint et al., 1996a). The numerical model was driven by stochastically modeled precipitation, modeled solar radiation and evapotranspiration, soil depth, and bulk bedrock properties and was calibrated to measurements of water content changes in neutron boreholes. The numerical model allowed for the investigation of infiltration under different climate scenarios. Under present-day climatic conditions, infiltration rates ranged from 0-80 mm/yr, with a mean of about 4.5 mm/yr across the general area of the repository (Plate 2). This model was a modified water-balance model (Flint et al., 1996a) and not based on Richards' equation and, therefore, did not require fracture-matrix equilibrium to initiate fracture flow.

Modeling with High Infiltration Rates: the Conversion to Fracture Flow

Lateral Diversion in the PTn

By 1995, most project hydrologists had accepted that the conceptual model of unsaturated zone flow must include fracture flow in the TSw and that numerical models must incorporate approaches other than the equivalent continuum model if they were to accommodate high fluxes through the TCw. However, some hydrologists still believed that lateral diversion in and above the PTn may reduce the volume of water that would penetrate the TSw. Although there was no significant or conclusive field evidence to support lateral diversion, a number of modeling exercises were done to test if the measured properties and geometry of the PTn could support or induce diversion (Ho, 1995; Altman et al., 1996; Moyer et al., 1996; Wilson, 1996). Quite typically in these simulations, lateral diversion could be attained but, consistent with an earlier study by Rulon et al. (1986), the percentage of water diverted was greatly reduced at high fluxes. At lower fluxes diverted water may have been an artifact of the model's simplified geometry, unrealistic hydrologic properties [extremely high air-entry pressure and high matrix permeability (Peters et al., 1984) are incompatible], idealized stratigraphic contacts as linear features or a misrepresentation of the gradational nature of the TCw-PTn transition (Moyer et al., 1996; Flint, 1998), or the downdip boundary conditions (e.g., the capillary properties assumed for the faults). In any case, on the basis of capillary barrier theory (Montazer and Wilson, 1984; Ross, 1990) and modeling results, lateral diversion could not be counted on to divert high fluxes from the repository horizon.

Evidence of Fast Fracture Flow

During this period, Fabryka-Martin et al. (1993, 1994), Fabryka-Martin and Liu (1995), and Yang et al. (1996) reported bomb-pulse concentrations of isotopes in surface-based boreholes; these pulses indicated possible fast pathways through the TCw and deep into the TSw. Measurements of ^{36}Cl and calculations of chloride mass balance supported the high rates of the spatially distributed estimates of infiltration and the existence of fracture flow. Secondary calcite deposits observed in fractures throughout the ESF indicated that considerable fracture flow had occurred (Paces et al., 1996). The results of several studies that had evaluated how water flows through fractures contradicted the traditional, but perhaps outdated, conceptual model of sheet flow in fractures (Rulon et al., 1986) and supported fast fracture flow by demonstrating channeling or fingering (Pruess and Tsang, 1990; Glass, 1993; Glass et al., 1994; Pruess, 1998). Nevertheless, modelers were still using either uniform low infiltration rates in their models (Wittwer et al., 1995) or spatially distributed rates that had been proportionally reduced to be consistent with previously low annual average infiltration values (Altman et al., 1996). Modelers justified their use of reduced distribution rates estimated from the infiltration maps of Hudson and Flint (1995) or Flint and Flint (1994) because their equivalent continuum models predicted unrealistic saturated matrix conditions and because they could not force enough water through the profiles to accommodate the high fluxes. Robinson et al. (1995) normalized the infiltration values from the Hudson and Flint (1995) map to a site-scale average of 1 mm/yr but noted that there was too much evidence for high infiltration rates and that the process modelers should now consider these higher rates. They proposed that nonequilibrium fracture flow could occur even if the matrix was not completely saturated. The method of Robinson et al. (1995) used the dual-permeability approach (Figure 2-10) to reduce the potential for saturation of the matrix under conditions of fracture flow by changing the parameters of the relative permeability curves of the fractures.

To reiterate the concepts that maintained long-term viability, it is pointed out here that the original proposed existence of nonequilibrium fracture flow came back full circle in the late 1990s to pre-Wang and Narasimhan (1985), who introduced the necessity of having a saturated matrix before fractures would flow. In the first conceptual model, Scott et al. (1983) suggested steady-state matrix saturation was necessary for fracture flow, but Montazer and Wilson (1984) maintained that a saturated matrix was not necessary to induce deep fracture flow, only wetted walls were required. Despite the recognition that fracture and matrix might not be in equilibrium, Montazer and Wilson (1984) suggested the use of the combined matrix-fracture moisture characteristic curve, or equivalent continuum approach, which necessitated high matrix saturations, except in cases of rapid wetting (Figure 2-7). This approach had survived for more than a decade.

Dual-Continuum Modeling Methods

Dual-permeability models (see Figure 2-10 for conceptual and numerical representations) were being used with the low infiltration rates estimated for Yucca Mountain in the early 1990s. To accommodate the high infiltration rates, Ho et al. (1995) decoupled fracture flow from matrix flow by reducing the simulated contact area between the fracture and matrix continua (dual-permeability model, Figure 2-10) by two orders of magnitude. Ho (1997) later suggested a reduction of at least four orders of magnitude in the contact area. This decoupling enabled the simulation of high rates of infiltration indicative of young isotopic ages and the fast transport of bomb-pulse nuclides that had been observed without causing the matrix to be unrealistically wet. In 1996, Wu et al. used the same technique presented by Robinson et al. (1995) to modify the equivalent continuum model by introducing different hydraulic properties for the fractures and the matrix. This was done to produce a dual-permeability-type approach that could accommodate 5 mm/yr of flux in two-dimensional cross sections without saturating the profiles, but also without decoupling the fracture-matrix interaction as Ho et al. (1995) had done.

Evidence supporting the high infiltration rates continued to accumulate. Analyses of thermal gradients in boreholes corroborated the estimated infiltration rates of 5-10 mm/yr for most of the repository (Bodvarsson and Bandurraga, 1996; Kwicklis and Rousseau, 1999). Estimates of 5-10 mm/yr also were calculated (Bodvarsson and Bandurraga, 1996) from analyses of perched water (Patterson, 1999).

Dual-continua modeling methods allowed the high fluxes to pass through the TSw, which, in turn, allowed simulations of young water in the CHn and in perched water. Model predictions of perched water at the vitric-zeolitic contact within the CHn or Prow Pass Tuff were supported by field observations (Figure 2-13a). An isopach surface (Figure 2-13b) of the top of the zeolitized units illustrates how water is envisioned as flowing across the top of the zeolites and into the water table either directly or through the faults (Bodvarsson and Bandurraga, 1996). Lateral flow at this contact also is supported by thermal gradient analyses, which indicate a decreased vertical percolation rate below this contact in the CHn and Prow Pass Tuff (Bodvarsson and Bandurraga, 1996). To account for the perched water, the site-scale model currently treats faults as permeability barriers to lateral flow, an assumption substantiated by pneumatic evidence (LeCain, 1997) and by the sharp drop in the potentiometric surface across the Solitario Canyon Fault (Tucci and Burkhardt, 1995). By 1996, the

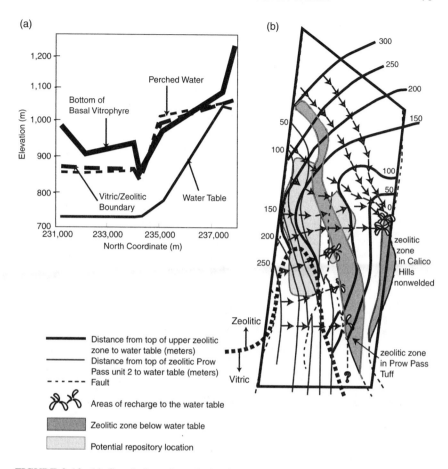

FIGURE 2-13 (a) Correlation of perched water with the vitric-zeolitic boundary. From Bodvarsson and Bandurraga (1996, Figure 7.4.1). (b) Contour map of the thickness of zeolites above the water table and below the base of the TSw, indicating possible flow paths and lateral diversion. From Bodvarsson and Bandurraga (1996, Figure 7.4.2).

conceptual model of hydrology for Yucca Mountain had evolved to include pervasive fracture flow in the TSw (Figure 2-14).

Development of Conceptual Model for Fast Pathways

Isotopic studies at the site began to support a conceptual model in which pervasive fracture flow and fast pathways through the unsaturated zone altered

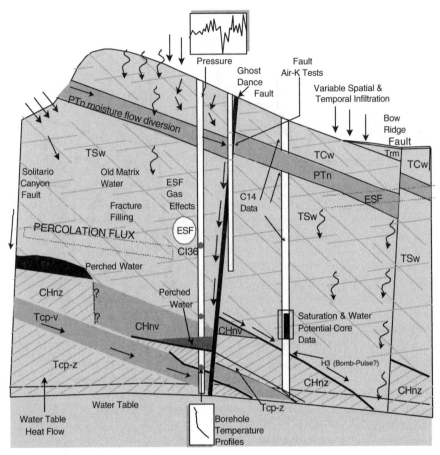

FIGURE 2-14   A schematic cross section through Yucca Mountain showing various conceptual model data and processes. From Bodvarsson and Bandurraga (1996, Figure 1.1.1).

the long-held conceptual model of flow in an arid environment (Winograd, 1981; Roseboom, 1983; Montazer and Wilson, 1984). Bomb-pulse concentrations of tritium and ^{36}Cl and high ^{14}C activities were detected in some of the several hundred unsaturated samples from boreholes and walls in the ESF (Yang et al., 1996, 1998; Fabryka-Martin et al., 1998). Tritium data also indicated that the groundwater was young at the top of the saturated zone in a fault zone, illustrating a mechanism for fast flow that may be important for local recharge. With the construction in 1995 of the ESF tunnel through the repository horizon, detailed in situ studies could be done to evaluate percolation flux through the TSw. An unexpectedly large number of fast pathways were discovered using feature-based and systematic sampling of porewater to measure [36]Cl (Figure 2-15), and a detailed conceptual model was developed to account for the distribution of the fast pathways (Fabryka-Martin et al., 1996, 1997; Sweetkind et al., 1997; Wolfsberg et al., 1999). Simulation results of this model indicate that the fastest pathways from the land surface to the repository horizon initiate as near-surface fracture flow beneath shallow soils that propagate along continuous fracture paths through the TCw, through faults in the PTn that allows water to bypass the rock matrix, and through continuous fracture paths in the TSw. Flow is required to remain in the fault only through the PTn. The concept of deep, rapid infiltration into fractured tuffs exposed at the surface or buried under shallow soils is supported by the presence of bomb-pulse [36]Cl and tritium at depth, as well as by low concentrations of chloride, which reflect fairly high infiltration rates (Fabryka-Martin et al., 1998).

Isotopic studies of the porewater salts in the underground ESF support a conceptual model for fast pathways in which faulting in the overlying PTn is the dominant factor controlling the spatial distribution of fast pathways to the repository horizon (Fabryka-Martin et al., 1997). Three conditions must be present to transmit bomb-pulse [36]Cl to the sampled depth (about 300 m below the surface) within 50 years. First, a continuous fracture path must extend from land surface, which requires the presence of faults that cut the PTn and increase its fracture conductivity and connectivity. Second, surface infiltration rates must be high enough to initiate and sustain at least a small component of fracture flow along the connected fracture path. Transport simulations indicate that the threshold rate

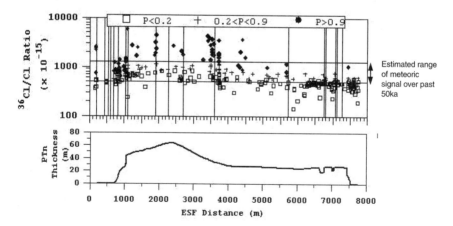

FIGURE 2-15 Distribution of mapped faults (vertical lines) in the exploratory studies facility (ESF) and [36]Cl/Cl measurements, with probability of being bomb pulse (P) indicated, together with thickness of the overlying PTn unit. From Wolfsberg et al. (1999).

is probably about 1 to 2 mm/yr (Wolfsberg et al., 1999). Third, the residence time of water in the soil cover must be less than 50 years, i.e., the soil thickness must be less than 3 m so that the bomb pulse is not retained within the soil profile. The ESF studies also support the theory that the capacity of the unfaulted PTn to buffer and redistribute infiltration appears to decrease in the south of the repository where the unit thins to 25 m (Wolfsberg et al., 1999).

Current Conceptual Model of Flow Through the Unsaturated Zone at Yucca Mountain

The current (2000) conceptual model of flow through the unsaturated zone at Yucca Mountain (Bodvarsson et al., 1998; Flint et al., in press) as presented here reflects those processes invoked and supported by the majority of research participants on the Yucca Mountain Project, as well as those concepts that are most consistent with most of the measured data and observations. The existing site-scale numerical flow models effectively encompass and integrate hydrologic processes that operate at multiple scales at Yucca Mountain. The ability of these numerical models to provide simulations that are consistent with a broad suite of characterization data and observations collected independently is substantial corroboration of the viability of the conceptual model on which the numerical models are based. The numerical gridding has become more detailed for the area of the potential repository (Figure 2-16) and currently contains approximately 10,000 surface grid nodes that extend vertically through approximately 28 hydrogeologic unit layers from the land surface to the water table (Bodvarsson et al., 1998).

The four most important features of the current conceptual model are (1) the existence of relatively high spatially and temporally variable infiltration rates that virtually eliminate (2) large-scale lateral diversion of water above and within the PTn and that force (3) the pervasive flow of water through fractures in densely welded tuff units, despite nonequilibrium water potentials between fractures and the adjacent matrix, and (4) vertical flow in the CHv and extensive lateral flow and perching of water at the zeolitic boundary abutting faults. This simplified version of the conceptual model is very similar to the earliest conceptual models for Yucca Mountain, particularly that of Scott et al. (1983). These four features basically control most of the remaining details of the hydrologic processes represented schematically in Figure 2-17.

The infiltration processes are governed primarily by the distribution and timing of precipitation, the properties of the surface soils and bedrock, and the components controlling evapotranspiration. Average annual rates of infiltration range from 0 to more than 80 mm/yr and average approximately 5-10 mm/yr across the repository block area, or 3 to 6 percent of the average annual precipitation of 170 mm/yr. Most water that becomes net infiltration infiltrates from the ridgetops and sideslopes where fractures are present and soils are shallow to depths below the effects of evapotranspiration. Net infiltration is negligible in

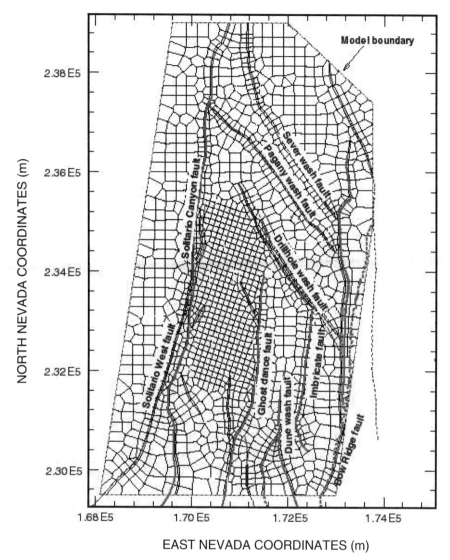

FIGURE 2-16 Plan view of the current (1999) site-scale unsaturated zone numerical model grid. From Bodvarsson et al. (1998).

deep soils because of the large storage component and evapotranspiration, except in the deep soils of channels fed by large volumes of runoff following extreme periods of precipitation.

Most of the infiltrating water passes quickly through the fractures of the TCw to be slowed during transition to matrix flow in the PTn except where

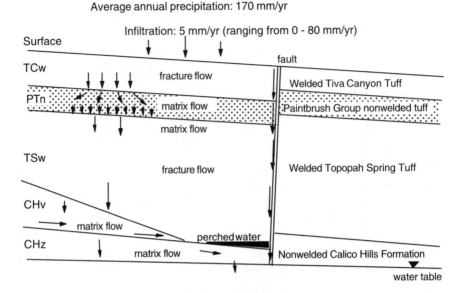

FIGURE 2-17 Current (2000) conceptual model of flow in the unsaturated zone at Yucca Mountain. From Flint et al. (in press, Figure 4b).

faults or broken zones disrupt the PTn, providing fast pathways for a small component of the flow. A small percentage of water in the TCw (less than 0.5 mm/yr) is lost to the atmosphere by way of upward vapor flow (barometric pumping; vapor diffusion; and convective, buoyancy-driven gas flow). Most flow is vertical and slow through the PTn matrix with possible local-scale lateral diversion just above the altered, nonwelded base of the Tiva Canyon Tuff, at linear contacts, or above low-permeability layers. Zones at the TCw-PTn and PTn-TSw contacts are nearly saturated but do not constitute perched layers. Water enters the TSw through faults or through localized broken zones at a low-conductivity vitrophyre at the top of the TSw and, to a lesser extent, through microfractures within the vitrophyre. The transition from highly porous tuffs to densely welded rocks occurs across a very short vertical distance, resulting in high saturations in this part of the flow system. Relatively pervasive broken-up areas that probably formed as the vitrophyre cooled provide ready access for entry of water into the underlying vapor-phase corroded nonlithophysal rocks of the TSw.

Once through the PTn, the bulk of the percolating water transitions back to vertical flow through the fractures and the matrix in the upper TSw and flows dominantly through fractures in the middle of the TSw. Fracture flow apparently occurs primarily under conditions of disequilibrium with the surrounding matrix when averaged for relatively large matrix blocks and may occur by channeling or focused flow. In many locations, particularly in the northern part of the site, the densely welded basal vitrophyre of the TSw coincides with the vitric-zeolitic boundary and therefore serves as a permeability barrier to vertical flow, which results in perched water at this stratigraphic location. Where the vitric-zeolitic boundary does not extend upward as far as the base of the vitrophyre, perching occurs locally at the vitric-zeolitic contact in the CHn. It is surmised that the sloping alteration boundary promotes lateral flow within the perched layers in which transport velocities may be quite high. This mechanism functions similarly in the vitric-zeolitic contacts within the Prow Pass Tuff.

The role of faults in the deep unsaturated zone is not yet fully understood. In the shallow unsaturated zone, faults are highly permeable to air flow and may be major conduits for rapid water flow in the PTn. Fast flow in the TSw may be more dispersed because of its high fracture permeability. Fast-flow pathways persist through the PTn where faults disrupt the generally matrix-flow-dominated nonwelded tuffs, but water moving along such paths probably is only a small fraction of the total flux because of matrix imbibition. Fault zones may have the capacity to conduct substantial volumes of water through the entire unsaturated zone to the water table; but without a supply from lateral flow, major faults may conduct little water. Faults may be locally impermeable to lateral flow, resulting in perching, where the fault has offset a permeable bed opposite a less permeable bed. In general, faults are vertical conduits for both air and water flow.

SUMMARY

This paper has described and traced the evolution of the conceptual model of hydrology for the unsaturated zone at Yucca Mountain. The study area is 145 km northwest of Las Vegas, Nevada, and covers approximately 45 km^2. The unsaturated zone is 500 m thick with the potential repository located 300 m below land surface. Precipitation ranges from less than 130 mm/yr at the lower elevations in the south to more than 280 mm/yr at the higher elevations in the north and averages 170 mm/yr across the potential repository. Yucca Mountain is in the Basin and Range physiographic province and consists of a Tertiary volcanic sequence that varies between 1 and 3 km in thickness with an unsaturated zone that varies from 500 m in thickness near the potential repository to more than 750 m thick in the north part of the study area. The sequence consists of alternating layers of welded and nonwelded tuffs ranging in age from approximately 11.6 million years to 13.5 million years. The layers have been downcut in fault-controlled washes and in erosion-controlled washes forming on uplifted fault blocks.

During the last 15 years, several iterations of conceptual models for Yucca Mountain have been developed. Most of the models have the same general nature but differ significantly in detail. The most persistent concepts assumed negligible infiltration rates, extensive lateral flow in the PTn, and virtually no fracture flow through the TSw (the potential repository horizon), resulting in extremely low vertical percolation rates and long (10^5 yr) travel times. Most of the conceptual models allowed some lateral flow in the underlying CHn for various reasons. The current conceptual model has relatively high infiltration rates (as much as 80 mm/ yr in some locations), fracture-dominated flow in the TCw, vertical matrix-dominated flow in the PTn (little lateral flow), fracture-dominated flow in the TSw, vertical matrix-dominated flow in the vitric rocks of the Calico Hills and Prow Pass, and extensive lateral flow above the zeolitic boundary in those units, all of which lead to much shorter ($<10^4$ yr) travel times through the system.

The faults are assumed to be permeable to water and air and to transmit bomb-pulse isotopes (fast-flow paths) through continuous, connected, fractured paths through the PTn, in which flow is otherwise slow because it is matrix-dominated. Once through the PTn, the water can continue in the fault zone or move into the fracture-dominated flow field of the TSw. Perched water bodies form when lateral flow along a zeolitic contact reaches a less permeable rock unit that has been offset by faulting. The fault will act as a conduit to slowly drain the perched water body in equilibrium with the large area inflow.

This paper has presented the most recent conceptual model of the unsaturated zone at Yucca Mountain but is by no means the final word. During the next several years as underground testing continues and more data are collected, analysis of and insights into the mechanisms of unsaturated flow likely will lead to a better understanding of the unsaturated zone at Yucca Mountain.

REFERENCES

Altman, S. J., B. W. Arnold, R. W. Barnard, G. E. Barr, C. K. Ho, S. A. McKenna, and R. R. Eaton, 1996. Flow Calculations for Yucca Mountain Groundwater Travel Time (GWTT-95). Albuquerque, N. Mex.: Sandia National Laboratories. SAND96-0819.

Anderson, L. A., 1981. Rock Property Analysis of Core Samples from the Calico Hills UE25a-3 Borehole, Nevada Test Site, Nevada. Denver, Colo.: U.S. Geological Survey Open-File Report 81-1337. 30 p.

Barnard, R. W., M. L. Wilson, H. A. Dockery, J. H. Gauthier, P. G. Kaplan, R. R. Eaton, F. W. Bingham, and T. H. Robey, 1992. TSPA 1991: An Initial Total-System Performance Assessment for Yucca Mountain. Albuquerque, N. Mex.: Sandia National Laboratories. SAND91-2795.

Bodvarsson, G. S., and T. M. Bandurraga, eds., 1996. Development and Calibration of the Three-Dimensional Site-Scale Unsaturated-Zone Model of Yucca Mountain, Nevada-Yucca Mountain Project Milestone OBO2. Berkeley, Calif.: Lawrence Berkeley National Laboratory Report LBNL-39315.

Bodvarsson, G. S., E. Sonnenthal, and Y-S. Wu, eds., 1998. Unsaturated Zone Flow and Transport Modeling of Yucca Mountain, FY98. Berkeley, Calif.: Lawrence Berkeley National Laboratory Milestone Report SP3CKJM4.

Botkin, D. B., R. A. Nisbet, S. Bicknell, C. Woodhouse, B. Bentley, and W. Ferren, 1991. Global climate change and California's natural ecosystems. In: Knox, J. B., and A. F. Scheuring, eds. Global Climate Change and California, Potential Impacts and Responses: Berkeley, Calif., University of California Press. p. 123-146.

Buesch, D. C., R. W. Spengler, T. C. Moyer, and J. K. Geslin, 1996. Proposed Stratigraphic Nomenclature and Macroscopic Identification of Lithostratigraphic Units of the Paintbrush Group Exposed at Yucca Mountain. Nevada: U.S. Geological Survey Open-File Report 94-469. 47 p.

Burger, P. A., and K. M. Scofield, 1994. Perched water occurrences at Yucca Mountain and their implications on the exploratory studies facility. Unpublished memorandum to the USGS Technical Project Officer.

Byers, F. M., Jr., W. J. Carr, P. P. Orkild, W. D. Quinlivan, and K. A. Sargent, 1976. Volcanic Suites and Related Cauldrons of the Timber Mountain-Oasis Valley Caldera Complex, Southern Nevada: U.S. Geological Survey Professional Paper 919. 70 p.

Carr, W. J., F. M. Byers, Jr., and P. P. Orkild, 1986. Stratigraphic and Volcano-Tectonic Relations of the Crater Flat Tuff and Some Older Volcanic Units, Nye County, Nevada: U.S. Geological Survey Professional Paper 1323. 28 p.

Fabryka-Martin, J. T., A. L. Flint, D. S. Sweetkind, A. V. Wolfsberg, S. S. Levy, G. J. C. Roemer, J. L. Roach, L. E. Wolfsberg, and M. C. Duff, 1997a. Evaluation of Flow and Transport Models of Yucca Mountain, Based on Chlorine-36 and Chloride Studies for FY97. Los Alamos, N. Mex.: Los Alamos National Laboratory, Yucca Mountain Project Milestone Report SP2224M3.

Fabryka-Martin, J. T., and B. Liu, 1995. Distribution of Chlorine-36 in UZ-14, UZ-16, Perched Water, and the ESF North Ramp, Yucca Mountain, Nevada. Los Alamos, N. Mex.: Los Alamos National Laboratory Milestone Report 3431.

Fabryka-Martin, J. T., S. J. Wightman, W. J. Murphy, M. P. Wickham, M. W. Caffee, G. J. Nimz, J. R. Southon, and P. Sharma, 1993. Distribution of chlorine-36 in the unsaturated zone at Yucca Mountain: an indicator of fast transport paths. In: FOCUS '93, Site Characterization and Model Validation, Las Vegas, Nev., Sept. 26-29, 1993. Proceedings. La Grange Park, Ill.: American Nuclear Society, pp. 58-68.

Fabryka-Martin, J. T., S. J. Wightman, B. A. Robinson, and E. W. Vestal, 1994. Infiltration Processes at Yucca Mountain Inferred from Chloride and Chlorine-36 Distributions. Los Alamos, N. Mex.: Los Alamos National Laboratory. Milestone Report 3417.

Fabryka-Martin, J. T., A. V. Wolfsberg, P. R. Dixon, S. Levy, J. Musgrave, and H. J. Turin, 1997b. Summary report of Chlorine-36 Studies: Sampling for Chlorine-36 in the Exploratory Studies Facility. Los Alamos, N. Mex.: Los Alamos National Laboratory. Report LA-13552-MS.

Fabryka-Martin, J. T., A. V. Wolfsberg, S. S. Levy, K. Campbell, P. Tseng, J. L. Roach, and L. E. Wolfsberg, 1998. Evaluation of Flow and Transport Models of Yucca Mountain, Based on Chlorine-36 and Chloride Studies for FY98. Los Alamos, N. Mex.: Los Alamos National Laboratory. Yucca Mountain Project Milestone Report SP32D5M3.

Flint, A. L., and L. E. Flint, 1994. Spatial Distribution of Potential Near Surface Moisture Flux at Yucca Mountain, Nevada. In: Proceedings of the International High-Level Radioactive Waste Conference, Las Vegas, Nev., May 22-26, 1994. La Grange Park, Ill.: American Nuclear Society, pp. 2352-2358.

Flint, A. L., J. A. Hevesi, and L. E. Flint, 1996a. Conceptual and Numerical Model of Infiltration for the Yucca Mountain Area, Nevada. Las Vegas, Nev.: U.S. Department of Energy Milestone Report 3GUI623M, September 1996, 210 p.

Flint, L. E., 1998. Characterization of Hydrogeologic Units Using Matrix Properties. U.S. Geological Survey Water-Resources Investigation Report 96-4342, 61 p.

Flint, L. E., and A. L. Flint, 1990. Preliminary Permeability and Moisture Retention of Nonwelded and Bedded Tuffs at Yucca Mountain, Nevada. U.S. Geological Survey Open-File Report 90-569, 57 p.

Flint, L. E., and A. L. Flint, 1995. Shallow Infiltration Processes at Yucca Mountain, Nevada. Neutron Logging Data 1984-93. U.S. Geological Survey Water-Resources Investigations Report 95-4035, 46 p.

Flint, L. E., A. L. Flint, and J. A. Hevesi, 1993. Shallow infiltration processes in arid watersheds at Yucca Mountain, Nevada,. In: Fifth International High Level Radioactive Waste Management Conference, Las Vegas, Nev., May 22-26. Proceedings. La Grange Park, Ill.: American Nuclear Society, pp. 2315-2322.

Flint, L. E., A. L. Flint, C. A. Rautman, and J. D. Istok, 1996b. Physical and Hydrologic Properties of Rock Outcrop Samples at Yucca Mountain, Nevada. U.S. Geological Survey Open-File Report 95-280, 70 p.

Flint, A. L., L. E. Flint, G. S. Bodvarsson, E. M. Kwicklis, and J. T. Fabryka-Martin, in press. The hydrology of Yucca Mountain. Reviews of Geophysics.

Gauthier, J. H., M. L. Wilson, and F. C. Lauffer, 1992. Estimating the consequences of significant fracture flow at Yucca Mountain. In: Third International Conference on High-Level Radioactive Waste Management, Las Vegas, Nev. Proceedings. La Grange Park, Ill.: American Nuclear Society, pp. 727-731.

Glass, R. J., 1993. Modeling gravity-driven fingering in rough-walled fractures using modified percolation theory. In: High Level Radioactive Waste Management Conference, Las Vegas, Nev. Proceedings. La Grange, Ill.: American Nuclear Society, v. 2, pp. 2042-2049.

Glass, R. J., A. L. Flint, V. C. Tidwell, W. Peplinski, Y. Castro, 1994. Fracture-matrix interaction in Topopah Spring tuff: Experiment and numerical simulation. In: International High-Level Radioactive Waste Conference, Las Vegas, Nev., May 22-26. Proceedings. La Grange Park, Ill.: American Nuclear Society, 9 p.

Grayson, D.K., 1993. The Desert's Past: A Natural Prehistory of the Great Basin. Washington, D.C.: Smithsonian Institution Press, 356 p.

Hammermeister, D. P., D. O. Blout, and J. C. McDaniel, 1985. Drilling and coring methods that minimize the disturbance of cuttings, core and rock formation in the unsaturated zone, Yucca Mountain, Nevada, in National Water Well Association Conference on Characterization and Monitoring of the Vadose (Unsaturated) Zone. Proceedings. Denver, Colo.: National Water Well Association, p. 507-541.

Hanks, T. C., I. C. Winograd, R. E. Anderson, T. E. Reilly, and E. P. Weeks, 1999. Yucca Mountain as a radioactive-waste repository at Yucca Mountain. A report to the Director: U.S. Geological Survey Circular 1184, 19 p.

Hevesi, J. A., and A. L. Flint, 1996. Geostatistical Model for Estimating Precipitation and Recharge in the Yucca Mountain Region, Nevada-California. U.S. Geological Survey Water-Resources Investigations Report 96-4123.

Hevesi, J. A., A. L. Flint, and J. D. Istok, 1992. Precipitation estimation in mountainous terrain using multivariate geostatistics. II. Isohyetal maps. Journal of Applied Meteorology 31(7): 677-688.

Ho, C. K., 1995. Assessing alternative conceptual models of fracture flow. In: TOUGH Workshop '95, K. Pruess, ed. Proceedings. Berkeley, Calif.: Lawrence Berkeley National Laboratory, Lawrence Berkeley Laboratory Report LBL-37200.

Ho, C. K., 1997. Models of fracture-matrix interactions during multiphase heat and mass flow in unsaturated fractured porous media. In: Sixth Symposium on Multiphase Transport in Porous Media. Dallas, Tex. 1997 ASME International Mechanical Engineering Congress and Exposition.

Ho, C. K., S. J. Altman, and B. W. Arnold, 1995. Alternative Conceptual Models and Codes for Unsaturated Flow in Fractured Tuff: Preliminary Assessments for GWTT-95, Yucca Mountain Site Characterization Project Report. Albuquerque, N. Mex.: Sandia National Laboratories, SAND95-1456.

Hudson, D. B., and A. L. Flint, 1995. Estimation of Shallow Infiltration and Presence of Fast Pathways for Shallow Infiltration in the Yucca Mountain Area, Nevada. Las Vegas, Nev.: U.S. Department of Energy Milestone Report 3GUI611M.

Istok, J. D., C. A. Rautman, L. E. Flint, and A. L. Flint, 1994. Spatial variability in hydrologic properties of a volcanic tuff. Groundwater 32: 751-760.

Klavetter, E. A., and R. R. Peters, 1986. Estimation of Hydrologic Properties of an Unsaturated Fractured Rock Mass. Albuquerque, N. Mex.: Sandia National Laboratories, SAND84-2642.

Klavetter, E. A., and R. R. Peters, 1987. An Evaluation of the Use of Mercury Porosimetry in Calculating Hydrologic Properties of Tuffs from Yucca Mountain, Nevada. Albuquerque, N. Mex.: Sandia National Laboratories, SAND86-0286, 74 p.

Kwicklis, E. M., and J. P. Rousseau, 1999. Analysis of percolation flux based on heat flow estimated in boreholes. In: Rousseau, J. P., E. M. Kwicklis, and D. C. Gillies, eds. Hydrogeology of the Unsaturated Zone, North Ramp Area of the Exploratory Studies Facility. Yucca Mountain, Nevada: U.S. Geological Survey Water-Resources Investigations Report 98-4050, 244 p.

LeCain, G. D., 1997. Air-Injection Testing in Vertical Boreholes in Welded and Nonwelded Tuff. Yucca Mountain, Nevada: U.S. Geological Survey Water-Resources Investigations Report 96-4262, 33 p.

Maxey, G. B., and T. E. Eakin, 1950. Ground Water in White River Valley, White Pine, Nye, and Lincoln Counties. Nevada. Nevada State Engineer, Water Resources Bulletin (8):59 p.

Montazer, P., and W. E. Wilson, 1984. Conceptual Hydrologic Model of Flow in the Unsaturated Zone, Yucca Mountain, Nevada. U.S. Geological Survey Water-Resources Investigation Report 84-4345, 55 p.

Moyer, T. C., and J. K. Geslin, 1995. Lithostratigraphy of the Calico Hills Formation and Prow Pass Tuff (Crater Flat Group) at Yucca Mountain, Nevada. U.S. Geological Survey Open-File Report 94-460, 59 p.

Moyer, T. C., J. K. Geslin, and L. E. Flint, 1996. Stratigraphic Relations and Hydrologic Properties of the Paintbrush Tuff Nonwelded (PTn) Hydrologic Unit, Yucca Mountain, Nevada. U.S. Geological Survey Open-File Report 95-397, 151 p.

Paces, J. B., L. A. Neymark, B. D. Marshall, J. F. Whelan, and Z. E. Peterman, 1996. Ages and Origins of Subsurface Secondary Minerals in the Exploratory Studies Facility (ESF). U.S. Geological Survey Milestone Report 3GQH450M.

Patterson, G. L., 1999. Occurrences of perched water in the vicinity of the Exploratory Studies Facility North Ramp. In: Rousseau, J. P., E. M. Kwicklis, and D. C. Gillies, eds. Hydrogeology of the unsaturated zone, North Ramp area of the Exploratory Studies Facility, Yucca Mountain, Nevada. U.S. Geological Survey Water-Resources Investigations Report 98-4050, Denver, Colo., 244 p.

Peters, R. R., and E. A. Klavetter, 1988. A continuum model for water movement in an unsaturated fractured rock mass. Water Resources Research 24(3):416-430.

Peters, R. R., E. A. Klavetter, I. J. Hall, S. C. Blair, P. R. Heller, and G. W. Gee, 1984. Fracture and Matrix Hydrogeologic Characteristics of Tuffaceous Materials from Yucca Mountain, Nye County, Nevada. Albuquerque, N. Mex.: Sandia National Laboratories, SAND84-1471, 108 p.

Pruess, K., 1998. On water seepage and fast preferential flow in heterogeneous, unsaturated rock fractures. Journal of Contaminant Hydrology 30(3-4):333-362.

Pruess, K., and Y. W. Tsang, 1990. On two-phase relative permeability and capillary pressure of rough-walled rock fractures. Water Resources Research 26(9):1915-1926.

Robinson, B. A., A. V. Wolfsberg, G. A. Zyvoloski, and C. W. Gable, 1995. An Unsaturated Zone Flow and Transport Model of Yucca Mountain. Los Alamos National Laboratory Yucca Mountain Project Milestone 3468.

Rockhold, M. L., B. Sagar, and M. P. Connelly, 1990. Multi-dimensional modeling of unsaturated flow in the vicinity of exploratory shafts and fault zones at Yucca Mountain, Nevada. In: First International High Level Radioactive Waste Management Conference, Las Vegas, Nev. Proceedings. La Grange Park, Ill.: American Nuclear Society, p. 153-162.

Roseboom, E. H., Jr., 1983. Disposal of high-level nuclear waste above the water table in arid regions. Alexandria, Va.: Geological Survey Circular 903, 21 p.

Ross, B., 1990. The diversion capacity of capillary barriers. Water Resources Research 26(10): 2625-2629.

Rousseau, J. P., E. M. Kwicklis, and D. C. Gillies, eds., 1998. Hydrogeology of the Unsaturated Zone, North Ramp Area of the Exploratory Studies Facility, Yucca Mountain, Nevada. U.S. Geological Water-Resources Investigations Report 98-4050, 244 p.

Rulon, J. J., G. S. Bodvarsson, and P. Montazer, 1986. Preliminary Numerical Simulations of Groundwater Flow in the Unsaturated Zone, Yucca Mountain, Nevada. Berkeley, Calif.: Lawrence Berkeley Laboratory, LBL 20553, 91 p.

Rush, F. E., W. Thordarson, and L. Bruckheimer, 1983. Geohydrologic and Drill-Hole Data for Test Well USW H-1, Adjacent to the Nevada Test Site, Nye County, Nevada. U.S. Geological Survey Open-File Report 83-141, 68 p.

Sass, J. H., and A. H. Lachenbruch, 1982. Preliminary Interpretation of Thermal Data from the Nevada Test Site. U.S. Geological Survey Open-File Report 82-973, 30 p.

Sawyer, D. A., R. J. Fleck, M. A. Lanphere, R. G. Warren, and D. E. Broxton, 1994. Episodic volcanism in the Miocene southwest Nevada volcanic field: Stratigraphic revisions, $^{40}Ar/^{39}Ar$ geochronologic framework, and implications for magmatic evolution. Geological Society of America Bulletin 106(10): 1304-1318.

Scott, R. B., and J. Bonk, 1984. Preliminary Geologic Map of Yucca Mountain with Geologic Sections, Nye County, Nevada. U.S. Geological Survey Open-File Report 84-494.

Scott, R. B., R. W. Spengler, S. Diehl, A. R. Lappin, and M. P. Chornack, 1983. Geologic character of tuffs in the unsaturated zone at Yucca Mountain, southern Nevada. In: Mercer, J. W., P. S. C. Rao, and I. W. Marine, eds. Role of the Unsaturated Zone in Radioactive and Hazardous Waste Disposal. Ann Arbor, Mich.: Ann Arbor Science, p. 289-335.

Sinnock, S., Y. T. Lin, and J. P. Brannen, 1984. Preliminary Bounds on the Expected Post-Closure Performance of the Yucca Mountain Repository Site, Southern Nevada. Albuquerque, N. Mex.: Sandia National Laboratories, SAND84-3918, 83 p.

Sinnock, S., Y. T. Lin. and J. P. Brannen, 1987. Preliminary bounds on the expected post-closure performance of the Yucca Mountain repository site, southern Nevada. Journal of Geophysical Research 92(B8): 7820-7842.

Snyder, D. B., and W. J. Carr, 1982. Preliminary Results of Gravity Investigations at Yucca Mountain and Vicinity, Southern Nye County, Nevada. U.S. Geological Survey Open-File Report 82-701, 36 p.

Striffler, P., G. O'Brien, T. Oliver, and P. A. Burger, 1996. Perched water characteristics and occurrences, Yucca Mountain, Nevada. Unpublished memorandum to the U.S. Geological Survey Yucca Mountain Project Technical Project Officer.

Sweetkind, D. S., J. T. Fabryka-Martin, A. L. Flint, C. J. Potter, and S. S. Levy, 1997. Evaluation of the Structural Significance of Bomb Pulse ^{36}Cl at Sample Locations in the Exploratory Studies Facility, Yucca Mountain, Nevada. Las Vegas, Nev.: U.S. Department of Energy Milestone Report SPG33M4.

Tucci, P., and D. J. Burkhardt, 1995. Potentiometric-Surface Map, 1993, Yucca Mountain and Vicinity, Nevada. U.S. Geological Survey Water-Resources Investigations Report 95-4149.

U.S. Department of Energy, 1984. Draft Environmental Assessment: Yucca Mountain Site, Nevada Research and Development Area, Nevada. Washington, D.C.: U.S. Department of Energy.

U.S. Department of Energy, 1992. Report of Early Site Suitability Evaluation of the Potential Repository Site at Yucca Mountain, Nevada. Washington, D.C.: U.S. Department of Energy.

Waddell, R. K., J. H. Robison, and R. K. Blankennagel, 1984. Hydrology of Yucca Mountain and Vicinity, Nevada-California: Investigative Results Through Mid-1983. U.S. Geological Survey Water-Resources Investigations Report 84-4267, 72 p.

Wang, J. S. Y., and T. N. Narasimhan, 1985. Hydrologic mechanisms governing fluid flow in a partially saturated, fractured, porous medium. Water Resources Research 21(12): 1861-1874.

Weeks, E. P., and W. E. Wilson, 1984. Preliminary Evaluation of Hydrologic Properties of Cores of Unsaturated Tuff, Test Well USW H-1, Yucca Mountain, Nevada. U.S. Geological Survey Water-Resources Investigations Report 84-4193, 30 p.

Whitfield, M. S., Jr., C. M. Cope, and C. L. Loskot, 1984. Borehole and Geohydrologic Data for Test Hole USW UZ-6, Yucca Mountain Area, Nye County, Nevada. U.S. Geological Survey Open-File Report 92-28, 41 p.

Wilson, M. L., 1996. Lateral diversion in the PTn unit: Capillary-barrier analysis. In: High Level Radioactive Waste Management Seventh Annual International Conference, Las Vegas, Nev., April 29-May 3. Proceedings. La Grange Park, Ill.: American Nuclear Society, SAND95-2186C, p. 111-113.

Winograd, I. J., 1971. Hydrogeology of ash-flow tuff: A preliminary statement. Water Resources Research 7(4): 994-1006.

Winograd, I. J., 1972. Near-surface storage of solidified high-level radioactive waste in thick (400-2,000 foot) unsaturated zones in the southwest. Geological Society of America, Abstracts with Programs 4: 708.

Winograd, I. J., 1974. Radioactive waste storage in the arid zone. EOS, Transactions of the American Geophysical Union 55: 884-894.

Winograd, I. J., 1981. Radioactive waste disposal in thick unsaturated zones. Science 212(4502): 1457-1464.

Winograd, I. J., and W. Thordarson, 1975. Hydrogeologic and Hydrochemical Framework, South-Central Great Basin, Nevada-California, With Special Reference to the Nevada Test Site. U.S. Geological Survey Professional Paper 712-C, 126 p.

Wittwer, C. S., G. S. Bodvarsson, M. P. Chornack, A. L. Flint, L. E. Flint, B. D. Lewis, R. W. Spengler, and C. A. Rautman, 1992. Design of a three-dimensional site-scale model for the unsaturated zone at Yucca Mountain, Nevada. In: International High-Level Radioactive Waste Conference, Las Vegas, Nev., April 12-16. Proceedings. La Grange Park, Ill.: American Nuclear Society, p. 263-271.

Wittwer, C. S., G. Chen, G. S. Bodvarsson, M. P. Chornack, A. L. Flint, L. E. Flint, E. M. Kwicklis, and R. W. Spengler, 1995. Preliminary Development of the LBL/USGS Three-Dimensional Site-Scale Model of Yucca Mountain, Nevada. Berkeley, Calif.: Lawrence Berkeley National Laboratory, LBL-37356, UC0814.

Wolfsberg, A. V., J. T. Fabryka-Martin, K. S. Campbell, S. S. Levy, and P. H. Tseng, 1999. Use of chlorine-36 and chloride data to evaluate fracture flow and transport models at Yucca Mountain. In: International Symposium: Dynamics of Fluids in Fractured Rocks: Concepts and Recent Advances, in Honor of Paul A. Witherspoon, February 10-12, 1999. Proceedings. Berkeley, Calif.: Lawrence Berkeley National Laboratory.

Wu, Y. S., S. Finsterle, and K. Pruess, 1996. Computer models and their development for the unsaturated zone model at Yucca Mountain, Chapter 4 in Development and Calibration of the Three-Dimensional Site-Scale Unsaturated-Zone Model of Yucca Mountain, Nevada. G. S. Bodvarsson and T. M. Bandurraga, eds. Berkeley, Calif.: Lawrence Berkeley National Laboratory Milestone OBO2, MOY-970317-04.

Yang, I. C., G. W. Rattray, and P. Yu, 1996. Interpretations of Chemical and Isotopic Data from Boreholes in the Unsaturated-Zone at Yucca Mountain, Nevada. U.S. Geological Survey Water-Resources Investigations Report 96-4058.

Yang, I. C., P. Yu, G. W. Rattray, and D. C. Thorstenson, 1998. Hydrogeochemical Investigations and Geochemical Modeling in Characterizing the Unsaturated Zone at Yucca Mountain, Nevada. U.S. Geological Survey Water-Resources Investigation Report 98-4132.

3

Conceptual Model of Vadose-Zone Transport in Fractured Weathered Shales

P.M. Jardine,[1] G.V. Wilson,[2] R.J. Luxmoore,[1] and J.P. Gwo[3]

ABSTRACT

The mobility of water and solutes in the vadose zone occurs through a complex continuum of pores that vary in size and shape. Fractured subsurface media that have weathered often depict the extreme case of pore-class heterogeneity because highly conductive voids surround low-permeability, high-porosity matrix blocks. In this chapter, the physical and chemical processes controlling water and solute transport in fractured, weathered shales are discussed. At the Oak Ridge National Laboratory (ORNL), the weathered shales are commonplace and are characterized by a highly interconnected fracture network with densities of 100-200 fractures per meter. The media are conducive to extreme preferential flow that results in physical, hydraulic, and geochemical nonequilibrium conditions between fractures and the surrounding soil matrix. This scenario is of significance with regard to contaminant fate and transport issues at ORNL, where the subsurface burial of toxic metals and radionuclides has occurred. A multiregion flow and transport concept has been adopted to describe the movement of water and solutes through the weathered shale. A variety of multiscale experimental and numerical endeavors have been undertaken to justify the multiregion concept. Experimental manipulations at the laboratory, intermediate, and field scales are designed to quantify the rates and mechanisms of intra- and inter-region mass transfer. These experimental techniques

---

[1] Environmental Sciences Division, Oak Ridge National Laboratory, Oak Ridge, Tennessee.
[2] Southern Nevada Science Center, Desert Research Institute, Las Vegas, Nevada.
[3] Center for Computational Sciences, Oak Ridge National Laboratory, Oak Ridge, Tennessee.

have significantly improved our conceptual understanding of time-dependent solute migration in fractured subsurface media and have also provided the necessary experimental constraints needed for accurate numerical quantification of the physical or geochemical nonequilibrium processes that control solute migration.

INTRODUCTION

Inceptisols, common to the southeastern United States, are weakly developed soils that have weathered from interbedded shale-limestone parent material. At the Oak Ridge National Laboratory (ORNL) in eastern Tennessee, the Department of Energy (DOE) has historically used these soils for the disposal of low-level radioactive waste in shallow-land burial trenches. Physical and chemical barriers were seldom used to impede waste migration since it was thought that the high cation exchange capacity (CEC) of the media (CEC ~20 cmol/kg) would significantly retard radionuclide mobility from the primary trench sources. Although the geochemistry of the media is sufficient for slowing radionuclide migration, the physical structure of the inceptisols coupled with large annual rainfall inputs (~1400 mm/y) have resulted in the formation of a large secondary contaminant source where radionuclides have been disseminated across a vast subsurface environment. The basic problem is that the soils are highly structured and conducive to rapid preferential flow coupled with significant matrix storage. This circumstance causes large physical and geochemical gradients between the various flow regimes of the media and typically results in nonequilibrium conditions during solute transport.

Because transport processes contributing to the formation of secondary contaminant sources are ill-defined, it is difficult to accurately assess the risk associated with the off-site migration of the contaminants and thus the need for remediation. To help resolve this dilemma, investigations were conducted to provide an improved conceptual understanding and predictive capability of solute transport processes in highly structured, heterogeneous subsurface environments that are complicated by fracture flow and matrix diffusion. The investigation involves multiscale experimental and numerical approaches to address coupled hydrological and geochemical processes controlling the fate and transport of contaminants in fractured vadose zone saprolites. Novel tracer techniques and experimental manipulation strategies using laboratory-, intermediate-, and field-scale experiments helped unravel how coupled transport processes affect the nature and extent of secondary contaminant sources. In so doing, a multiregion flow and transport concept is developed and numerically implemented.

SELECT CHARACTERISTICS OF WEATHERED, FRACTURED SHALE

The subsurface materials used in the disposal of low-level radioactive waste at ORNL are acidic inceptisols that have been weathered from interbedded shale-limestone sequences within the Conasauga Formation. The limestone has weathered to massive clay lenses devoid of carbonate, and the more resistant shale has weathered to an extensively fractured saprolite. Fractures are highly interconnected with densities in the range of 200 fractures per meter (Dreier et al., 1987). Fold-related fractures in the saprolite are observable in the field and consist of (a) fractures along bedding planes, (b) two sets of orthogonal extensional fractures that are perpendicular to bedding planes, and (c) shear fractures. The extensional fractures are either parallel or perpendicular to the strike of the bedding planes and form an orthogonal fracture network with the bedding-plane fractures. Bedding-plane fractures dominate the fracture network of the media and have aperture spacing of <0.05 mm. Cross-cutting extensional fractures are less numerous but have larger apertures ranging from 0.2-0.5 mm (G. R. Moline, pers. comm.). Many of the fracture surfaces contain secondary deposits (e.g., clays, Mn and Fe oxides) suggesting that they are hydrologically active. Fracture orientation and connectivity can give rise to extensive preferential flow within the media (Solomon et al., 1992). Fractures surround low-permeability, high-porosity matrix blocks that have water contents ranging from 30 to 50 percent. Functional relationships between water content (θ), pressure head (h), and hydraulic conductivity (K) suggest that the soil behaves as a three-region media consisting of macropores, mesopores, and micropores (Wilson et al., 1992). Macropores and mesopores are conceptualized as primary and secondary fractures, and micropores as the soil matrix. The three-region conceptualization of the fractured, weathered shales is supported by field measurements of Wilson and Luxmoore (1988) that showed mesopore convergent flow into macropores with subsequent bypass of the soil matrix. The occurrence of preferential flow in these media is common due to large annual rainfall inputs—as much as 50 percent of the infiltrating precipitation results in groundwater and surface water recharge (10 and 40 percent, respectively). The condition of variably saturated preferential flow suggests that hydraulic nonequilibrium (i.e., hydraulic gradient between flow regions) and physical nonequilibrium (i.e., concentration gradient) exist during storm-driven solute transport.

The fractured, weathered shales are slightly acidic (e.g., pH = 4.5 to 6.0) with typical cation exchange capacities of 15-20 cmol/kg. The <2 μm clay fraction is predominately illite with lesser quantities of 2:1 interstratified material and vermiculite. Carbonates are completely weathered from the upper several meters of soil, and most of the fracture pathways and matrix blocks are coated with amorphous Fe- and Mn-oxides. The diverse mineralogy results in a highly reactive solid phase that can significantly alter the geochemical behavior and transport of solutes. In an effort to enhance our conceptual understanding of and predictive capability for solute transport processes in these unsaturated fractured soils, a multiscale experimental and numerical approach was used to quantify the rates and mechanisms of pore class interactions.

LABORATORY-SCALE ASSESSMENT OF TRANSPORT IN FRACTURED WEATHERED SHALES

Undisturbed columns (typically 15 cm diameter × 40 cm length) are used in tracer transport experiments to assess the interaction of hydrology, geochemistry, and microbiology on the fate and transport of nonreactive and reactive solutes. The primary purpose of research at this scale was to quantify transport mechanisms that are operative at the field scale, but difficult to quantify at these larger scales. At the column scale, several techniques were employed to assess nonequilibrium processes that result from the large difference in hydraulic conductivity of fractures versus matrix blocks. The techniques include (1) controlling flow-path dynamics with manipulations of pore-water flux and soil-water tension, (2) isolating diffusion and slow geochemical processes with flow interruption, (3) using multiple tracers with different diffusion coefficients, and (4) using multiple tracers with grossly different sizes.

Controlling Flow-Path Dynamics

Variations in Pore-Water Flux

A relatively simple technique for confirming and quantifying physical non-equilibrium in soil systems involves displacement experiments performed at a variety of experimental fluxes using a single representative tracer. Alteration of the experimental flux or specific discharge through a soil system perturbs the rate of approach toward equilibrium by changing the hydraulic or concentration gradient. In heterogeneous systems that exhibit a large distribution of pore sizes, an increase in the overall pore-water flux should result in greater system non-equilibrium due to a decrease in solute residence time within the porous media. This is usually the case, as solute movement into the matrix is a combination of advective and diffusive processes, and is typically the rate-limiting step as the system approaches equilibrium. This condition can be observed in Figure 3-1, which shows the breakthrough of a nonreactive Br⁻ tracer at several different pore-water fluxes through an undisturbed column of weathered, fractured shale. As is typical of heterogeneous media, tracer displacement was characterized by an initial rapid solute breakthrough followed by extended tailing to longer times. In this system, the fracture network of the weathered shale controlled the advective transport of solutes, which was coupled with diffusion into the surrounding matrix blocks. The largest and smallest flux experiments were conducted over periods of 0.25 days and 94 days, respectively, with the relative amount of tracer mass remaining in the column at the end of each pulse ranging from 22 percent (fast flux) to 38 percent (slow flux). These results indicate that the system became increasingly removed from equilibrium as the pore-water flux was increased. At

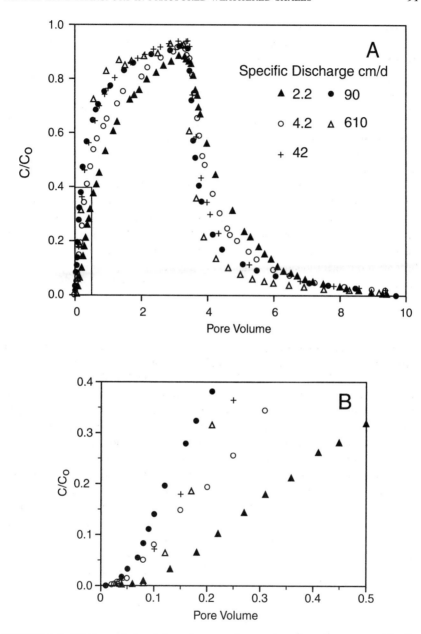

FIGURE 3-1 (A) Bromide breakthrough curves for a series of steady-state specific discharges in an undisturbed column of fractured, weathered shale. The largest and slowest flux experiments were conducted over periods of 6 and 2,200 h respectively. The rectangle in the lower corner outlines the expanded portion of the plot shown in (B). From R. O'Brien (1994, ORNL, unpublished data).

the conditions of faster flux, the tracer residence times in the mobile fracture regions were significantly decreased, and thus not as much mass was lost to the matrix. However, mass loss to the matrix as a function of pore-water flux was highly nonlinear, suggesting that the average rate of mass transfer from fractures to the matrix is greater at larger fluxes. The velocity dependence of mass-transfer processes in various porous media and soils has been shown by others (Akratanakul et al., 1983; Nkedi-Kizza et al., 1983: Jensen, 1984; Schulin et al., 1987; Anamosa et al., 1990; Kookana et al., 1993; Reedy et al., 1996).

Controlling flow-path dynamics through variations in pore-water flux is a relatively simple technique for assessing the significance of physical nonequilibrium in soil systems. However, when used by itself the technique is semiquantitative as it is difficult to know how system variables change in response to flux variations (e.g., are the proportions of advective flow paths constant with changes in flux?). When this technique is combined with the other manipulative experimental strategies (e.g. multiple tracers, flow interruption) described below, it can become a powerful means of quantifying physical nonequilibrium processes in structured media (see, for example, Hu and Brusseau, 1995; Reedy et al., 1996).

Variations in Pressure Head

Controlling flow-path dynamics by manipulation of the soil water content with pressure-head variations is an excellent technique to assess nonequilibrium processes (Seyfried and Rao, 1987; Jardine et al., 1993a). The basic concept of the technique is to collect water and solutes from select sets of pore classes in order to determine how each set contributes to the bulk flow and transport processes that are observed for the whole system. In heterogeneous systems, a decrease in pressure head (more negative) will cause larger pores, such as fractures, to drain and become nonconductive during solute transport. Because advective flow processes tend to dominate in large pore regimes, a decrease in pressure head, which will restrict flow and transport to smaller pores, will limit the disparity of solute concentrations among pore groups. By minimizing the concentration gradient in the system, the extent of physical nonequilibrium is decreased. Figure 3-2 conveys this concept by showing the breakthrough curves of a nonreactive Br⁻ tracer at three different pressure heads in an undisturbed column of the weathered, fractured saprolite from the Oak Ridge Reservation (ORR). The increasing asymmetry of the breakthrough curves with increasing saturation (less negative pressure head) is indicative of enhanced preferential flow coupled with mass loss into the matrix. As the soil becomes increasingly unsaturated, breakthrough curve tailing becomes less significant because of a decrease in the participation of larger pores (fractures) involved in the transport process. These findings suggest that mass-transfer limitations (nonequilibrium conditions) become less significant for these unsaturated conditions because fracture flow has been eliminated. An interesting finding of these studies was that the application

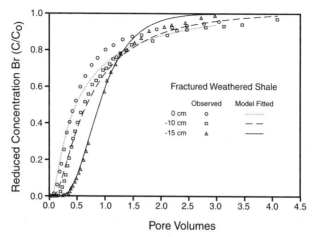

FIGURE 3-2 Breakthrough curves for a nonreactive Br⁻ tracer as a function of pressure head (h) in an undisturbed column of fractured, weathered shale. For conditions where h = 0 cm, transport occurred under saturated flow and the entire fracture network was conductive. When h = –10 cm the primary fracture network became nonconductive, and when h = –15 cm primary fractures and a portion of the secondary fractures became nonconductive. The model-fitted curves used the classical convective-dispersive model with optimization of the dispersion coefficient to the observed data. Modified from Jardine et al. (1993a), with permission.

of –10- and –15-cm pressure heads resulted in 5- and 40-fold decreases in the mean pore-water flux, respectively, with relatively little change in soil water content relative to saturated conditions (0.55 cm^3/cm^3 at h = 0 to 0.51 cm^3/cm^3 at h = –15 cm). This suggests that most of the water flux may be channeled through pores that hold water with tensions <10 cm (primary fractures, macropores) even though their surface area and contribution to the total system porosity is very small (Wilson and Luxmoore, 1988; Wilson et al., 1989). These results differ from the findings of Krupp and Elrick (1968), who investigated Cl⁻ breakthrough in variably saturated glass bead media. In this study, breakthrough asymmetry was more pronounced during unsaturated conditions relative to saturated flow. This observation resulted from the uniformity of the media (i.e., single pore class) and the use of a constant volumetric flow rate that caused a wide pore-water velocity distribution to develop during unsaturated flow (Wilson and Gelhar, 1974).

Reactive solute transport is also dramatically influenced by water content changes and pore regime connectivity. With enhanced preferential flow, important geochemical reactions such as sorption, oxidation/reduction, complexation/dissociation, and precipitation/dissolution become increasingly limited due to a decrease in residence time of pore water with the soil matrix. Jardine et al. (1988,

1993a, 1993b) found that the reactivity of divalent contaminants and chelated radionuclides increased dramatically with a slight decrease in pressure head from 0 to −10 cm. This was the result of a decrease in preferential bypass when fractures were empty and a greater contact time between the high surface area matrix and secondary fracture network.

Flow Interruption

Another useful technique for isolating diffusion or slow time-dependent geochemical reactions involves flow interruption during a portion of a tracer displacement experiment. The technique involves inhibiting the flow process during an experiment for a designated period of time and allowing a new physical or chemical equilibrium state to be approached. When physical nonequilibrium processes are significant in a soil system, the flow-interruption method will cause an observed concentration perturbation for a conservative tracer when flow is resumed. Interrupting flow during tracer injection will result in a decrease in tracer concentration when flow is resumed, whereas interrupting flow during tracer displacement (washout) will result in an increase in tracer concentration when flow is resumed. The concentration perturbations that are observed after flow interrupts are indicative of solute diffusion between pore regions of hetero-geneous media. Conditions of preferential flow create concentration gradients between pore domains (physical nonequilibrium), resulting in diffusive mass transfer between the regions. Therefore, during injection, tracer concentrations within advection-dominated flow paths (i.e., fractures, macropores) are higher than those within the matrix. Upon flow interruption, the relative concentration decrease that is observed indicates that solute diffusion is occurring from larger, more conductive pores into the smaller pores. During tracer displacement or washout, the concentrations within the preferred flow paths are lower than those within the matrix. Thus, solute diffusion is occurring from smaller pores into larger pores, and a concentration increase is observed when flow interruption has been imposed.

The utility of the flow-interrupt method for confirming and quantifying physical nonequilibrium can be observed in Figure 3-3, which shows Br^- breakthrough curves at two fluxes in an undisturbed column of fractured, weathered shale from the ORR. The observed concentration perturbations on the ascending and de-scending limbs of the breakthrough curves are the result of prolonged flow inter-rupt and the system approaching a new state of physical equilibrium. The concen-tration perturbations that are induced by flow interruption are significantly more pronounced at larger fluxes (Figure 3-3b). This is because the system is further removed from equilibrium at the larger fluxes, as a greater concentration gradient exists between advection-dominated flow paths and the soil matrix.

The flow-interrupt method has also been used during reactive tracer studies that focus on the determination of rate-limiting geochemical reactions (i.e., geo-

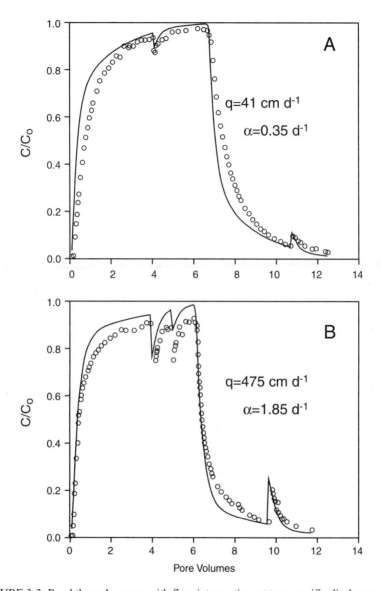

FIGURE 3-3 Breakthrough curves with flow interruption, at two specific discharges, for a nonreactive Br$^-$ tracer in an undisturbed column of fractured, weathered shale. Flow interruption was initiated for 7 days after (A) approximately 4 and 11 pore volumes of tracer were displaced at a flux of 41 cm/d, and (B) approximately 4, 5, and 10 pore volumes of tracer were displaced at a flux of 475 cm/d. The solid lines represent simulations using a two-region model with optimization of the mass transfer coefficient, α, that accounts for mass exchange between different pore regions. Modified from Reedy et al. (1996), with permission.

chemical nonequilibrium) such as sorption, precipitation, transformation, and complexation (Murali and Aylmore, 1980; Hutzler et al., 1986; Brusseau et al., 1989; Ma and Selim, 1994; Mayes et al., 2000). Mayes et al. (2000) provided the first application of the flow-interruption technique for quantifying geochemical nonequilibrium of reactive contaminant and chelated radionuclides in fractured, weathered shales. The flow-interruption process was very effective at decoupling rate-limiting redox and dissociation reactions versus time-dependent solid phase desorption reactions.

Multiple Tracers with Different Diffusion Coefficients

A powerful technique for quantifying physical nonequilibrium in structured subsurface media is the simultaneous use of multiple tracers with different diffusion coefficients. In general, when the technique is used to quantify physical nonequilibrium processes in soils and rock, two or more conservative tracers that have different diffusion coefficients are simultaneously displaced through the porous media. Tracers such as Br^-, Cl^-, fluorobenzoates, ^3H_2O, and, for water-saturated conditions, dissolved gases such as He, Ne, SF_6, and Kr, are frequently suitable for assessing physical nonequilibrium processes in structured media using the multiple-tracer technique (Carter et al., 1959; Raven et al., 1988; Bowman and Gibbens, 1992; Wilson and MacKay, 1993; Gupta et al., 1994; Jaynes, 1994; Linderfelt and Wilson, 1994; Clark et al., 1996; Sanford et al., 1996; Sanford and Solomon, 1998; Jardine et al., 1999). Colloidal tracers that are size-excluded from the matrix porosity of the media are not included in this type of experimental technique. When physical nonequilibrium processes are significant in porous media, tracers with larger molecular diffusion coefficients will be preferentially lost from advective flow paths (i.e., fractures) due to more rapid diffusion into the surrounding solid phase matrix. Likewise, tracers with smaller molecular diffusion coefficients (e.g., larger molecules) will remain in the advective flow paths for longer times due to slower diffusion into the matrix porosity. When advective processes are dominant in a system and matrix diffusion is negligible, multiple tracer breakthrough profiles will not differ considerably (see Brusseau, 1993, for an example).

The utility of multiple tracers for quantifying physical nonequilibrium processes in structured media can be observed in Figure 3-4, where the simultaneous transport of two nonreactive tracers, Br^- and pentafluorobenzoic acid (PFBA), was investigated in large undisturbed columns of fractured, weathered shale at two different pore-water fluxes. The molecular diffusion coefficient for PFBA is 40 percent smaller than the diffusion coefficient for Br^- (Bowman, 1984). Differences in the breakthrough curves for these solutes can be attributed to differences in the rates of tracer diffusion into the soil matrix. Because the PFBA diffused more slowly into the weathered shale matrix, its breakthrough at the column exit was initially more rapid than Br^- but required longer times to approach equilibrium ($C/C_0 = 1$) (Figure 3-4 a, b). Thus, Br⁻ had a larger mass loss to the matrix at any given time and exhibited a more retarded breakthrough relative to PFBA. However, Br⁻ will reach equilibrium more rapidly than PFBA and the tracer breakthrough curves will eventually cross at longer times. In contrast, the mobility of these two nonreactive tracers would be identical in a column of unstructured media because pore class heterogeneity would be minimal, thus limiting the significance of physical nonequilibrium during transport (see Brusseau, 1993, for an example).

Shropshire (1995) investigated the transport of an inorganic anion (Br⁻) and a dissolved gas tracer (He) in saturated columns of fractured weathered shale from the ORR. The breakthrough and displacement (washout) of He was significantly more sluggish relative to Br⁻. This resulted from the fact that the molecular diffusion coefficient for He was about three times greater than that for Br⁻. Thus, the He tracer was preferentially lost to the matrix relative to Br⁻. Using a discrete fracture flow model, Shropshire (1995) found that the multiple-tracer technique was an excellent method for approximating fracture aperture and network geometry in the weathered shale media.

Multiple Tracers with Grossly Different Sizes

Another sensitive technique for confirming and quantifying physical nonequilibrium in heterogeneous soil and rock systems is the use of multiple tracers with distinctly different sizes. Specifically, this technique uses both dissolved solutes and colloidal tracers so that flow-path accessibility can be controlled. Viruses, bacteria, fluorescent microspheres, DNA-labeled microspheres, radiolabeled Fe oxide particles, and synthetic polymers have all been used as colloidal tracers in various subsurface media (Barraclough and Nye, 1979; Gerba et al., 1981; Smith et al., 1985; Bales et al., 1989; Harvey et al., 1989, 1993, 1995; Toran and Palumbo, 1991; McKay et al., 1993a, 1993b; Hinsby et al., 1996; Reimus, 1996; Yang et al., 1996). Colloidal particles are typically large enough to be excluded from the matrix porosity of soils and geologic material. If they are not severely retarded by the porous media, colloidal particles serve as excellent tracers for quantifying advective flow velocities in systems conducive to preferential flow. When colloidal tracers are coupled with dissolved solutes that can interact with the matrix porosity, a unique technique emerges for assessing physical nonequilibrium processes in subsurface media.

The utility of using multiple tracers of different sizes can be seen in Figure 3-5, which shows the simultaneous injection of two strains of bacteriophage (PRD-1 and MS-2) and two dissolved solutes (Br⁻ and PFBA) into a column of fractured, weathered shale in order to assess physical nonequilibrium processes. The bacteriophage travel times were much more rapid than those of the dissolved solutes, and the bacteriophage strains exhibited significantly less total dispersion in their transport, as evidenced by steeper breakthrough characteristics relative to

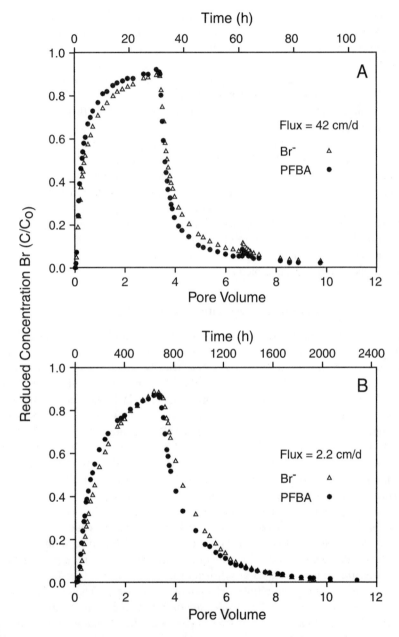

FIGURE 3-4 Breakthrough curves for the simultaneous injection of two nonreactive tracers, Br⁻ and PFBA, at a flux of (A) 42 cm/d and (B) 2.2 cm/d in an undisturbed column of fractured, weathered shale. The free water diffusion coefficient for Br⁻ is 40 percent larger than that of PFBA. From R. O'Brien (1994, ORNL, unpublished data).

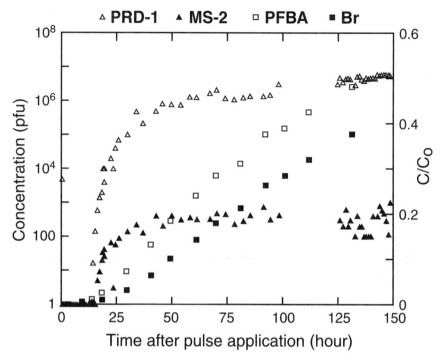

FIGURE 3-5 Effluent concentrations of two bacteriophage strains (PRD-1 and MS-2, with a mean size of 0.062 and 0.026 µm, respectively) and reduced concentrations of the dissolved tracers PFBA and Br⁻, which were simultaneously injected at 2.2 cm/d into an undisturbed column of fractured, weathered shale. Note the different concentration axes used for each of the two tracer types. From R. O'Brien (1994, ORNL, unpublished data).

PFBA and Br⁻. The larger bacteriophagies were preferentially transported through a smaller range of flow paths and were minimally affected by diffusion into the matrix porosity. The dissolved tracers, on the other hand, were influenced by diffusive mass-transfer processes between fractures and the matrix (Figure 3-5). Besides providing visual evidence of physical nonequilibrium processes in structured media, the experiments provided advective flow velocities that were used to parameterize numerical models designed to simulate the observed data.

Significant research regarding colloidal tracer transport in the fractured, weathered shales has been conducted by the research group led by Dr. Larry McKay at the University of Tennessee. This team of researchers has investigated the influence of coupled hydrologic and geochemical conditions on the fate and transport of bacteriophage and microspheres in undisturbed columns of the fractured saprolites. Studies by Harton (1996) have shown that the retention of bacteriophage to the solid phase is extremely sensitive to flow rate, where essentially no retention was observed at rapid pore-water fluxes and a gradient of 1.0, versus near-complete retention at slower fluxes and a gradient of 0.01. Cumbie and McKay (1999) further showed that for any given pore-water flux, the transport characteristics of microspheres in the fractured saprolites were optimized in the size range of 0.5-1.0 μm. Smaller particles down to 0.05 μm tended to adhere to the fracture walls and enter the soil matrix, and larger particles up to 4.25 μm tended to clog advective flow paths due to straining and gravitational settling. These results were corroborated by Haun (1998), who also found that microsphere retention mechanisms were enhanced by increasing aqueous phase ionic strength and the valence state of cotransported cations. Studies by G. R. Moline (pers. comm.) at ORNL have also used fluorescent microspheres to interrogate the fracture network and flow-path heterogeneity of the weathered saprolites. These results were coupled with computed tomography imaging of the undisturbed core specimens, and the fracture network patterns were used to simulate tracer displacement experiments conducted on the undisturbed cores.

INTERMEDIATE-SCALE ASSESSMENT OF TRANSPORT IN FRACTURED WEATHERED, SHALES

A logical progression from laboratory-scale undisturbed columns is the use of intermediate-scale in situ pedons for assessing the interaction of coupled processes on the fate and transport of solutes in the fractured, weathered shales (Figure 3-6). This research scale, unlike the column scale, encompasses more macroscopic structural features common to the field (e.g., dip of bedding planes, more continuous fracture network, convergent flow processes), yet allows for a certain degree of experimental control as the pedon can be hydrologically isolated from the surrounding environment. The pedon is an undisturbed block of soil (2m × 2m × 3m deep) with three excavated sides refilled with compacted soil and a concrete wall with access ports placed in good contact with the front soil face. The pedon was instrumented with a variety of solution samplers designed to monitor water and solutes through various pore regimes as a function of depth. Fritted glass plate lysimeters of varying porosity were held under different tensions to derive solutions from various pore regimes. Coarse fritted glass samplers were held at zero tension and collected free-flowing advective pore water (primary fractures); medium porosity frits were held at 20 cm tension and collected pore water from mesopores or secondary fractures; and fine frits were held at 250 cm tension and collected pore water from the soil matrix. Numerous tracer experiments have been conducted at this facility and others on the ORR using both nonreactive and reactive tracers with various steady-state infiltration rates or transient flow conditions driven by storm events (Jardine et al., 1989, 1990; Wilson, unpublished data). The purpose of the experiments was to determine transport properties and mass transfer rates for the various pore regimes. An example of tracer mobility (Br$^-$) through the soil for two different infiltration

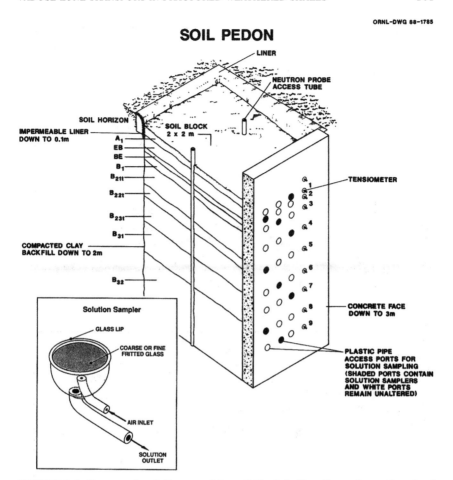

FIGURE 3-6 Cross-sectional diagram of the soil block facility (2 m × 2 m × 3 m deep) that was used for tracer transport investigations in fractured, weathered shale. The inset illustrates the fritted glass solution samplers that were installed laterally within the soil as a function of depth. Each depth interval contained a coarse, medium, and fine frit sampler, each of which was held at a different tension for the purpose of extracting pore water from primary fractures, secondary fractures, and the soil matrix, respectively. Modified from Geoderma 46, Jardine, P. M., G. V. Wilson, and R. J. Luxmoore, Unsaturated solute transport through a forest soil during rainstorm events, pp. 103-118, 1990, with permission from Elsevier Science.

rates can be seen in Figure 3-7. At an infiltration rate of 30 cm/d, Br⁻ is transported exclusively through medium and small pore regimes indicative of secondary fractures and the soil matrix, respectively. Flow through the large pore regimes, indicative of primary fractures, is essentially excluded since the imposed

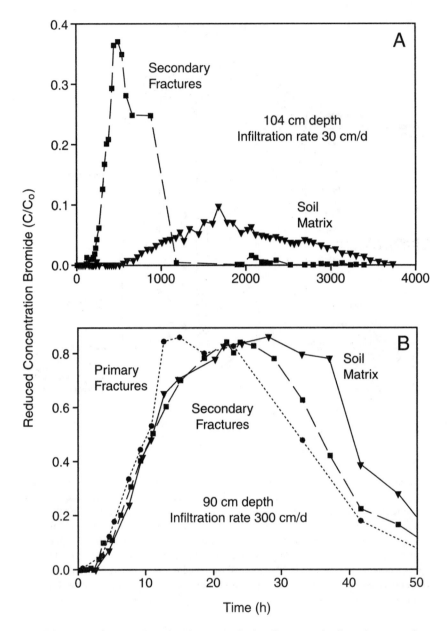

FIGURE 3-7 Breakthrough curves for a nonreactive Br⁻ tracer in discrete pore regimes during two infiltration experiments on the fractured, weathered shale soil block. At an infiltration rate of 30 cm/d (A) only secondary fractures and the soil matrix are conductive, whereas at an infiltration rate of 300 cm/d (B) both primary and secondary fractures and the soil matrix are conductive.

infiltration velocity is not sufficient to accommodate their conductivity. Larger infiltration rates (e.g., 300 cm/d), representative of larger-scale convergent flow processes, did produce flow through the primary fractures, and thus three distinct breakthrough curves can be observed for tracer movement through primary and secondary fractures, as well as the soil matrix. Large infiltration rates, however, create local-scale perched water tables within the soil, allowing small pore regime samplers to extract a portion of the larger pore water. This is why the ascending limb of the three breakthrough curves at 300 cm/d exhibit diminutive differences. Nevertheless, the distinction of tracer migration through the different pore regions allows for semi-quantitative estimates of pore-flow velocities, dispersion coefficients, and mass transfer rates between the pore classes at a scale one step closer to the realities of the field scale.

When the primary fracture network is nonconductive (e.g., 30 cm/d) solute residence times are sufficient to allow significant interaction of reactive solutes with the secondary fracture walls and the soil matrix. Near-equilibrium conditions were observed for Sr^{2+} transport through the pedon at an infiltration rate of 30 cm/d. In contrast, when primary fractures are conductive, reactive tracers can bypass the soil matrix and migrate in the same way as a nonreactive tracer (e.g., Jardine et al., 1988, 1993a). This scenario can be observed in undisturbed laboratory columns as well, where it was found that batch adsorption isotherms were appropriate for describing the transport of reactive solutes during unsaturated conditions (primary fractures nonconductive and preferential flow restricted to secondary fractures) but were inappropriate for describing solute transport during saturated conditions (Jardine et al., 1988, 1993a). During unsaturated conditions in the undisturbed columns, solute residence times (~10 d) were sufficient to establish geochemical equilibrium between the secondary fracture network and the soil matrix. These results are consistent with those of Reedy et al. (1996), who used a flow-interruption technique to show that ~4 days was sufficient to establish equilibrium conditions between nonreactive solutes in the fractures and nonreactive solutes in the soil matrix. Both the column and pedon studies have shown that the rate of mass transfer between fractures and the soil matrix is relatively rapid, and thus it can be inferred that the soil matrix remains an integral part of the entire system during storm events and is not necessarily excluded from transport processes when preferential flow occurs in fractures.

FIELD-SCALE ASSESSMENT OF TRANSPORT IN FRACTURED, WEATHERED SHALES

Waste migration issues in fractured, weathered shales are field-scale problems that are complicated by large-scale media heterogeneities that cannot be replicated at the laboratory or intermediate scale. In order to validate our conceptual understanding of vadose-zone transport that was derived from laboratory and pedon-scale observations, field-scale solute fate and transport experiments must be performed. At ORNL, a field facility has been constructed for assessing storm-driven solute mobility in the vadose zone of the fractured saprolites (Luxmoore and Abner, 1987). The experimental subwatershed drains an area of 0.63 hectares from an elevation of 275-258 m. The landscape is forested with hardwoods, and the soil profile is 0.5-3 m thick depending on elevation. The facility contains a buried line source for tracer release to simulate leakage of trench waste from contaminated areas and is equipped with an elaborate array of water and solute monitoring devices. Lateral subsurface flow is intercepted by a 2.5 m deep by 16 m long trench that has been excavated across the outflow region of the sub-watershed (Figure 3-8). Six massive stainless steel pans with 10-cm lips have been pressed back into the soil face to capture subsurface drainage from different portions of the landscape. Subsurface drainage from the pans, as well as overland storm flow and drainage under the pans, is routed into tipping bucket rain gauges situated in two H-flumes that are equipped with ultrasonic sensors for measuring water levels. The tipping buckets and ultrasonic sensors are equipped with computer data acquisition, allowing for real-time monitoring of tracer fluxes during storm events. Besides the ability to capture subsurface drainage, the field facility is equipped with numerous multilevel solution samplers, tensiometers, and piezometers, used to assess perched water table dynamics during storm events. The field facility has been well characterized with respect to spatial variability of surface and subsurface hydraulic conductivities, basic geochemical properties, and mineralogical analyses (Watson and Luxmoore, 1986; Jardine et al., 1988, 1993a, b, 1998; Wilson and Luxmoore, 1988; Wilson et al., 1989, 1992, 1993, 1998; Luxmoore et al., 1990; Kooner et al., 1995; Reedy et al., 1996).

In an effort to address field-scale fate and transport processes during transient flow in the fractured weathered shale media, Wilson et al. (1993) released a Br⁻ tracer from the ridgetop buried-line source during a storm event and monitored its mobility through the subsurface for nearly 8 months. During the release they observed that a small portion of the total Br⁻ mass (~5 percent) migrated very rapidly through the hillslope via lateral storm flow with subsequent export through the weirs that were located 70 m from the line source. The first arrival of Br⁻ at the weirs was 3 h after the initiation of the release, which illustrates the incredibly large hydraulic conductivity of the primary fracture network during hillslope convergent flow. These results are consistent with the measurements of Wilson and Luxmoore (1988) that showed 85 percent of ponded water flux went through primary fractures that constitute only 0.1 percent of the total soil porosity. Subsequent storm events over the 8-month period resulted in the export of ~20 percent of the injected Br⁻ mass (e.g., Figure 3-9a). While the actual tracer release revealed a rapid transport through the fracture network of the soil, the mass transfer into the low-permeability matrix was significant since >50 percent of the applied tracer was found to reside within matrix porosity of the soil primarily at a depth of 1-1.5 m (Figure 3-9b), which is synonymous with where lateral storm flow occurs through the hillslope. Large-scale structural heterogeneities of the sub-

ORNL DWG 90M-4998

SURFACE- AND SUBSURFACE-FLOW WEIR AND MONITORING CELLAR

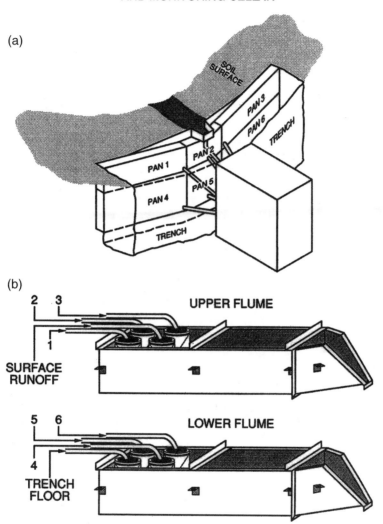

FIGURE 3-8 Schematic illustration of the subsurface weirs that intercept lateral storm flow from the subcatchment that is used for field-scale tracer injection experiments in the fractured, weathered shale soil: (a) shows the six stainless steel pans pressed against the trench face for collecting free lateral flow, and (b) the two H-flumes that contain tipping bucket rain gauges and ultrasonic sensors for continuous monitoring of subsurface flow. Reprinted from Journal of Hydrology, 145, Wilson, G. V., P. M. Jardine, J. D. O'Dell, and M. Collineau, Field-scale transport from a buried line source in variable saturated soil, pp. 83-109, 1993, with permission from Elsevier Science.

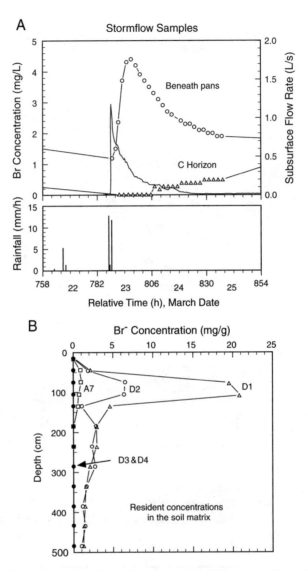

FIGURE 3-9 Storm flow and soil matrix tracer results following a release of Br⁻ at the field-scale tracer injection facility in the fractured, weathered shale soil: (a) shows an example of a subsurface flow hydrograph (solid line) for the lower flume that resulted from a storm event, with the corresponding Br⁻ concentrations from the C-horizon and from flow beneath the pans; and (b) Br⁻ residence concentrations from the soil matrix as a function of depth for six sampling locations downslope from the line source where tracer was released. Modified from Journal of Hydrology, 145, Wilson, G. V., P. M. Jardine, J. D. O'Dell, and M. Collineau, Field-scale transport from a buried line source in variable saturated soil, pp. 83-109, 1993, with permission from Elsevier Science.

surface media controlled solute mobility, since tracer was preferentially trans-ported along bedding planes that dipped in an opposing direction to the hillslope topography. The interconnected nature of bedding plane parallel fractures with cross-cutting extensional fractures results in continuous preferential flow paths that can accommodate large storm-flow inputs. The sheer magnitude of bedding plane parallel fractures versus strike parallel extensional fractures tends to drive solutes along the dip of the bedding planes (G. R. Moline, pers. comm.).

Strong evidence indicating matrix-fracture interactions during transient storm flow can be inferred from tracer breakthrough patterns at the subsurface weirs. Storm events that followed the release of tracer resulted in delayed tracer break-through pulses relative to the subsurface flow hydrograph (Figure 3-9a). Stable isotope and solute chemistry analysis revealed that subsurface flow was predomi-nately new water at peak flow and almost exclusively old water during the de-scending limb of the subsurface hydrograph. These results suggested that the storm-driven export of Br^- through the weirs was the result of tracer mass transfer from the soil matrix into the fracture network with subsequent mobility through the hillslope. The field-scale endeavors provided an improved conceptual under-standing of how transient hydrodynamics and media structure controlled the migration and storage of contaminants in the subsurface. The field-scale findings were consistent with the multiregion conceptual framework established with labo-ratory- and intermediate-scale observations of flow and transport through these heterogeneous systems.

MULTIREGION CONCEPTUAL FRAMEWORK OF VADOSE ZONE TRANSPORT IN FRACTURED MEDIA

The multiscale experimental endeavors have provided a conceptual frame-work for a multiregion flow and transport mechanism that controls solute mobil-ity in the fractured, weathered shales. Experimentally we are able to distinguish three pore-size classes (primary fractures, secondary fractures, soil matrix); thus a representative elemental volume (REV) at any point in the soil consists of three regions, each with its own flow and transport parameters (Figure 3-10). Intra-region mass transfer is described by flow from a physical point to a neighboring one and interregion mass transfer between the various pore regimes.

This transfer is controlled by both advective and diffusive processes, where hydraulic gradients, caused by differences in fluid velocity in different-sized pores, drive advective mass transfer and concentration gradients drive diffusive processes. This concept is illustrated by the following example.

Clean rainwater infiltrating into secondary fractures and converging into primary fractures would bypass an initially contaminant-rich soil matrix. Thus, physical nonequilibrium—that is, concentration gradients—would exist between regions and cause diffusive transfer from the matrix to the mobile regions. How-ever, pressure gradients would also exist initially from the secondary and primary

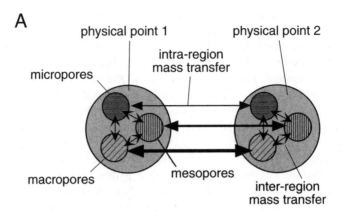

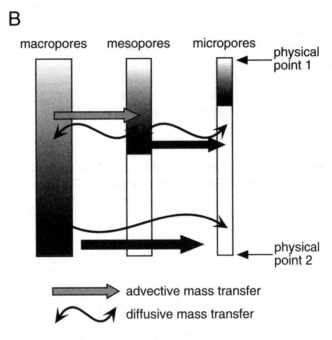

FIGURE 3-10 Triple-porosity, triple-permeability example of the multiregion flow and transport concept. (A) The REV(large circle) at two physical points consist of three pore regions. Intra-region flow and transport is indicated by lines between large circles (REVs), and inter-region transfer is depicted by lines within each large circle (REV). (B) Both advective and diffusive mass transfer may occur in parallel or counter to each other. From Gwo, J. P., P. M. Jardine, G. V. Wilson, and G. T. Yeh, 1996. Using a multiregion model to study the effects of advective and diffusive mass transfer on local physical nonequilibrium and solute mobility in a structured soil. Water Resources Research 32: 561-570. Copyright by American Geophysical Union.

fracture regions toward the soil matrix. Thus, diffusion of contaminants from the matrix would initially be counter to the interregion advective transfer (Figure 3-10b). The resulting export of contaminants out of the profile would be slow, followed by a gradual increase as advective mass transfer between regions subsides. In this example, primary fractures would serve to provide relatively clean water to aquifers and streams. In the absence of primary fractures, the initiation of outflow may require a longer period; however, the eventual export of contaminants would be much greater due to the increased displacement of this contaminant-rich matrix water. From this example, advection as well as diffusion of solutes between pore regions are relevant processes in solute transport under field conditions.

Another process inherent in multiregion flow and transport in structured systems is the time-dependent nature of both the advective and diffusive mass transfer rates between the various pore domains. Time-dependent mass transfer rates take into account changes in concentration gradients as solute mass is transferred between pore regimes. Time-dependent rate coefficients also indirectly account for variabilities in matrix block sizes.

The concept of multiregion flow and transport in structured media has been numerically implemented by Gwo et al. (1991), with the mathematical formulation of the code described in Gwo et al. (1994, 1995a, b, 1996) and the simulation of experimental data illustrated in Gwo et al. (1998, 1999). A two-region formulation of the model has been used to describe multiple tracer transport through undisturbed soil columns of fractured, weathered shale. The structure of the porous media was embedded into the two-region model in which the interregion mass flux was characterized by mesoscale mass-transfer coefficients. Predictions were significantly more accurate than those obtained using a simple single-fracture conceptual model that represents a least-information scenario. Modeling results suggested that mesoscale spreading of tracer in structured porous media may be largely attributed to interregion mass transfer. The multiregion model has also been used to numerically simulate storm-driven solute transport experiments conducted at the watershed scale within the fractured saprolites. Comparison of numerical results with the field data indicated that multiregion, preferential flow occurs under partially saturated conditions that can be confirmed theoretically, and that mass transfer between pore regions is an important process influencing contaminant movement in the subsurface.

SUMMARY

In a manner similar to that of Solomon et al. (1992), the basic concepts of groundwater flow and transport in the vadose zone of fractured, weathered shales on the ORR are summarized below.

Flow and transport through the fractured saprolites is a multiporosity scenario where an interconnected network of highly conductive primary and secondary fractures surrounds low-permeability, high-porosity matrix blocks.

Flow and transport are driven by storm events with significant water flow occurring through a 1- to 2-m-thick storm-flow zone that develops as a result of perched water tables. Rapidly infiltrating storm water in the upper soil horizons becomes perched on lower conductivity, high bulk density saprolites, promoting subsurface lateral flow through hillslopes. The flow is transient and preferentially follows the dip of bedding planes, with most storm flow being discharged into local streams.

The majority of storm flow occurs in primary and secondary fractures, and most water storage occurs in the soil matrix. Fractures can accommodate large rainfall inputs (85 percent of convergent storm flow) but only constitute 0.1 percent of the total soil porosity. The soil matrix, on the other hand, has low permeabilities but porosities as high as 50 percent.

The fractured, weathered shales are conducive to extreme preferential flow, which results in hydraulic, physical, and geochemical nonequilibrium between fractures and the surrounding soil matrix. Differences in fluid velocities in different-sized pores create hydraulic gradients that drive time-dependent interregion advective mass transfer. Solute concentration differences between the various sized pores creates concentration gradients that drive time-dependent inter-region diffusive mass transfer. Storm-enhanced preferential flow in these weathered shales disrupts geochemical equilibrium between the solid, liquid, and gas phases.

The rapid transport of small amounts of mass occurs through large pores (fractures), but because of the prevalence of nonequilibrium conditions, the majority of mass resides in small pores (soil matrix), which greatly retards bulk mass migration rates.

ACKNOWLEDGMENTS

This research was supported by the Environmental Technology Partnership program of the Office of Biological and Environmental Research, U.S. Department of Energy. The authors thank Mr. Paul Bayer, contract officer for DOE's ETP program, for financially supporting this research. Oak Ridge National Laboratory is managed by University of Tennessee-Battelle, LLC, for the U.S. Department of Energy, under contract DE-AC05-00OR22725. Publication No. 5057, Environmental Sciences Division, ORNL.

REFERENCES

Akratanakul, S., L. Boersma, and G. O. Klock, 1983. Sorption processes in soils as influenced by pore water velocity. 2. Experimental results. Soil Science 135: 331-341.

Anamosa, P. R., P. Nkedi-Kizza, W. G. Blue, and J. B. Sartain, 1990. Water movement through an aggregated, gravelly oxisol from Cameroon. Geoderma 46: 263-281.

Bales, R. C., C. P. Gerba, G. H. Grondin, and S. L. Jensen, 1989. Bacteriophage transport in sandy soil and fractured tuff. Appl. Environ. Microbiol. 55: 2061-2067.

Barraclough, D., and P. H. Nye, 1979. The effect of molecular size on diffusion characteristics in soil. Journal of Soil Science 30: 29-42.

Bowman, R., 1984. Evaluation of some new tracers for soil water studies. Soil Sci. Soc. Am. J. 48: 987-993.

Bowman, R. S., and J. F. Gibbens, 1992. Difluorobenzoates as nonreactive tracers in soil and ground-water. Ground Water 30: 8-14.

Brusseau, M. L., 1993. The influence of solute size, pore water velocity, and intraparticle porosity on solute dispersion and transport in soil. Water Resources Research 29: 1071-1080.

Brusseau, M. L., P. S. C. Rao, R. E. Jessup, and J. M. Davidson, 1989. Flow interruption: A method for investigating sorption nonequilibrium. Journal of Contaminant Hydrology 4: 223-240.

Carter, R. C., W. J. Kaufman, G. T. Orlob, and D. K. Todd, 1959. Helium as a ground-water tracer. Journal of Geophysical Research 64: 2433-2439.

Clark, J. F., P. Schlosser, M. Stute, and H. J. Simpson, 1996. SF_6-^3He tracer release experiment: A new method of determining longitudinal dispersion coefficients in large rivers. Environmental Science and Technology 30: 1527-1532.

Cumbie, D. H., and L. D. McKay, 1999. Influence of diameter on particle transport in a fractured shale saprolite. Journal of Contaminant Hydrology 37(1-2): 139-157.

Dreier, R. B., D. K. Solomon, and C. M. Beaudoin, 1987. Fracture characterization in the unsaturated zone of a shallow land burial facility. In: Flow and Transport Through Unsaturated Rock. D. D. Evans and T. J. Nicholson, eds. Geophysical Monograph 42, Washington, D.C.: American Geophysical Union, pp. 51-59.

Gerba, C. P., S. M. Goyal, I. Cech, and G. F. Bogdan, 1981. Quantitative assessment of the adsorptive behavior of viruses to soils. Environmental Science and Technology 15: 940-944.

Gupta, S. K., L. S. Lau, and P. S. Moravcik, 1994. Ground-water tracing with injected helium. Ground Water 32: 96-102.

Gwo, J. P., G. T. Yeh, and G. V. Wilson, 1991. Proceedings of the International Conference on Transport and Mass Exchange Processes in Sand and Gravel Aquifers. 2. Field and Modeling Studies. Ottawa, Canada, pp. 578-589.

Gwo, J. P., P. M. Jardine, G. T. Yeh, and G. V. Wilson, 1994. MURF user's guide: A finite element model of multiple-pore-region flow through variably saturated subsurface media. Oak Ridge National Laboratory, ORNL/GWPO-011.

Gwo, J. P., P. M. Jardine, G. V. Wilson, and G. T. Yeh, 1995a. A multiple-pore-region concept to modeling mass transfer in subsurface media. Journal of Hydrology 164: 217-237.

Gwo, J. P., P. M. Jardine, G. T. Yeh, and G. V. Wilson, 1995b. MURT user's guide: A finite element model of multiple-pore-region transport through variably saturated subsurface media. Oak Ridge National Laboratory, ORNL/GWPO-015.

Gwo, J. P., P. M. Jardine, G. V. Wilson, and G. T. Yeh, 1996. Using a multiregion model to study the effects of advective and diffusive mass transfer on local physical nonequilibrium and solute mobility in a structured soil. Water Resources Research 32: 561-570.

Gwo, J. P., R. O'Brien, and P. M. Jardine, 1998. Mass transfer in structured porous media: Embedding mesoscale structure and microscale hydrodynamics in a two-region model. Journal of Hydrology 208: 204-222.

Gwo, J. P., G. V. Wilson, P. M. Jardine, and E. F. D'Azevedo, 1999. Modeling subsurface contaminant reactions and transport at the watershed scale. In: Assessment of Non-Point Source Pollution in the Vadose Zone, D. L. Corwin, K. Loague, and T. R. Ellsworth, eds. Geophysical Monograph Series 108: 31-43.

Harton, A. D., 1996. Influence of Flow Rate on Transport of Bacteriophage in a Column of Highly Weathered and Fractured Shale. M.S. thesis, University of Tennessee.

Harvey, R. W., L. H. George, R. L. Smith, and D. L. LeBlanc, 1989. Transport of microspheres and indigenous bacteria through a sandy aquifer: Results of natural- and forced-gradient tracer experiments. Sci. Technol. 23: 51-56.

Harvey, R. W., N. E. Kinner, D. MacDonald, D. W. Metge, and A. Bunn, 1993. Role of physical heterogeneity in the interpretation of small-scale laboratory and field observations of bacteria, microbial-sized microsphere, and bromide transport through aquifer sediments. Water Resources Research 29: 2713-2721.

Harvey, R. W., N. E. Kinner, D. MacDonald, D. W. Metge, and A. Bunn, 1995. Transport behavior of groundwater protozoa and protozoan-sized microspheres in sandy aquifer sediments. Applied and Environmental Microbiology, January: 209-217.

Haun, D. D., 1998. Influence of Ionic Strength and Cation Valence on Transport of Colloid-Sized Microspheres in Fractured Shale Saprolite. M.S. thesis, University of Tennessee.

Hinsby, K., L. D. McKay, P. Jørgensen, M. Lenczewski, and C. P. Gerba, 1996. Fracture aperture measurements and migration of solutes, viruses and immiscible creosote in a column of clay till. Ground Water 34: 1065-1075.

Hu, Q., and M. L. Brusseau, 1995. Effect of solute size on transport in structured porous media. Water Resources Research 31: 1637-1646.

Hutzler, N. J., J. C. Crittenden, J. S. Gierke, and A. S. Johnson, 1986. Transport of organic compounds with saturated groundwater flow: Experimental results. Water Resources Research 22: 285-295.

Jardine, P. M., G. V. Wilson, and R. J. Luxmoore, 1988. Modeling the transport of inorganic ions through undisturbed soil columns from two contrasting watersheds. Soil Sci. Soc. Am. J. 52: 1252-1259.

Jardine, P. M., G. V. Wilson, R. J. Luxmoore, and J. F. McCarthy, 1989. Transport of inorganic and natural organic tracers through an isolated pedon in a forested watershed. Soil Sci. Soc. Am. J. 53: 317-323.

Jardine, P. M., G. V. Wilson, and R. J. Luxmoore, 1990. Unsaturated solute transport through a forest soil during rain storm events. Geoderma 46:103-118.

Jardine, P. M., G. K. Jacobs, and G. V. Wilson, 1993a. Unsaturated transport processes in undisturbed heterogeneous porous media. I. Inorganic contaminants. Soil Sci. Soc. Am. J. 57: 945-953.

Jardine, P. M., G. K. Jacobs, and J. D. O'Dell, 1993b. Unsaturated transport processes in undisturbed heterogeneous porous media. II. Co-contaminants. Soil Sci. Soc. Am. J. 57: 954-962.

Jardine, P. M., R. O'Brien, G. V. Wilson, and J. P. Gwo, 1998. Experimental techniques for confirming and quantifying physical nonequilibrium processes in soils. In: Physical Nonequilibrium in Soils: Modeling and Applications. H. M. Selim and L. Ma, eds. Chelsea, Michigan: Ann Arbor Press, Inc., pp. 243-271.

Jardine, P. M., W. E. Sanford, J. P. Gwo, O. C. Reedy, D. S. Hicks, R. J. Riggs, and W. B. Bailey, 1999. Quantifying diffusive mass transfer in fractured shale bedrock. Water Resources Research 35: 2015-2030.

Jaynes, D. B., 1994. Evaluation of fluorobenzoate tracers in surface soils. Ground Water 32: 532-538.

Jensen, J. R., 1984. Potassium dynamics in soil during steady flow. Soil Science 138:285-293.

Kookana, R. S., R. D. Schuller, and L. A. G. Aylmore, 1993. Simulation of simazine transport through soil columns using time-dependent sorption data measured under flow conditions. Journal of Contaminant Hydrology 14: 93-115.

Kooner, Z. S., P. M. Jardine, S. Feldman, 1995. Competitive surface complexation reactions of SO_4^{2-} and natural organic carbon on soil. Journal of Environmental Quality 24: 656-662.

Krupp, H. K., and D. E. Elrick, 1968. Miscible displacement in an unsaturated glass bead medium. Water Resourc. Res. 4: 809-815.

Linderfelt, W. R., and J. L. Wilson, 1994. Field study of capture zones in a shallow sand aquifer. In: Transport and Reactive Processes in Aquifers. Dracos and Stauffer, eds. Rotterdam: Balkema.

Luxmoore, R. J., and C. H. Abner, 1987. Field facilities for subsurface transport research. DOE/ER-0329. Washington, D.C.: U.S. Department of Energy. 32 pp.

Luxmoore, R. J., P. M. Jardine, G. V. Wilson, J. R. Jones, and L. W. Zelazny, 1990. Physical and chemical controls of preferred path flow through a forested hillslope. Geoderma 46: 139-154.

Ma, L., and H. M. Selim, 1994. Predicting the transport of atrazine in soils: Second-order and multireaction approaches. Water Resources Research 30: 3489-3498.

Mayes, M. A., P. M. Jardine, I. L. Larsen, S. C. Brooks, and S. E. Fendorf, 2000. Multispecies transport of metal-EDTA complexes and chromate through undisturbed columns of weathered fractured saprolite. Journal of Contaminant Hydrology 45: 243-265.

McKay, L. D., J. A. Cherry, R. C. Bales, M. T. Yahya, and C. P. Gerba, 1993a. A field example of bacteriophage as tracers of fracture flow. Environmental Science and Technology 27: 1075-1079.

McKay, L. D., R. W. Gillham, and J. A. Cherry, 1993b. Field experiments in a fractured clay till. 2. Solute and colloid transport. Water Resources Research 29: 3879-3890.

Murali, V., and L. A. G. Aylmore, 1980. No-flow equilibration and adsorption dynamics during ionic transport in soils. Nature 283: 467-469.

Nkedi-Kizza, P., J. W. Biggar, M. Th. van Genuchten, P. J. Wierenga, H. M. Selim, J. D. Davidson, and D. R. Nielsen, 1983. Modeling tritium and chloride 36 transport through an aggregated oxisol. Water Resources Research 19: 691-700.

Raven, K. G., K. S. Novakowski, and P. A. Lapcevic, 1988. Interpretation of field tracer tests of a single fracture using a transient solute storage model. Water Resources Research 24: 2019-2032.

Reedy, O. C., P. M. Jardine, G. V. Wilson, and H. M. Selim, 1996. Quantifying the diffusive mass transfer of nonreactive solutes in columns of fractured saprolite using flow interruption. Soil Sci. Soc. Am. J. 60: 1376-1384.

Reimus, P. W., 1996. The Use of Synthetic Colloids in Tracer Transport Experiments in Saturated Rock Fractures. Ph.D. dissertation, University of New Mexico, LA-13004-T.

Sanford, W. E., R. G. Shropshire, and D. K. Solomon, 1996. Dissolved gas tracers in groundwater: Simplified injection, sampling, and analysis. Water Resources Research 32: 1635-1642.

Sanford, W. E., and D. K. Solomon, 1998. Site characterization and containment assessment with dissolved gases. Journal of Environmental Engineering 124: 572-574.

Schulin, R., P. J. Wierenga, H. Flühler, and J. Leuenberger, 1987. Solute transport through a stony soil. Soil Sci. Soc. Am. J. 51: 36-42.

Seyfried, M. S., and P. S. C. Rao, 1987. Solute transport in undisturbed columns of an aggregated tropical soil: Preferential flow effects. Soil Sci. Am. J. 51: 1434-1444.

Shropshire, R. G., 1995. Dual-Tracers: A Tool for Studying Matrix Diffusion and Fracture Parameters. M.S. thesis, University of Waterloo, Ontario, Canada.

Smith, M. S., G. W. Thomas, R. E. White, and D. Ritonga, 1985. Transport of Escherichia coli through intact and disturbed soil columns. Journal of Environmental Quality 14: 87-91.

Solomon, D. K., G. K. Moore, L. E. Toran, R. B. Dreier, and W. M. McMaster, 1992. A hydrologic framework for the Oak Ridge Reservation. ORNL/TM-12026.

Toran, L., and A. V. Palumbo, 1991. Colloid transport through fractured and unfractured laboratory sand columns. Journal of Contaminant Hydrology 9: 289-303.

Watson, K. W., and R. J. Luxmoore, 1986. Estimating macroporosity in a forest watershed by use of a tension infiltrometer. Soil Sci. Soc. Am. J 50: 578-582.

Wilson, G. V., and R. J. Luxmoore, 1988. Infiltration, macroporosity, and mesoporosity distributions on two forested watersheds. Soil Sci. Soc. Am. J. 52: 329-335.

Wilson, G. V., J. M. Alfonsi, and P. M. Jardine, 1989. Spatial variability of saturated hydraulic conductivity of the subsoil of two forested watersheds. Soil Sci. Soc. Am. J. 53: 679-685.

Wilson, G. V., P. M. Jardine, and J. P. Gwo, 1992. Modeling the hydraulic properties of a multiregion soil. Soil Sci. Soc. Am. J. 56: 1731-1737.

Wilson, G. V., P. M. Jardine, J. D. O'Dell, and M. Collineau, 1993. Field-scale transport from a buried line source in variable saturated soil. Journal of Hydrology 145: 83-109.

Wilson, G. V., J. P. Gwo, P. M. Jardine, and R. J. Luxmoore, 1998. Hydraulic and physical non-equilibrium effects on multi-region flow and transport. In: Physical Nonequilibrium in Soils: Modeling and Application. H. M. Selim and L. Ma, eds. Chelsea, Michigan: Ann Arbor Press, Inc., pp. 37-61.

Wilson, J. L., and L. W. Gelhar, 1974. Dispersive mixing in a partially saturated porous medium. Water Resources and Hydrodynamics, Department of Civil Engineering, Massachusetts Institute of Technology, Boston, Massachusetts, Report No. 191.

Wilson, R. D., and D. M. MacKay, 1993. The use of sulfur hexafluoride as a conservative tracer in saturated sandy media. Ground Water 31: 719-724.

Yang, Z., R. S. Burlage, W. E. Sanford, and G. R. Moline, 1996. DNA-labeled silica microspheres for groundwater tracing and colloid transport studies. In: Proceedings, 212th National Meeting, American Chemical Society, Orlando, Fla., Aug. 25-30.

4

Evaluation of Conceptual and Quantitative Models of Fluid Flow and Chemical Transport in Fractured Media

Brian Berkowitz,[1] Ronit Nativ,[2] and Eilon Adar[3]

ABSTRACT

Patterns of fluid flow and chemical transport in heterogeneous porous and fractured media tend to be highly complex. Models of flow and transport processes in these systems must account for the inherent heterogeneity and scaling properties of porous rocks, as well as the typically sparse and uncertain field data that can be obtained to characterize a geological formation. Recent introduction of advanced theoretical and experimental techniques is providing new insight into our understanding of these highly non-uniform patterns. Statistical network models can be used to characterize the structure of pore and fracture systems, and to define power laws that quantify flow and transport processes within them. In a similar spirit, a random walk formalism can quantify anomalous (non-Gaussian) patterns of chemical transport that are frequently observed in laboratory and field experiments. Analyses of transport of reactive chemicals, including effects of precipitation and dissolution, as well as changes in fracture morphology, also demonstrate highly non-uniform behavior. In the context of these results, we discuss some conceptual and quantitative models of fluid flow and chemical transport in the fractured vadose zone, and the laboratory and field data that are required to evaluate them.

[1] Department of Environmental Sciences and Energy Research, Weizmann Institute of Science, Rehovot, Israel
[2] Department of Soils and Water, The Hebrew University of Jerusalem, Rehovot, Israel
[3] Blaustein Institute for Desert Research, Ben Gurion University of the Negev, Israel

INTRODUCTION

Major efforts have been devoted over the last two decades to the development of realistic theoretical models capable of simulating flow and transport processes in fractured and heterogeneous porous formations. These efforts have led to significant understanding of the dynamics of flow and transport processes in these systems. However, predictive capabilities related to real fractured and heterogeneous media remain limited. This is due, in part, to the very complex nature of fracture networks in the subsurface, and to the virtual impossibility of obtaining detailed structural, hydraulic, and geochemical characterizations of fractures in situ, in diverse geological and lithological settings. As a result, studies must often rely on extrapolation of exposed features and indirect measurements, together with subjective considerations, to generate a statistical characterization of fracture systems (e.g., Berkowitz and Adler, 1998).

These analyses are demanding because fractures exist in a broad range of geological formations and rock types, and are produced under a variety of geological and environmental processes. As a result of the variability in rock properties and structures, as well as the variety in fracturing mechanisms, fracture sizes range from microfissures of the order of microns to major faults of the order of kilometers, while fracture network patterns range from relatively regular polygonal arrangements to apparently random distributions. The hydraulic and transport properties of these formations vary considerably, being dependent largely on the degree of fracture interconnection, aperture variations in the fractures, and chemical characteristics of the fractures and host rock.

With regard to fluid flow and chemical transport processes in fractured vadose zones, our ability to predict actual flow and transport behavior—and even our understanding of the basic dynamics—is severely limited, and in some cases rudimentary. Several factors force development of specific conceptual pictures and models tailored to dealing with these processes: (1) fracture walls are rough; (2) fractures often contain filling material; (3) this filling material and the walls themselves are subject to processes of chemical reaction, dissolution, precipitation, mineralization, and/or particle detachment and trapping; (4) fluid migration in partially saturated conditions occurs in preferential paths and channels, with a complex interplay between the fractures and the porous host rock; and (5) fluid and chemical migration patterns are often temporally and spatially unstable. Quantitative models of fluid flow and chemical transport in the fractured vadose zone must therefore be predicated on conceptual pictures of fractured, heterogeneous, and otherwise "disordered" porous media, which account for a variety of non-uniform flow and transport behaviors.

In this chapter, we consider conceptual pictures and quantitative models of fluid flow and chemical transport in fractured and heterogeneous porous media relevant to the fractured vadose zone. To this end, we build on our more established understanding of these processes in saturated systems. Other models specific to fluid flow and chemical transport in unsaturated fractured media, such as film flow and fracture-matrix interactions, are discussed in other chapters in this volume. Given the wide variety of flow and transport processes, and the different perspectives that arise as a function of the actual problem of interest, the quantitative models that we examine are problem-specific. Clearly, there is no single generic model that is appropriate, and so as a consequence, a range of probabilistic and statistical approaches must be introduced.

In the next section, we discuss conceptual pictures and modeling approaches for fluid flow and chemical transport. We consider these processes at the scale of individual fractures and in fracture networks. Several experimental efforts are then examined, and in light of these results, we address the issue of how to evaluate conceptual and quantitative models at the laboratory and field scales.

CONCEPTUAL PICTURES AND MODELING APPROACHES FOR FLUID FLOW AND CHEMICAL TRANSPORT

Single Fractures

Fluid Flow in Saturated Fractures

Many laboratory experiments (e.g., Raven and Gale, 1985; Pyrak-Nolte et al., 1987; Durham and Bonner, 1994) and field studies (e.g., Novakowski et al., 1985, 1995; Rasmuson and Neretnieks, 1986; Raven et al., 1988) indicate that the classical view of a rock fracture as a pair of smooth, parallel plates is not adequate to fully quantify fluid flow. However, though more advanced conceptual models have been introduced in recent years, an alternative model for flow in fractures has yet to be generally accepted. Moreover, some very basic aspects of the theoretical problem remain largely unstudied.

A critical, intimately related issue is the validity of the local cubic law. To date, the majority of theoretical and simulation studies of single-fracture flow (Iwai, 1976; Brown, 1987; Moreno et al., 1988; Thompson and Brown, 1991; David, 1993; Unger and Mase, 1993; Amadei and Illangasekare, 1994) have postulated that local flow magnitudes are well described by the Reynolds equation. From this equation, local flow magnitudes are proportional to the cube of the local aperture; hence the name "local cubic law" (LCL). The LCL, a key feature in conceptual models, states that the volumetric flow through a fracture varies as the cube of the fracture aperture. Thus definition of fracture aperture, and application of the LCL, strongly influence quantitative analyses and interpretation of laboratory and field measurements on flow in fractures.

The Reynolds equation originates from lubrication theory, which was formulated for narrow void spaces between artificially smoothed surfaces. However, the roughness of rock fracture surfaces is usually more significant and irregular, and has been shown to possess self-affine fractal properties, with a roughness, or

Hurst exponent $H \cong 0.8$ (Brown and Scholz, 1985; Poon et al., 1992; Schmittbuhl et al., 1995). A typical self-affine fracture surface with H = 0.8 is shown in Figure 4-1. Moreover, computer simulations (Mourzenko et al., 1995; Brown et al., 1995) have compared predictions from the Stokes and Reynolds equations, and suggest that the Reynolds equation is of limited validity for rough fractures. The adequacy of the LCL for rough geometries remains, essentially, an open question.

A detailed review of these studies is beyond the scope of this report. Here we focus on a new approach that we have proposed recently, which addresses the above issues. Oron and Berkowitz (1998) used a two-dimensional order-of-magnitude analysis of the Navier-Stokes equations for a general rough-wall geometry to examine the LCL assumption. This study hypothesized that LCL behavior should not be assumed to be correct or incorrect a priori for the entire fracture; rather, it should be studied on a local basis. As a result, three conditions can be defined for the applicability of LCL flow, as a leading-order approximation, in a local fracture segment with parallel or nonparallel walls. Consider a local cross section of some fracture segment of length L, with a typical half-aperture B. We define a geometric aspect ratio parameter, $\delta \equiv B/L$, and a nondimensional local roughness parameter $\varepsilon \equiv \max(\sigma_u/B, \sigma_l/B) \ll 1$, where σ_u and σ_l are the root-mean-square roughnesses (deviations from the means) of the upper and lower walls, respectively, averaged over the length L. Oron and Berkowitz (1998) demonstrated that the LCL is an adequate first-order approxi-

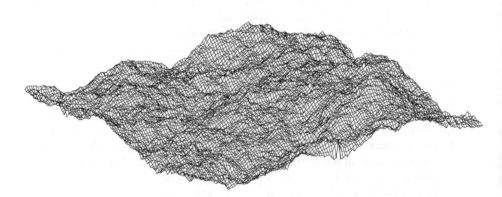

FIGURE 4-1 A typical self-affine fracture surface, generated numerically on a 128×128 grid with $H = 0.8$. After Oron, A. P., and B. Berkowitz, 1998. Flow in rock fractures: The local cubic law assumption reexamined. Water Resources Research 34: 2811-2825. Copyright by American Geophysical Union.

mation of the flow under three conditions: (1) the local segment must be long enough relative to its aperture, so that $\delta^2 \ll 1$; (2) over this segment length, $\varepsilon \ll 1$; and (3) inertial effects are limited via $Re \cdot \max(\delta,\varepsilon) \ll 1$, where Re is the Reynolds number in the segment.

These conditions define dynamic-geometric conditions for LCL flow, which are a function of fluid velocity, relative to the fracture aperture and the (relative) wall roughness. From these conditions it follows that the cubic law aperture should be measured as an average over a distance L, normal to the mean trend of the walls, and not with point-by-point methods as suggested by, for example, Ge (1997) or Mourzenko et al. (1995). An illustration of various methods for defining the aperture is given in Figure 4-2. The orientation of measurement over L is set by the general orientation of the segment. Under the above conditions, the LCL, even though it is only a first-order approximation, provides second-order accuracy for estimates of local discharge and of mean fluid velocity. The LCL may also be adapted to pathways with nonparallel walls, as long as the mean path half-angle is moderate. Moreover, the aspect-ratio condition (1) sets a finite resolution in the x-y plane to all analyses based on the LCL. Also, these conditions suggest that even when Re is as low as ~1-10, transition to non-LCL (nonlaminar) flow may occur.

Extending to the third dimension, in addition to defining apertures over segment lengths, Oron and Berkowitz (1998) find that the geometry of the contact regions between fracture walls influences flow paths more significantly than might be expected from consideration of only the nominal area fraction of these contacts. Moreover, this effect is enhanced by the presence of non-LCL regions around these contacts. Accounting for the self-affine nature of fracture walls, typical (fractal) fracture wall contact areas generated numerically are illustrated in Figure 4-3. We find a wide variability among the hydraulic conductivity maps, in terms of the shape of the regions in which LCL flow exists, due to the varying distribution and morphology of contact and non-LCL regions in the fractures. While contact ratios of 0.1-0.2 are usually assumed to have a negligible effect, our calculations suggest that contact ratios as low as 0.03-0.05 can be significant. Analysis of computer-generated fractures with self-affine walls demonstrates a nonlinear increase in contact area, and a faster-than-cubic decrease in the overall hydraulic conductivity, with decreasing fracture aperture. These results are in accordance with existing experimental data on flow in fractures (Durham and Bonner, 1994; see also the summary in Figure 14 of Mourzenko et al., 1997).

We conclude this section by considering the fundamental ill-posedness of this problem. In view of the complexity of local flow behavior, a global cubic law need not exist for every fracture; the cubic law results from the parabolic velocity profile in perfect Poiseuille flow, and not from any dimensional considerations. Flow, or permeability, versus aperture curves are derived indirectly, on the basis of displacement measurements on the perimeter of laboratory fracture samples. At best, one can measure the mechanical aperture, that is, the apparent, average,

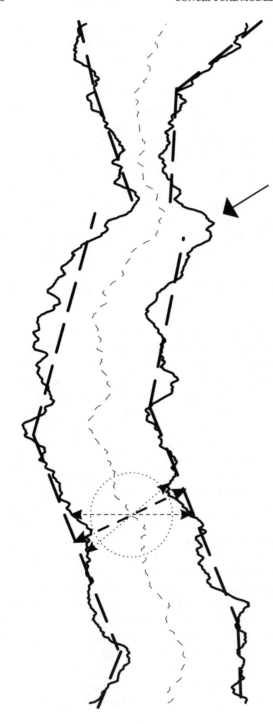

FIGURE 4-2 A demonstration of the various methods for measuring the LCL aperture. Shown are the vertical aperture (short dashes), the normal-to-local-centerline aperture of Ge (1997) (dots), and the ball aperture of Mourzenko et al. (1995). The approach of Oron and Berkowitz (1998) indicates that the cubic law aperture is not a point-by-point property but an average over a segment; the bold dashes on the pathway walls indicate a possible division into cubic law segments. The location marked with an arrow is inadequate for local cubic law because of a rapidly varying wall geometry. The rightmost, diagonal segment has a diverging angle that makes this region marginal for cubic law flow. After Oron, A. P., and B. Berkowitz, 1998. Flow in rock fractures: The local cubic law assumption reexamined. Water Resources Research 34: 2811-2825. Copyright by American Geophysical Union.

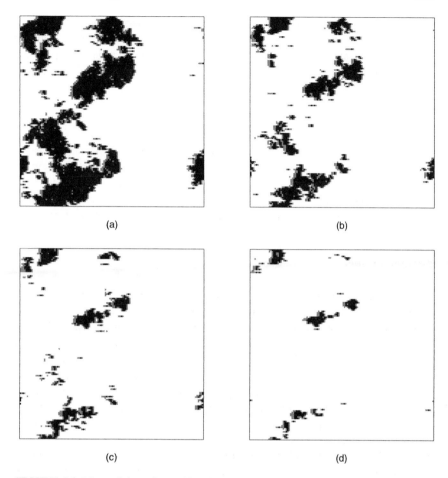

(a)   (b)

(c)   (d)

FIGURE 4-3 Maps of the estimated local hydraulic conductivity for a numerically generated fracture with self-affine walls, with $H = 0.8$ and a grid size of 128×128, for flow in
the x-direction (left to right). The length of each side of the map is equivalent to 6.4 mm.
Black indicates fracture wall contact, while grey indicates regions where LCL is not
applicable; LCL flow exists only in the white regions. Average fracture apertures are (a)
100 μm, (b) 150 μm, (c) 175 μm, and (d) 200 μm. After Oron, A. P., and B. Berkowitz,
1998. Flow in rock fractures: The local cubic law assumption reexamined. Water Resources Research 34: 2811-2825. Copyright by American Geophysical Union.

separation between the fracture walls. In general, we do not have exact information on the (varying) void space within the sample itself, as it changes under
different applied stresses, as well as following geochemical reactions (see the
next section and the section on Field-Scale Analysis of Fluid Flow and Chemical
Transport in the Fractured Vadose Zone), and we have only relative displacement

measurements rather than the actual aperture values. Any conversion of these measurements to local apertures, and definition of an effective aperture, is based on assumptions and convenience. As a consequence, although natural, the question of flow versus aperture, and validity of a cubic law, may be the wrong one to ask about rough fractures.

Chemical Transport Processes in Saturated Fractures

In addition to rough fracture walls, the presence of filling material further influences the hydraulic properties of a fracture. In general, fracture wall roughness and fracture infilling evolve over time, as a function of a variety of dynamic mechanical and geochemical processes (e.g., Steefel and Lichtner, 1994). There is evidence that these changes may also take place over very short time intervals. It is critical that we understand these processes, which can occur naturally or be induced artificially (e.g., by injection of grouting in efforts to seal underground repository walls).

The highly complex interactions between the mass transport mechanisms and the changing properties of a fractured porous medium make quantitative analysis of such systems very difficult. Precipitation and dissolution can significantly modify the physical and chemical properties of fractured media. Physical changes in fracture aperture, tortuosity, and thus effective mass diffusivities and permeabilities are coupled to the subsequent fluid flow and solute transport and the precipitation and dissolution reactions. As such, a variety of closure patterns can be observed in the field, which range from fully filled (mineralized) fractures to those with pockets or irregular layers on the walls. Chemical changes in the fractures, such as sorption and reaction capacities, also have a significant effect on chemical transport and fracture wall evolution.

As a first step to quantifying precipitation/dissolution processes in fractures, we first consider modeling of reactive transport. A considerable body of literature deals with the transport of conservative contaminants in individual fractures. In contrast, however, little attention has been devoted to the analysis of the transport of reactive contaminants in fractures. The existing literature has been reviewed by Berkowitz and Zhou (1996), who developed a model to quantify the transport of reactive contaminants in a simplified fracture. The model, which accounts for advection, molecular diffusion, and interphase mass transfer between the aqueous phase and the fracture walls, demonstrates that the solute transport is controlled by the interplay between the Damköhler number (characterizing the effect of reaction relative to that of molecular diffusion on solute transport) and the Peclet number (characterizing the effect of advection relative to that of molecular diffusion on solute transport).

Clearly, sorption reactions in groundwater systems involve many kinds of chemical constituents and different chemical and physical processes. Thus no single model can successfully describe all kinds of sorption reactions. Berkowitz and Zhou (1996) considered several surface reaction models, including irreversible first-order kinetics, instantaneous reversible reactions, and reversible first-order kinetics. They determined criteria under which one can model reactive solutes as nonreactive, and reversible reactions as irreversible, in terms of time evolution of concentration distributions. They also gave a criterion for applicability of the local equilibrium assumption and subsequent use of a retardation coefficient.

Numerical models describing precipitation and dissolution in porous media and fractures were presented by Sallès et al. (1993) and Békri et al. (1995), who investigated a variety of idealized geometries that included parallel plate channels and channels with opposite protrusions on each wall. Similar numerical models for precipitation in deterministic and stochastic fractures were presented by Mourzenko et al. (1996) and Békri et al. (1997). Numerical models for precipitation and dissolution in a constant aperture fracture and the surrounding evolving porous matrix were given by Savage and Rochelle (1993) and Steefel and Lichtner (1994). Realistic parameters were used to simulate a fracture originating from a cement-bearing nuclear waste repository. These studies examined effects on the surrounding medium but did not study the evolution of the fracture geometry itself. A model for the evolution of fracture network maze patterns, for fractures with initially large apertures, was presented by Palmer (1975, 1991), while Dreybrodt (1996) analyzed the time needed for the development of mature karst channels. A numerical model to investigate the minimum hydrogeochemical conditions for fracture dissolution in limestone was given by Groves and Howard (1994a). Their results indicate that the selective enlargement of fracture flow passages may be strongly influenced by their initial sizes. Groves and Howard (1994b) and Howard and Groves (1995) then developed numerical models for fracture dissolution in limestone for laminar and turbulent flow. Their results show that for laminar flow, passage enlargement taking place early in the conduit network development is highly selective, while the transition to turbulent flow often results in more general passage enlargement.

Dijk and Berkowitz (1998) examined the evolution of fracture aperture over the entire length of the fracture, as a function of time, due to solute precipitation and dissolution. Irreversible first-order kinetic surface reactions were considered. The effect of the relevant transport mechanisms, the fracture geometries, and the initial and boundary conditions on the evolution of the fracture properties, fluid flow, and solute transport for geological systems and timescales were investigated. Calculations were based on a wide range of parameter values estimated from data available in the literature. Results showed the evolution of the solute transport and fracture geometry as a function of the Damköhler and Peclet numbers. Figure 4-4 illustrates typical simulated fracture mineralization patterns. Fracture closure times are found to be of the order of days to millions of years, for half-life reaction times (Langmuir and Mahoney, 1984) of the order of seconds to years, and for fluid residence times of the order of minutes to days. These closure times are compatible with typical hydrogeological time scales.

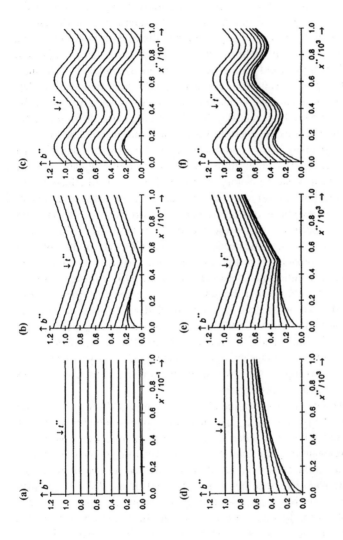

FIGURE 4-4 Simulated fracture mineralization patterns for three fractures and two rates of flow and chemical reaction. Predicted dimensionless half-aperture b^{**} as a function of dimensionless location within the fracture, x^{**}, and of dimensionless time t^{**}. For initially flat (a), (d); linearly constricted (b), (e); and sinusoidal (c), (f) fracture walls. Line of symmetry (fracture center) in each fracture is given by the x-axis. Cases (a), (b), (c): large flow rate relative to rate of reaction; cases (d), (e), (f): low flow rate relative to rate of reaction. Modified from Dijk, P. E., and B. Berkowitz, 1998. Precipitation and dissolution of reactive solutes in fractures. Water Resources Research 34(3): 457-470. Copyright by American Geophysical Union.

Fluid Flow and Chemical Transport in Partially Saturated Fracture Systems

The conceptual pictures and quantitative models of fluid flow and chemical transport in the fully saturated fractures that we have discussed above are not directly applicable to fluid flow and chemical transport in the fractured vadose zone. However, our established understanding of these processes in saturated systems forms a valuable knowledge base from which we can work.

In the context of fluid flow and chemical transport in partially saturated fractured systems, we can consider flow in initially unsaturated (or partially saturated) single fractures, accounting only for the fracture itself, or accounting for the fracture together with the porous medium in which it is embedded. However, even at this stage, we must define what, precisely, we mean by infiltration, or fluid flow and chemical transport, in unsaturated (or partially saturated) fractures. Consider a fracture that is, initially, completely unsaturated. As shown by, for example, Di Pietro et al. (1994) and Di Pietro (1996), using numerical simulations with lattice gas automata, the pattern of infiltration and fluid distribution is largely a function of the rate of infiltration (relative to the dimensions of the fracture). In other words, for high rates of fluid entering the fracture, the fracture becomes fully saturated, and we need consider only the transient advance of the wetting front as partially saturated (e.g., Nicholl et al., 1992, 1993a, 1993b). For slower rates of fluid infiltration, film flows can develop along the fracture walls. The significant advance of fluid along such films, even with thicknesses of the order of 1 μm, has been demonstrated by Tokunaga and Wan (1997). Similar observations about saturation patterns can be made for mixed fracture/porous matrix systems (e.g., Glass and Norton, 1992). Models that account for film flow and fracture-matrix interactions are discussed in detail in other chapters in this volume, and we therefore do not dwell on them here.

A critical feature in these cases is that medium heterogeneity and irregular distributions of fluid lead to the development of preferential flow paths. Because fluid flow patterns may be highly ramified and sensitive to small-scale details, use of simulation models that yield only averaged, "effective" behavior may be of highly limited value, and will in many instances yield meaningless results. We can, however, conceptualize the fracture/porous matrix system as a heterogeneous porous medium, particularly if the fracture has very rough walls, and/or contains filling material. Several modeling approaches can then be considered, depending upon whether the emphasis is on (1) the steady-state fluid flow and transient chemical transport through a partially saturated domain, or (2) the transient evolution of either the advancing front of fluid or contaminant, or the actual flow paths.

In the former case, flow and transport models for heterogeneous, fully saturated domains can be applied simply by restricting the conducting, saturated portion of the domain to account for the air phase (e.g., Birkholzer and Tsang, 1997). Alternatively, in the latter case, we can introduce a variety of pore network and other statistical models (such as percolation theory variants) in order to focus on an advancing wetting front or a transient infiltration pattern (Blunt and Scher, 1995), or use a "dripping faucet" analogy (Shaw, 1984) to characterize the often irregular (chaotic) patterns of inflow and outflow in such fractures. Simple growth models that mimic the essential physics of growth processes are known for specific instances, such as viscous fingering (Paterson, 1984; Lenormand et al., 1988), capillary fingering (Wilkinson and Willemsen, 1983; Wilkinson, 1986; Lenormand et al., 1988), and gravity fingering (Meakin et al., 1992; Glass and Yarrington, 1996). Other generalized growth models that attempt to mimic the full range of behavior are the subject of current investigations (Ewing and Berkowitz, 1998; Glass et al., 1998). Such models account for a wide range of viscous, gravity, and capillary forces, all in the presence of medium heterogeneities.

The literature devoted to description of these various models is vast, and is beyond the scope of this analysis. The interested reader can find detailed descriptions of these models, and extensive references to the relevant literature, in, for example, Blunt and Scher (1995), Berkowitz and Ewing (1998), Ewing and Berkowitz (1998), and Glass et al. (1998).

Fracture Networks

We consider now conceptual and quantitative models of fluid flow and chemical transport in fracture networks. These models are influenced directly by the flow and transport mechanisms and behaviors outlined above for single fractures.

Network Structure and Fluid Flow

The complex and usually non-Gaussian nature of distributions of fluid fluxes and migrating contaminants in fracture networks is well-known in the literature. In particular, field and laboratory experiments in natural fractures have demonstrated strong evidence of channeling and highly preferential flow paths in individual fractures and in fracture networks (e.g., Neretnieks et al., 1982; Neretnieks, 1993 and references therein). Field data, for example, from a large-scale investigation of fracture flow in a granite uranium mine at Fanay-Augères, France, show a difference of four orders of magnitude between the largest and smallest injection flow rates, despite very good fracture connectivity. Cacas et al. (1990a, 1990b) concluded that the high degree of heterogeneity and channeling is due to a broad distribution of fracture conductivities, and that it overwhelmingly governs fluid flow and transport behavior.

In order to study these phenomena, two-dimensional (2D) (e.g., Schwartz et al., 1983; Smith and Schwartz, 1984; Charlaix et al., 1987; David, 1993; Berkowitz and Scher, 1997, 1998) and three-dimensional (3D) (e.g., Andersson and Dverstorp, 1987; Billaux et al., 1989; Cacas et al., 1990a, 1990b; Nordqvist et al., 1996) models have considered aspects of the influences of fracture connectivity and fracture conductivity distributions, both in fracture planes and at the scale of a full fracture network. Tsang and Tsang (1987), Tsang et al. (1988), Moreno and Neretnieks (1993), and others have used similar models and also replaced the heterogeneities by systems of statistically equivalent, variable aperture, one-dimensional channels. The major emphasis of these studies has been on analysis of contaminant breakthrough curves.

Another approach to modeling flow and transport in fracture networks, which is able to capture distinct preferential flow paths and channeling, is based on percolation theory (e.g., Englman et al., 1983; Robinson, 1983, 1984; Hestir and Long, 1990; Balberg et al., 1991; Berkowitz and Balberg, 1993; Berkowitz, 1995). In this framework, and near the percolation threshold—the point at which the fractures are "just connected" across the entire domain—network structures display channeling patterns and transport properties that are quantifiable by power law relationships. The proximity to the threshold often depends on the scale of measurement, and in some cases the backbone substructure is obscured due to the presence of many disconnected fractures. However, there is considerable evidence from field measurements and theoretical considerations that many field-scale fracture networks are indeed near the percolation threshold (Chelidze, 1982; Guéguen et al., 1991; Crampin and Zatsepin, 1996; Renshaw, 1996).

The principal feature of disordered networks near the percolation threshold is the distribution of critical single bonds that connect large "blobs" of fractures. Flow through the percolating cluster occurs only along the backbone, that is, the hydraulically conducting portion of the percolating cluster. Many of the open fracture segments in a percolation-generated domain do not conduct fluid, since either they do not belong to the percolating cluster, or they form dangling branches on the backbone analogous to dead-end pores.

The analogy between real fracture networks and percolation systems is well known. Real fracture networks can, at least in some cases, be mapped to percolation lattices and random domains (e.g., Winterfeld et al., 1981; Jerauld et al., 1984a, 1984b; Hestir and Long, 1990) and, in general, the same percolation scaling properties hold for lattices and continuum systems. Numerical simulations of 3D fracture networks can map fractures to a regular orthogonal array of platelets (Englman and Jaeger, 1990). Alternatively, flow channels that exist within the planes of real fractures (Neretnieks, 1993) can be mapped to networks of tubes.

In general, models accounting for fracture-system complexity have thus been based on the incorporation of fracture planes containing aperture distributions, or idealized flow paths in 3D connecting the centers of intersecting disks, or geometrically sparse networks near the percolation threshold. Most recently, Margolin et al. (1998) developed a lattice model that includes all of these key elements in a single framework. This approach allows a simple characterization of a range of geometrical structures with a range of aperture distributions, in both 2D and 3D fracture systems. The model uses anisotropic percolation to generate random interconnecting bonds (representing channels within fractures), and accounts for aperture variation within fracture planes and among fractures.

Margolin et al. (1998) examined the interplay and relative importance of the key factors that lead to formation of channeled flow paths in fracture networks. Specifically, they considered (1) structural effects, including network density and anisotropy, and (2) hydraulic effects, governed by aperture variability. The presence of rough-walled fractures and filling material can also be incorporated in the aperture variability. Analysis of flow in these lattice networks demonstrates that either of the elements—fracture geometry and aperture variability—can give rise to channeled flow, and that the interplay between them is especially important. Figure 4-5 illustrates typical preferential flow patterns in this system. The bonds represent channels within fracture planes. As the degree of aperture variability increases, the preferential paths become sparser and more sharply defined. Thus, in practical terms, it is clear that parameter values that are measured or inferred for use in discrete fracture network models may have only limited physical significance. A critical observation for our analysis is that the effect of large aperture variability renders even dense networks poor conductors, so that from the point of view of fluid flow, the networks behave as though near the percolation threshold. Margolin et al. (1998) developed a scaling relationship that quantifies the dependence of effective hydraulic conductivity on aperture variability and on the network structure and fracture element density; this result is also used to derive an explanation for the field-length dependence of permeability frequently observed in fractured formations.

In light of the above, it seems reasonable to expect that in many field-scale situations, the portion of the fracture network that actively contributes to fluid flow and chemical transport is, functionally, near a percolation threshold, due either to structural characteristics and/or aperture variability. Analogously, in partially saturated systems, only a small fraction of fracture areas actually conduct fluid (as discussed earlier), and we therefore suggest that fluid flows in partially saturated fracture networks may also exhibit flow patterns similar to those shown in Figure 4-5. Moreover, the presence of film flow may permit preferential (or percolation-like) fluid flow and chemical transport even in fractured porous media with very low bulk saturation. In these contexts, dynamics of fluid flow and chemical transport in fracture networks can also be quantified using the modeling approaches outlined earlier.

Chemical Transport

The issue of how to quantify the migration of contaminants in fractured and fractured-porous geological formations has received overwhelming attention during the last three decades, and a vast literature dealing with the subject has

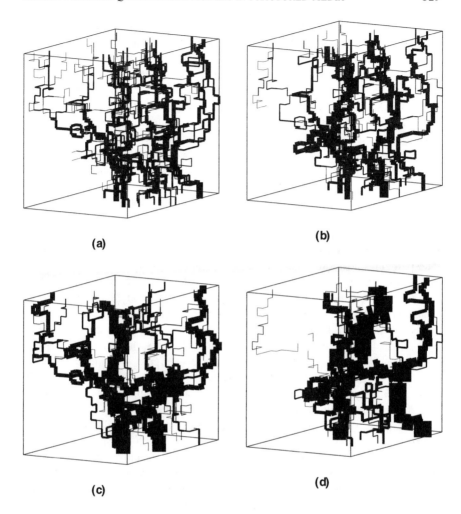

(a)

(b)

(c)

(d)

FIGURE 4-5 Illustration of channeled flow in fracture networks. Typical spatial distributions of (normalized) bond discharges in three-dimensional lattice networks, for a range of bond aperture variability (defined by b, the standard deviation of the log aperture). The lattice is relatively well connected and consists of 30 percent of the bonds (located randomly) in a full orthogonal lattice. (a) $b = 0.0$, (b) $b = 0.5$, (c) $b = 1.0$, (d) $b = 2.0$. The thickness of the bonds is proportional to the relative discharges they carry; only those bonds carrying at least 1 percent of the total volumetric flow through the network are shown. After Margolin, G., B. Berkowitz, and H. Scher, 1998. Structure, flow, and generalized conductivity scaling in fracture networks. Water Resources Research 34(9): 2103-2121. Copyright by American Geophysical Union.

developed. And yet, our ability to quantitatively predict contaminant migration in these systems remains severely limited, and questions remain about how best to model such transport. Discrete fracture models require detailed information on structural and hydraulic properties of fractures, and are hampered by heavy computational requirements. On the other hand, use of the advection-dispersion equation (ADE), and stochastic theories that generalize the ADE by substituting a more detailed stochastic process for the Fickian assumption, have not successfully quantified observed behavior (e.g., Dagan and Neuman, 1997).

It is beyond the scope of this report to review all of the relevant literature. Rather, we shall briefly discuss a new framework that we have recently proposed (Berkowitz and Scher, 1995, 1997, 1998; Berkowitz et al., 2000) in which frequently observed anomalous (non-Gaussian) chemical transport can be quantified: if the variation in the velocity field is sufficiently large, a highly discrete picture of the contaminant (particle) motion can be developed. The particle motion is approximated as a series of discrete steps, each having a different velocity. The particle's position is determined at the points of transition between these steps. A continuous time random walk (CTRW) can be introduced to naturally account for the cumulative effects of a sequence of transitions. The transitions are characterized by $\psi(\mathbf{s}, t)$, the probability rate for a displacement $\mathbf{s}$ with a difference of arrival times of t. In the CTRW approach, one can determine the evolution of the particle distribution (plume), $P(\mathbf{s}, t)$, for a general $\psi(\mathbf{s}, t)$, so there is no a priori need to consider the moments of $P(\mathbf{s}, t)$. The challenge is to map the important aspects of the particle motion in the medium onto a $\psi(\mathbf{s}, t)$. A mapping that results in a $\psi(\mathbf{s}, t)$ with an algebraic low velocity tail for large time leads to the description of anomalous transport (e.g., non-Gaussian plumes).

We have applied the CTRW to an explicit, discrete medium—a random fracture network. We analyzed a series of Monte Carlo transport simulations based on synthetic random fracture networks (Berkowitz and Scher, 1997, 1998). The spatial and temporal behavior of the contaminant advance, based on particle tracking simulations in these networks, is shown in Figure 4-6a, while theoretical concentration profiles based on CTRW are shown in Figure 4-6b. The CTRW curves are characterized by a peak that lies close to the origin, with a forward tail that spreads in response to the flow field. As time progresses, the distribution approaches a step function, increasingly uniform in space, with the residual position of the peak indicated by the sharp drop near the origin.

The shapes of the spreading pulse $P(x, t)$ shown in Figure 4-6 are qualitatively similar to those found from the particle tracking simulations in these fracture networks. Despite the statistical noise in these results due to vertical averaging and a relatively small number of realizations, the CTRW theory reproduces the development of highly skewed particle plumes. Significantly, the CTRW solution also captures the variation of both the movement of the center of mass of the particles and the standard deviation of particle location around this center of mass.

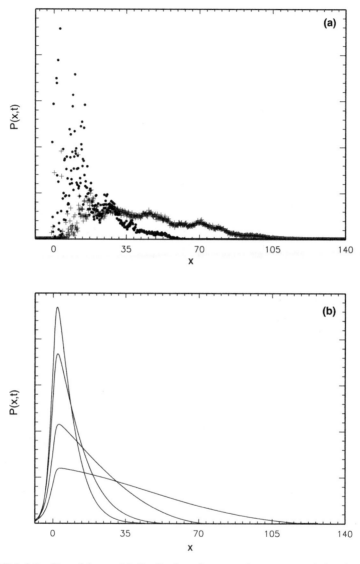

FIGURE 4-6 Profiles of the spatial distribution of a contaminant at two relative times (t = 1, points; t = 2.5, crosses), determined from numerical simulations of particle transport in a two-dimensional, random fracture network. The concentration profile, $P(x, t)$ is the vertical average (along the y-axis) of $P(s, t)$, in arbitrary units; particle injection point is at $x = 0$. (b) Profile of the spatial distribution of a contaminant, $P(x, t)$, determined from the CTRW solution, at four relative times t = 1, 2.5, 10, and 37.5. The concentration profile, $P(x, t)$, is the vertical average (along the y-axis) of $P(s, t)$, in arbitrary units; particle injection point is at $x = 0$. After Berkowitz, B., and H. Scher, Phys. Rev. Lett. 79(20), 1997, pp. 4038-4041. Copyright by the American Physical Society.

As in the previous sections, we suggest that this framework, developed for fully saturated systems, may also be suitable for describing chemical transport in partially saturated fractured media. In partially saturated systems, only a small portion of the domain actually conducts fluid and chemicals, and as suggested in the section on Network Structure and Fluid Flow, fluid flows in partially saturated fracture networks may also exhibit flow patterns similar to those shown in Figure 4-6. Thus, it may be possible to modify the transition probability $\psi(\mathbf{s}, t)$ to capture the broad distribution of fluid velocities (from fast film flows to stagnant regions, as well as fracture-matrix interactions). Alternatively, and especially for slower flow regimes, statistical growth models such as outlined in the section on Fluid Flow and Chemical Transport in Partially Saturated Fracture Systems may be applicable.

EVALUATION OF CONCEPTUAL AND QUANTITATIVE MODELS: FIELD AND LABORATORY DATA

From the survey in the previous section, it is clear that a variety of theoretical models must be invoked in order to quantify the broad spectrum of fluid flow and chemical transport behaviors. Similarly, experiments at a variety of laboratory and field scales, under both fully and partially saturated conditions, are required in our efforts to understand and predict these behaviors. In this context, we discuss below several recent, representative experimental studies.

Laboratory-Scale Study of Single Fractures

As discussed in the section on Single Fractures, the roughness of fracture walls, and dynamic changes in the roughness (as well as filling material) can significantly influence the hydraulic properties of a fracture. Fluid flow in rough-walled rock fractures is complex even under laminar flow regimes, and measuring flow in such fractures is prohibitively difficult. As a result of the paucity of experimental data, existing models are based largely on unvalidated assumptions and simplifications. Dijk et al. (1999) use nuclear magnetic resonance imaging (NMRI) to directly measure flow patterns in natural rock fractures, and to examine the effects of fracture wall morphology. The investigation focuses on a qualitative and quantitative description of the fracture surface geometry, water flow velocities and flow rates, flow and stagnant regions, critical aperture and velocity paths, and flow pattern stability and reproducibility. In particular, the effects of a sharp step discontinuity of the fracture walls are studied.

Four sets of 2D slices through a 3D velocity image, including a pair of (zero fluid velocity) calibration tubes, are displayed in Plate 3. The 3D velocity images presented here show the velocity components in the principal (z) flow direction. A variety of flow features, including irregular distributions of high- and low-velocity zones, and the effect of the fracture discontinuity can be detected easily.

The significant backflow region in the vicinity of the fracture discontinuity is indicated by the arrow. It should be noted that in the vicinity of rough fracture walls, eddies and turbulent regions may develop, while at some distance from the walls, fully laminar flow is expected to occur.

From the velocity data and theoretical considerations, the applicability of the local cubic law (LCL) is examined. As a result of the complex 3D geometry, velocity profiles are generally parabolic, as assumed by the LCL, but often highly asymmetric with respect to the fracture walls. These asymmetric velocity profiles are clustered together, with significant correlations; they are not just local random phenomena. Analysis of the z-direction velocity component in a single cross-sectional plane (Dijk et al., 1999) suggests that the effects of the measured asymmetry on volumetric flow rates are insignificant. However, analysis of the complete 3D distribution of the z-direction velocity components (Dijk and Berkowitz, 1999) demonstrates that the actual flow rate is significantly less than that predicted by the LCL. It is expected, moreover, that the asymmetries will have a significant influence on chemical solute transport, dissolution, and precipitation. The features discussed in these studies emphasize the strong heterogeneity and the highly 3D nature of the flow patterns in natural rock fractures, even for relatively small Reynolds numbers, and the need for 3D flow analyses.

Fracture wall morphology may change over time as a result of geochemical and mechanical processes. The evolution of unsaturated fracture walls was investigated by Weisbrod et al. (1998, 1999), who used high-resolution laser scanning to measure topographical variations of fractured chalk surfaces. The chalk surfaces were either (1) immersed in tap water and then air-dried over two wetting cycles, for 10 min and for 14 h (Weisbrod et al., 1998), or (2) installed in customized flow cells and exposed to short flow events (8, 9, and 24 h) of synthetic rainwater, followed by long drying periods (Weisbrod et al., 1999). Both coated and uncoated fracture surfaces were investigated in the latter study. The surfaces were found to erode by up to 0.352 mm following the two cycles of immersion in tap water and by up to 0.313 mm following the three cycles of synthetic rainwater flow. Erosion was more pronounced in the coated surface than in the uncoated surface, reflecting its mechanical instability. Figure 4-7 shows the changes in one of the (uncoated) fracture walls after each of the wetting cycles, along with the changes in topography along a single cross-sectional profile of a fracture wall. The changes in wall morphology can be related to a non-uniform release of particles from the fracture walls. The erosion thickness was found to be strongly correlated to the thickness of a layer calculated from the total accumulated mass of particles and soluble salts released from the surface. An important result is that processes of precipitation and dissolution, as well as particle deposition, weathering, and clay swelling, may be significant in relatively soft rock formations. Moreover, these processes may be enhanced in the fractured vadose zone because of the transient nature of water infiltration and water retention.

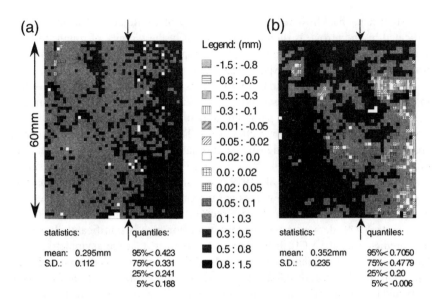

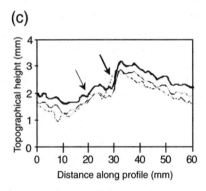

FIGURE 4-7 Map of the topographic height difference of a fracture wall following (a) a 10-min wetting cycle and (b) a 14-h wetting cycle. Note that some areas, particularly along the right edge, were uplifted after the 14-h wetting cycle. Basic statistics characterizing the differences are also shown. Each map contains ~2500 data points. Positive and negative values indicate decreases and increases in topographical height, respectively. Figure 4-7c shows the topographical height of three profiles along a cross section of the fracture wall sample, indicated by the arrows on (a) and (b). The bold line depicts the initial topographical height of the sample. The solid and dashed lines were obtained after submerging the core in tap water for 10-min and 14-h, respectively. After Weisbrod, N., R. Nativ, D. Ronen, and E. Adar, 1998. On the variability of fracture surfaces in unsaturated chalk. Water Resources Research 34: 1881-1887. Copyright by American Geophysical Union.

Chemical Transport

We introduced a CTRW framework (see above) to quantify anomalous (non-Gaussian) chemical transport in fractured and heterogeneous porous media. We have used these solutions (Berkowitz and Scher, 1998) to suggest the occurrence of anomalous transport in natural, heterogeneous porous media, by interpreting field data from a large-scale field study performed at the Columbus Air Force Base (Adams and Gelhar, 1992). At this site, a highly heterogeneous alluvial aquifer, bromide was injected as a pulse and traced over a 20-month period by sampling from an extensive 3D well network. The tracer plume that evolved was remarkably asymmetric and could not be described by classical Gaussian models (Adams and Gelhar, 1992). Significantly, the CTRW captured the variation of both the movement of the center of mass of the particles and the standard deviation of particle location around this center of mass in the direction of flow, determined from the field data. Here, in agreement with CTRW theory, interpretation of the moments suggested that both the mean and standard deviation of the plume position scale with time (t) as $t^{0.6}$. Details of the CTRW solutions and full analyses of these data, including discussion of how the parameters for the CTRW model can be obtained from site data, are given in Berkowitz and Scher (1998) and Berkowitz et al. (2000).

We have also recently used the CTRW framework to evaluate the first passage time distributions, or breakthrough curves, and compared them to measurements from a tracer migration experiment in a heterogeneous sandbox model (Berkowitz et al., 2000). We find that the curves fit the range of measured data remarkably well, and in particular match both the early time behavior and the late time (long tail) concentration breakthroughs. In contrast, these data could not be adequately modeled using the (Gaussian) ADE.

Clearly, long tail or otherwise skewed concentration profiles can, in some cases, be reconciled with the ADE by, for example, imposing a trend in the mean velocity. However, such trends can only be considered if they are indicated by site-specific field measurements. On the other hand, the CTRW theory would appear to be a promising general framework that allows analysis of transport in formations containing complex heterogeneities, at large scale, which are not amenable to analysis using classical advection-dispersion theory. While much research remains to be done in this direction, the CTRW seems to represent a potentially valuable tool in the assessment of dispersive processes in heterogeneous porous media.

Our analysis of laboratory and field data indicates that the observed anomalous (non-Gaussian) behavior results from subtle features in the flow fields. These distributions are present in (steady-state) partially saturated porous media, and incorporate mixtures of fast and slow paths, over long and short distances. We suggest, therefore, that similar behaviors will be found in partially saturated systems, as a function of the degree of saturation and of the medium itself.

Field-Scale Analysis of Fluid Flow and Chemical Transport in the Fractured Vadose Zone

A small number of studies have attempted to directly measure fracture flow in laboratory set-ups (as discussed above), but extrapolation of such results to field conditions is not straightforward. The few studies in which fracture flow and transport were investigated on a field scale determined these parameters indirectly, that is, via the matrix surrounding the preferential pathways. Measurements of matrix chemical and isotopic compositions (Komor and Emerson, 1994; Nativ et al., 1995; Wood et al., 1997), or matrix water content and permeability (Scanlon, 1991, 1992) were used to infer preferential flow and transport in the embedded fractures. Dahan et al. (1998) presented a methodology designed to address the need for direct quantitative measurements of flow and transport in fractures. These measurements provide insight into field-scale heterogeneity that strongly affects the measured properties.

The methodology and the experimental set-up described here were developed and applied in a study located in the northern part of the Negev Desert, Israel. A large chemical industrial complex and the National Site for Treatment and Isolation of Hazardous Wastes have been operating there since 1975. The underlying geological infrastructure in the region consists of an Eocene chalk formation partly covered by thin layers of sand and loess. Because of the aridity of the area (180 mm/yr rainfall) and the low permeability of the chalk matrix (~2 millidarcy), the vadose zone was considered to be a natural barrier to potential groundwater contamination resulting from the industrial activities on the land surface. However, the high concentration of industrial pollutants found in the local groundwater (Nativ and Nissim, 1992), as well as other factors such as seasonal fluctuations in groundwater levels, water table responses to flood events, and the occurrence of tritium in the groundwater (Nativ and Nissim, 1992; Rophe et al., 1992), indicate that the groundwater is rapidly affected by on-surface activities. Evidently, infiltrating water and contaminants bypass the low-permeability chalk matrix via preferential flow paths, possibly through the numerous vertical to subvertical fractures intersecting the chalk matrix (characterized by Bahat, 1988). Indeed, preferential water flow and solute transport across the fractured chalk in the vadose zone was demonstrated using chemical and isotopic tracers (Nativ et al., 1995).

The experimental set-up described here (see Figure 4-8) was designed to measure directly and to assess the spatial distribution of flow and transport along fractures from water ponded at the land surface (Dahan et al., 1998). The experimental site was located in an ephemeral wash. Twenty-one percolation ponds were installed sequentially on top of a freshly cleaned rock ledge, along 5.3 m of a vertical fracture opening, dividing it into 21 equal sections. The area of each pond was 25 × 40 cm and its depth was 35 cm. One meter below the surface of the rock ledge, a 25-cm-diameter, 4.3-m-long horizontal borehole was cored along

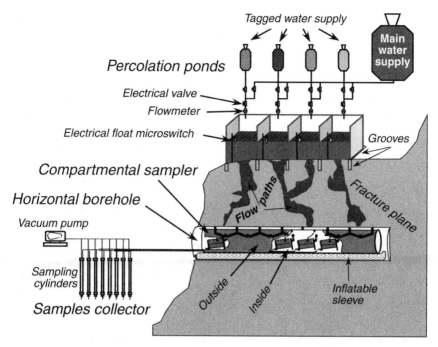

FIGURE 4-8 Schematic cross section of a measurement system designed to determine water fluxes and tracer migration rates in a field set-up through a single natural fracture in the vadose zone. The system components are a set of individually tagged percolation ponds, located above a rock ledge intersected by a vertical fracture; a horizontal borehole cored into the same fracture at a depth below the ledge and the percolation ponds; a compartmental sampler introduced into the borehole to collect the tagged water drained from the various ponds; and a sample collector to retrieve the fluids stored in the compartmental sampler. After Dahan et al. (1998). Reprinted by permission of Ground Water. Copyright 1998.

the fracture plane. A compartmental sampler composed of 21 cells (each 20 cm long) was introduced into the borehole. Thus packers attached to the top of the sampler divided the fracture opening at the borehole ceiling into 21 separate sections. The use of various tracers in the percolation ponds, and their detection in the drained effluents, enables delineation of flow trajectories connecting the ponds at the fracture inlet to the sampler cells in the horizontal borehole below.

A five-day (119-h) experiment was held in May 1997. The percolation ponds were filled to a constant water head, and the flow rate out of each pond was automatically measured and recorded as soon as water was introduced into the ponds. Simultaneously, the flow rate of the drained effluents collected by the sampler in the horizontal borehole was measured and the water was sampled for tracer concentration. The active flow paths along the fracture plane were resolved by two tracer tests. In each test, water in the percolation ponds was tagged, for 6 h and 4.5 h during the first and second tests, respectively. The first tracer test (test A) began 43 h after the beginning of the experiment, and the second test (test B) started 29 h later. The flow trajectories were defined by relating the tracers found in the various sampler cells in the borehole to their source ponds on the surface where they were applied. Details of the entire experiment are given in Dahan et al. (1999).

Flow rates into the upper opening of the fracture from each of the percolation ponds and into the lower opening in each of the sampler cells are presented in Figure 4-9. The temporal flow rates reflect the variability in fluid flow from each of the ponds to each of the sampler cells. The high variability in maximum flow rates observed at the inlet and outlet sides of the fracture indicates that only about 50 percent of the length of the fracture opening allowed any appreciable flow, whereas less than about 20 percent of the fracture opening transmitted high flow rates. Significantly, the temporal variations in flow rates—with no clear pattern—persisted throughout the duration of the five-day experiment; a steady flow pattern was not reached in any of the ponds or sampler cells during this period. Both abrupt and gradual temporal variations were observed.

Turbidity measurements in the drained effluents reflect the transport of suspended material through the fracture. Abrupt changes in turbidity (with maximum values of about 1500 nephelometric turbidity units [NTU]) were observed in samples from most cells throughout the duration of the experiment. These fluctuations reflect the unstable structure of the fracture void and its filling material (as noted by Weisbrod et al., 1998, 1999), and are linked to flow rate changes and stability.

Flow trajectories through the fracture—between the percolation ponds and the sampler cells—were estimated on the basis of the tracer tests, as well as by correlating fluxes between ponds and cells. The connecting lines in Figure 4-10 represent the relative contribution from each pond to each sampler cell. This relative contribution was determined by the percentage of the maximum relative concentration of a tracer derived from a pond and observed in the sampler cell. Not surprisingly, the flow trajectories between the ponds and the sampler cells— separated by only 1 m—were often not vertical, but rather shifted horizontally along the fracture plane.

Two important—and perhaps unexpected—observations can be made. First, the paths of tagged water from different source ponds often crossed without any observed mixing. For example, tagged water from pond 8 moving to sampler cells 4 and 5 crossed the zone where tagged water from ponds 3, 5, and 7 drained downward (Figure 4-10), and yet no indication of tracer from these ponds was found in sampler cells 4 and 5. This finding suggests that flow paths consist of discrete and unconnected channels within the main fracture plane. These flow paths could arise through the complex structure of the consolidated and uncon-

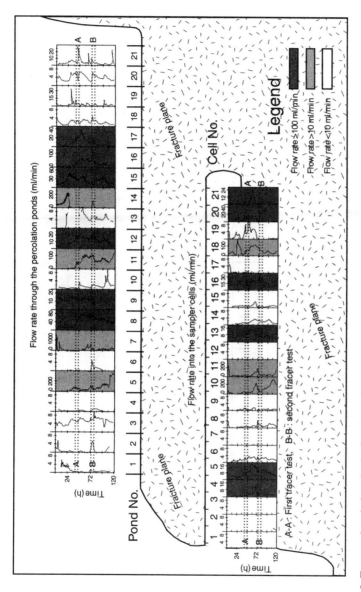

FIGURE 4-9 Temporal variations in the flow rate of each of the percolation ponds (above) and the sampler cells (below). The figure displays the exact position of the ponds with respect to the sampler cells. The flow-rate scale is variable according to the maximum measured flow rate in each individual pond/cell. The duration of the two tracer tests A and B is shown by dotted lines (A-A and B-B, respectively) on both graphs. After Dahan, O., R. Nativ, E. Adar, B. Berkowitz, and Z. Ronen, 1999. Field observation of flow in a fracture intersecting unsaturated chalk. Water Resources Research 35(11): 3315-3326. Copyright by American Geophysical Union.

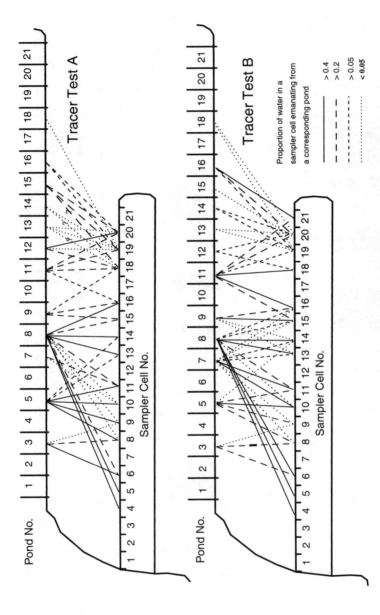

FIGURE 4-10 Flow trajectories connecting the source ponds on land surface to the sampler cells in the borehole, as defined by the two tracer tests. The line styles represent the relative water contribution of a certain pond to its corresponding sampler cell. After Dahan, O., R. Nativ, E. Adar, B. Berkowitz, and Z. Ronen, 1999. Field observation of flow in a fracture intersecting unsaturated chalk. Water Resources Research 35(11): 3315-3326. Copyright by American Geophysical Union.

solidated filling material with the fracture void, micro-dissolution channels along the fracture walls, or intersecting small fractures.

Second, the relative contributions of fluid from each pond to each sampler cell changed significantly during the experiment. The temporal variations in flow paths during the two tracer tests (Figure 4-10) are remarkable—evidently due to dissolution of filling material and walls, as well as to particle mobilization, migration, and entrapment, as evidenced by the turbidity measurements. Analysis of the data also indicated that the instability of the flow regime observed on a small scale at each individual fracture segment did not decrease on larger scales, for several combined segments or for the entire flow domain.

The observations of Dahan et al. (1998) suggest that large portions of the fracture plane play only a minor role, with small sections controlling most of the flow activity, probably within a few major discrete flow paths. Close examination of the fracture opening along the bottom of the ponds and along the borehole ceiling indicated that the hydraulically active sections of the fracture are related to dissolution channels along its plane. These dissolution channels were found to be closely related to fracture plane crossings between the main fracture plane and secondary fracture planes.

CONCLUDING REMARKS

We have considered aspects of fluid flow and chemical transport in the fractured vadose zone by building on our more established (though by no means complete) understanding of these processes in saturated systems. Other relevant experiments and models specific to unsaturated fractured media, such as film flow and fracture-matrix interactions, are discussed in other chapters in this volume.

Fluid flow and chemical transport behavior in fractured porous formations is highly complex. Models of flow and transport in these systems must account for the inherent heterogeneity and scaling properties of porous rocks, as well as the typically sparse and uncertain field data that can be obtained to characterize a geological formation. In the fractured vadose zone, key physical parameters that must be considered include, over a range of scales, the geometrical and hydraulic characteristics of individual fractures, their interconnection within networks, and their interplay with the porous host rock. Analysis of chemical transport must also include parameters that describe chemical reaction, dissolution, and precipitation, and particle detachment and trapping, all of which lead to spatial and temporal variations in flow and transport patterns. These parameters are also fundamental to flow and transport models in saturated, highly heterogeneous formations, and similar conceptual frameworks and quantitative tools can often, though not always, be applied. For example, the finding that 3D flow paths within a fracture plane can consist of discrete and unconnected channels supports our suggestion that fractures in the vadose zone may be treated as highly heterogeneous porous media.

From the modeling point of view, it is important to distinguish between predictions of large-scale (average) and small-scale flow and transport behavior. Here we stress the importance of capturing essential fluid flow and chemical transport behavior, such as highly ramified fluid flow patterns; such behavior is often extremely sensitive to small-scale medium details. Large-scale (average) models are clearly irrelevant for analysis of small-scale fluid velocity and chemical concentration distributions, or, for example, for prediction of first arrival times of a contaminant. But significantly, models that attempt to apply effective, or homogenized, parameters to describe even the average flow and transport behavior in heterogeneous media often fail. Careful measurement in the laboratory and the field has indeed demonstrated the dominant influence of small-scale heterogeneity. Thus, we consider alternative modeling approaches, depending on whether the emphasis is on steady-state fluid flow and transient chemical transport through a partially saturated domain, or on transient evolution of either the advancing front of fluid or contaminant, or of the actual flow paths.

It is essential that we obtain additional field and laboratory data in order to determine the applicability of these various models. To date, there exist only a very limited number of large-scale laboratory and field experiments in the fractured vadose zone. A major difficulty lies in actually measuring the parameters of interest, particularly under unsaturated conditions. The other key difficulty is that we must account for high degrees of uncertainty, and for the interplay of flow and transport dynamics over a range of spatial and temporal scales. For example, if fluid flow in the fractured vadose zone follows a path analogous to a percolation system near the threshold, then the likelihood of measuring and actually delineating this path is very small. On the other hand, theoretical investigations of fluid flow and chemical transport, predicated on available experimental evidence, have yielded tremendous insight into the dynamics of flow and transport in the fractured vadose zone. The use of probabilistic and statistical tools allows us to characterize these dynamics, and to place quantitative bounds on fluid flow and chemical transport processes. New experiments with measurements that are sensitive over a range of scales to the geometrical, hydraulic, and chemical heterogeneity of fractured vadose zones are required to further refine these models.

ACKNOWLEDGMENT

Brian Berkowitz thanks the European Commission Environment and Climate Program (Contract No. ENV4-CT97-0456) for partial support.

REFERENCES

Adams, E. E., and L. W. Gelhar, 1992. Field study of dispersion in a heterogeneous aquifer. 2. Spatial moment analysis. Water Resources Research 28(12): 3293-3308.

Amadei, B., and T. Illangasekare, 1994. A mathematical model for flow and solute transport in non-homogeneous and anisotropic rock fractures. Int. J. Rock Mech. Min. Sci. & Geomech. Abs. 31: 719-731.

Andersson, J., and B. Dverstorp, 1987. Conditional simulations of fluid flow in three-dimensional networks of discrete fractures. Water Resources Research 23(10): 1876-1886.

Bahat, D., 1988. Early single layer and late multi-layer joints in the lower Eocene chalk near Beer Sheva, Israel. Annales Tectonicae ii: 3-1.

Balberg I., B. Berkowitz, and G. E. Drachsler, 1991. Application of a percolation model to flow in fractured hard rocks. Journal of Geophysical Research 96(B6): 10015-10021.

Békri, S., J. F. Thovert, and P. M. Adler, 1995. Dissolution of porous media. Chem. Eng. Sci. 50(15): 2765-2791.

Békri, S., J. F. Thovert, and P. M. Adler, 1997. Dissolution and precipitation in fractures. Eng. Geol. 48: 283-308.

Berkowitz, B., 1995. Analysis of fracture network connectivity using percolation theory. Math. Geology 27(4): 467-483.

Berkowitz, B., and P. M. Adler, 1998. Stereological analysis of fracture network structure in geological formations. Journal of Geophysical Research 103(B7): 15,339-15,360.

Berkowitz, B., and I. Balberg, 1993. Percolation theory and its application to groundwater hydrology. Water Resources Research 29(4): 775-794.

Berkowitz, B., and H. Scher, 1995. On characterization of anomalous dispersion in porous and fractured media. Water Resources Research 31(6): 1461-1466.

Berkowitz, B., and J. Zhou, 1996. Reactive solute transport in a single fracture, Water Resources Research 32(4): 901-913.

Berkowitz, B., and H. Scher, 1997. Anomalous transport in random fracture networks, Phys. Rev. Lett. 79(20): 4038-4041.

Berkowitz, B., and R. P. Ewing, 1998. Percolation theory and network modeling applications in soil physics. Surveys in Geophysics 19(1): 23-72.

Berkowitz, B., and H. Scher, 1998. Theory of anomalous contaminant migration in random fracture networks. Phys. Rev. E 57(5): 5858-5869.

Berkowitz, B., H. Scher, and S. E. Silliman, 2000. Anomalous transport in laboratory-scale, heterogeneous porous media. Water Resources Research 36(1):149-158 (correction: 36(5): 1371).

Billaux, D., J. P. Chiles, K. Hestir, and J. Long, 1989. Three-dimensional statistical modelling of a fractured rock mass: An example from the Fanay-Augères mine. Int. J. Rock Mech. Min. Sci. & Geomech. Abstr. 26(3/4): 281-299.

Birkholzer, J., and C.-F. Tsang, 1997. Solute channeling in unsaturated heterogeneous porous media. Water Resources Research 33(10): 2221-2238.

Blunt, M., and H. Scher, 1995. Pore-level modeling of wetting. Phys. Rev. E 52: 6387-6403.

Brown, S. R., 1987. Fluid flow through rock joints: The effect of surface roughness. Journal of Geophysical Research 92B: 1337-1347.

Brown, S. R., and C. H. Scholz, 1985. Broad bandwidth study of the topography of natural rock surfaces. Journal of Geophysical Research 90B: 12,575-12,582.

Brown, S. R., H. W. Stockman, and S. J. Reeves, 1995. Applicability of the Reynolds equation for modeling fluid flow between rough surfaces. Geophysical Research Letters 22: 2537-2540.

Cacas, M. C., E. Ledoux, G. de Marsily, B. Tillie, A. Barbreau, E. Durand, B. Feuga, and P. Peaudecerf, 1990a. Modeling fracture flow with a stochastic discrete fracture network: Calibration and validation. 1. The flow model. Water Resources Research 26(3): 479-489.

Cacas, M. C., E. Ledoux, G. de Marsily, A. Barbreau, P. Calmels, B. Gaillard and R. Margritta, 1990b. Modeling fracture flow with a stochastic discrete fracture network: Calibration and validation. 2. The transport model. Water Resources Research 26(3): 491-500.

Charlaix, E., E. Guyon, and S. Roux, 1987. Permeability of a random array of fracture of widely varying apertures. Transp. Porous Media 2: 31-43.

Chelidze, T. L., 1982. Percolation and fracture. Phys. Earth Planet. Int. 28: 93-101.

Crampin, S., and S. V. Zatsepin, 1996. Rock, dead or alive? Theory prompts new understanding of fluid-rock interaction and shear-wave splitting. EOS, Transactions, American Geophysical Union 77(30): 281.

Dagan, G. and S. P. Neuman (eds.), 1997. Subsurface Flow and Transport: A Stochastic Approach. Cambridge: Cambridge University Press.

Dahan, O., R. Nativ, E. Adar, and B. Berkowitz, 1998. A measurement system to determine water flux and solute transport through fractures in the unsaturated zone. Ground Water 36: 444-449.

Dahan, O., R. Nativ, E. Adar, B. Berkowitz, and Z. Ronen, 1999. Field observation of flow in a fracture intersecting unsaturated chalk. Water Resources Research 35(11): 3315-3326.

David, C., 1993. Geometry of flow paths for fluid transport in rocks. Journal of Geophysical Research 98(B7): 12,267-12,278.

Di Pietro, L. B., 1996. Application of a lattice-gas numerical algorithm to modelling water transport in fractured porous media. Transp. Porous Media 22: 307-325.

Di Pietro, L. B., A. Melayah, and S. Zaleski, 1994. Modeling water infiltration in unsaturated porous media by interacting lattice gas-cellular automata. Water Resources Research 30(10): 2785-2792.

Dijk, P. E., and B. Berkowitz, 1998. Precipitation and dissolution of reactive solutes in fractures. Water Resources Research 34(3): 457-470.

Dijk, P. E., and B. Berkowitz, 1999. Three-dimensional flow measurements in rock fractures. Water Resources Research 35(12): 3955-3960.

Dijk, P. E., B. Berkowitz, and P. Bendel, 1999. Investigation of flow in water-saturated rock fractures using nuclear magnetic resonance imaging (NMRI). Water Resources Research 35(2): 347-360.

Dreybrodt, W., 1996. Principles of early development of karst conduits under natural and man-made conditions revealed by mathematical analysis of numerical models. Water Resources Research 32(9): 2923-2935.

Durham, W. B., and B. P. Bonner, 1994. Self-propping and fluid flow in slightly offset joints at high effective pressures. Journal of Geophysical Research 99(B5): 9391-9399.

Englman, R., and Z. Jaeger, 1990. Fracture in concrete due to percolating cracks and pores. Physica A 168: 655-671.

Englman, R., Y. Gur, and Z. Jaeger, 1983. Fluid flow through a crack network in rocks. J. Appl. Mech. 50: 707-711.

Ewing, R. P., and B. Berkowitz, 1998. A generalized growth model for simulating initial migration of dense non-aqueous phase liquids. Water Resources Research 34(4): 611-622.

Ge, S., 1997. A governing equation for fluid flow in rough fractures. Water Resources Research 33(1): 53-61.

Glass, R. J., and D. L. Norton, 1992. Wetted-region structure in horizontal unsaturated fractures: Water entry through the surrounding porous matrix. In: Proceedings, International High Level Radioactive Waste Management Conference, Las Vegas, Nevada, April.

Glass, R. J., and L. Yarrington, 1996. Simulation of gravity fingering in porous media using a modified invasion percolation model. Geoderma 70: 231-252.

Glass, R. J., M. J. Nicholl, and L. Yarrington, 1998. A modified invasion percolation model for low-capillary-number immiscible displacements in horizontal rough-walled fractures: Influence of local in-plane curvature. Water Resources Research 34(12): 3215-3234.

Groves, C. G., and A. D. Howard, 1994a. Minimum hydrochemical conditions allowing limestone cave development. Water Resources Research 30(3): 607-615.

Groves, C. G., and A. D. Howard, 1994b. Early development of karst systems. 1. Preferential flow path enlargement under laminar flow. Water Resources Research 30(10):2837-2846.

Guéguen, Y., C. David, and P. Gavrilenko, 1991. Percolation and fluid transport in the crust. Geophysical Research Letters 18(5): 931-934.

Hestir, K., and J. C. S. Long, 1990. Analytical expressions for the permeability of random two-dimensional Poisson fracture networks based on regular lattice percolation and equivalent media theories. Journal of Geophysical Research 95(B13): 21565-21581.

Howard, A. D., and C. G. Groves, 1995. Early development of karst systems. 2. Turbulent flow. Water Resources Research 31(1): 19-26.

Iwai, K., 1976. Fundamental Studies of Fluid Flow Through a Single Fracture. Ph.D. dissertation, University of California, Berkeley.

Jerauld, G. R., J. C. Hatfield, L. E. Scriven, and H. T. Davis, 1984a. Percolation and conduction on Voronoi and triangular networks: A case study in topological disorder. J. Phys. C17: 1519-1529.

Jerauld, G. R., L. E. Scriven, and H. T. Davis, 1984b. Percolation and conduction on the 3D Voronoi and regular networks: A second case study in topological disorder. J. Phys. C17: 3429-3439.

Komor, S. C., and D. G. Emerson, 1994. Movement of water, solute and stable isotopes in unsaturated zones of two sand plains in the upper Midwest. Water Resources Research 30: 253-267.

Langmuir, D., and J. Mahoney, 1984. Chemical equilibrium and kinetics of geochemical processes in groundwater studies. In: First Canadian/American Conference in Hydrogeology. B. Hitchon and E. I. Wallick, eds. Dublin, Ohio: National Water Well Association, pp. 69-75.

Lenormand, R., E. Touboul, and C. Zarcone, 1988. Numerical models and experiments on immiscible displacements in porous media. J. Fluid Mech. 189: 165-187.

Margolin, G., B. Berkowitz, and H. Scher, 1998. Structure, flow, and generalized conductivity scaling in fracture networks. Water Resources Research 34(9): 2103-2121.

Meakin, P., J. Feder, V. Frette, and T. Jøssang, 1992. Invasion percolation in a destabilizing gradient. Phys. Rev. A 46: 3357-3368.

Moreno, L., and I. Neretnieks, 1993. Fluid flow and solute transport in a network of channels. Journal of Contaminant Hydrology 14: 163-192.

Moreno, L., Y. W. Tsang, C. F. Tsang, F. V. Hale, and I. Neretnieks, 1988. Flow and tracer transport in a single fracture: A stochastic model and its relation to some field observations. Water Resources Research 24: 2033-2048.

Mourzenko, V. V., J.-F. Thovert, and P. M. Adler, 1995. Permeability of a single fracture: Validity of the Reynolds equation. J. Phys II France 5: 465-482.

Mourzenko, V. V., S. Békri, J. F. Thovert, and P. M. Adler, 1996. Deposition in fractures. Chem. Eng. Comm. 148(15): 431-464.

Mourzenko, V. V., O. Galamay, J.-F. Thovert, and P. M. Adler, 1997. Fracture deformation and influence on permeability. Phys. Rev. E56(3): 3167-3184.

Nativ, R., and I. Nissim, 1992. Characterization of a desert aquitard: Hydrologic and hydrochemical consideration. Ground Water 30: 598-606.

Nativ, R., E. Adar, O. Dahan, and M. Geyh, 1995. Water recharge and solute transport through the vadose zone of fractured chalk under desert conditions. Water Resources Research 31:253-261.

Neretnieks, I., 1993. Solute transport in fractured rock: Applications to radionuclide waste repositories. In: Flow and Contaminant Transport in Fractured Rock. J. Bear, C.-F. Tsang, and G. de Marsily, eds. New York: Academic Press, Inc., pp. 39-127.

Neretnieks, I., T. Eriksen, and P. Tähtinen, 1982. Tracer movement in a single fissure in granitic rock: Some experimental results and their interpretation. Water Resources Research 18(4): 849-858.

Nicholl, M. J., R. J. Glass, and H. A. Nguyen, 1992. Gravity-driven fingering in unsaturated fractures. In: Proceedings, International High Level Radioactive Waste Management Conference, Las Vegas, Nevada, April.

Nicholl, M. J., R. J. Glass, and H. A. Nguyen, 1993a. Small-scale behavior of single gravity-driven fingers in an initially dry fracture. In: Proceedings, International High Level Radioactive Waste Management Conference, Las Vegas, Nevada, April.

Nicholl, M. J., R. J. Glass, and H. A. Nguyen, 1993b. Wetting front instability in an initially wet unsaturated fracture. In: Proceedings, International High Level Radioactive Waste Management Conference, Las Vegas, Nevada, April.

Nordqvist, A. W., Y. W. Tsang, C.-F. Tsang, B. Dverstorp, and J. Andersson, 1996. Effects of high variance of fracture transmissivity on transport and sorption at different scales in a discrete model for fractured rocks. Journal of Contaminant Hydrology 22: 39-66.

Novakowski, K. S., G. V. Evans, D. A. Lever, and K. G. Raven, 1985. A field example of measuring hydrodynamic dispersion in a single fracture. Water Resources Research 21: 1165-1174.

Novakowski, K. S., P. A. Lapcevic, J. Voralek, and G. Bickerton, 1995. Preliminary interpretation of tracer experiments conducted in a discrete rock fracture under conditions of natural flow. Geophysical Research Letters 22: 1417-1420.

Oron, A. P., and B. Berkowitz, 1998. Flow in rock fractures: The local cubic law assumption reexamined. Water Resources Research 34: 2811-2825.

Palmer, A. N., 1975. The origin of maze caves. National Speleological Society 37: 56-76.

Palmer, A. N., 1991. The origin and morphology of limestone caves. Geological Society of America Bulletin 103: 1-21.

Paterson, L., 1984. Diffusion-limited aggregation and two-fluid displacements in porous media. Phys. Rev. Lett. 52: 1621-1624.

Poon, C.Y., R. S. Sayles, and T. A. Jones, 1992. Surface measurement and fractal characterization of naturally fractured rocks. J. Phys. D 25: 1269-1275.

Pyrak-Nolte, L. J., L. R. Meyer, N. G. W. Cook, and P. A. Witherspoon, 1987. Hydraulic and mechanical properties of natural fractures in low permeable rock. Proc. 6th Int. Cong. Rock Mech. 225-231.

Rasmuson, A., and I. Neretnieks, 1986. Radionuclide transport in fast channels in crystalline rock. Water Resources Research 22: 1247-1256.

Raven, K. G., K. S. Novakowski, and P. A. Lapcevic, 1988. Interpretation of field tracer tests of a single fracture using a transient solute storage model. Water Resources Research 24: 2019-2032.

Raven, K. G., and J. E. Gale, 1985. Water flow in a natural rock fracture as a function of stress and sample size. Int. J. Rock Mech. Min. Sci. & Geomech. Abstr. 22(4): 251-261.

Renshaw, C. E., 1996. Influence of subcritical fracture growth on the connectivity of fracture networks. Water Resources Research 32(6): 1519-1530.

Robinson, P. C., 1983. Connectivity of fracture systems: A percolation theory approach. J. Phys. A 16(3): 605-614.

Robinson, P. C., 1984. Numerical calculations of critical densities for lines and planes. J. Phys. A 17(4): 2823-2830.

Rophe, B., B. Berkowitz, M. Magaritz, and D. Ronen, 1992. Analysis of subsurface flow and anisotropy in a fractured aquitard using transient water level data. Water Resources Research 28: 199-207.

Sallès, J., J. F. Thovert, and P. M. Adler, 1993. Deposition in porous media and clogging. Chem. Eng. Sci. 48: 2839-2858.

Savage, D., and C. A. Rochelle, 1993. Modeling reactions between cement pore fluids and rock: Implications for porosity change. Journal of Contaminant Hydrology 13: 365-378.

Scanlon, B. R., 1991. Evaluation of moisture flux from chloride data in desert soils. Journal of Hydrology 128: 137-156.

Scanlon, B. R., 1992. Evaluation of liquid vapor water flow in desert soils based on chlorine-36 and tritium tracers and nonisothermal flow simulations. Water Resources Research 28: 285-297.

Schmittbuhl, J., F. Schmitt, and C. H. Scholz, 1995. Scaling invariance of crack surfaces. Journal of Geophysical Research 100B: 593-597.

Schwartz, F. W., L. Smith, and A. S. Crowe, 1983. A stochastic analysis of macroscopic dispersion in fractured media. Water Resources Research 19(5): 1253-1265.

Smith, L., and F. W. Schwartz, 1984. An analysis of the influence of fracture geometry on mass transport in fractured media. Water Resources Research 20(9): 1241-1252.

Shaw, R., 1984. The Dripping Faucet as a Model Chaotic System. Santa Cruz, Calif.: Aerial Press.

Steefel, C. I., and P. C. Lichtner, 1994. Diffusion and reaction in rock matrix bordering a hyper-alkaline fluid-filled fracture. Geochemica et Cosmochimica Acta 58: 3595-3612.

Thompson, M. E., and S. R. Brown, 1991. The effect of anisotropic surface roughness on flow and transport in fractures. Journal of Geophysical Research 96B: 21,923-21,932.

Tokunaga, T. K., and J. Wan, 1997. Water film flow along fracture surfaces of porous rock. Water Resources Research 33: 1287-1295.

Tsang, Y. W., and C. F. Tsang, 1987. Channel model of flow through fractured media. Water Resources Research 23(3): 467-479.

Tsang, Y. W., C. F. Tsang, I. Neretnieks, and L. Moreno, 1988. Flow and tracer transport in fractured media: A variable aperture channel model and its properties. Water Resources Research 24(12): 2049-2060.

Unger, A. J. A., and C. W. Mase, 1993. Numerical study of the hydromechanical behavior of two rough fracture surfaces in contact. Water Resources Research 29: 2101-2114.

Weisbrod, N., R. Nativ, D. Ronen, and E. Adar, 1998. On the variability of fracture surfaces in unsaturated chalk. Water Resources Research 34: 1881-1887.

Weisbrod, N., R. Nativ, E. Adar, and D. Ronen, 1999. The impact of intermittent rainwater and wastewater flow on coated and uncoated fractures in chalk. Water Resources Research 35(11): 3211-3222.

Wilkinson, D., 1986. Percolation effects in immiscible displacement. Phys. Rev. A 34: 1380-1391.

Wilkinson, D., and J. F. Willemsen, 1983. Invasion percolation: a new form of percolation theory. J. Phys. A: Math. Gen. 16: 3365-3376.

Winterfeld, P. H., L. E. Scriven, and H. T. Davis, 1981. Percolation and conductivity of random two-dimensional composites. J. Phys. C 14: 2361-2376.

Wood, W. W., K. A. Rainwater, and D. B. Thompson, 1997. Quantifying macropore recharge: Examples from a semi-arid area. Ground Water 35(6): 1097-1106.

5

Uniform and Preferential Flow Mechanisms in the Vadose Zone

Jan M.H. Hendrickx[1] and Markus Flury[2]

ABSTRACT

The two major flow mechanisms in the vadose zone are uniform flow and preferential flow. Both types of flow occur often simultaneously, but have considerably different consequences for water flow and chemical leaching. The objectives of this paper are to describe and classify flow mechanisms in the subsurface and to present illustrative field and laboratory studies. Since preferential flow occurs at a number of scales, scale is used as the primary classification criterion. Three distinctive scales are recognized on the basis of three different conceptual and physical models for water flow in the vadose zone: pore scale, Darcian scale, and areal scale. A common example of pore-scale preferential flow is saturated and unsaturated flow through macropores and fractures. At the Darcian scale we observe flow through stony soils, unstable flow occurring in water repellent and wettable homogeneous soils, unstable flow in layered soil profiles, preferential flow induced by variability in soil hydraulic properties, and flow through displacement faults. At the areal scale, surface depressions and discontinuous layers with lower or higher permeabilities can cause preferential flow. The paper concludes with a short section on measurement techniques for preferential flow and with guidelines for the formulation of conceptual models for the vadose zone.

---

[1] Department of Earth and Environmental Science, New Mexico Tech, Socorro
[2] Department of Crop and Soil Sciences, Washington State University, Pullman

INTRODUCTION

It has long been recognized that water flow in soils can either be uniform (Green and Ampt, 1911) or non-uniform (Lawes et al., 1881). *Uniform flow* leads to stable wetting fronts that are parallel to the soil surface; non-uniform flow results in irregular wetting. As a direct consequence of these irregular flow patterns, water moves faster and with increased quantity at certain locations in the vadose zone than at others. This non-uniform movement of water and dissolved solutes is commonly denoted *preferential flow*.

The term preferential flow neither distinguishes between the causes of the non-uniform flow pattern nor differentiates between the types of patterns. As such the term preferential flow comprises all phenomena where water and solutes move along certain pathways, while bypassing a fraction of the porous matrix. The reasons for the non-uniform flow patterns are manifold, and several identified mechanisms have coined an own term: *Macropore flow* is preferential water movement along root channels, earthworm burrows, fissures, or cracks. It occurs predominantly in fine-textured soils or media with a pronounced structure. Water bypasses the denser and less-permeable soil matrix by using the pathway of least resistance through macropores. *Unstable flow* is often observed in coarse-textured materials, and may be induced by textural layering, water repellency, air entrapment, or continuous non-ponding infiltration. As in the case of macropore flow, a considerable portion of the porous matrix is bypassed by the infiltrating water. *Funnel flow* refers to the lateral redirection and funneling of water caused by textural boundaries. Water again moves along the pathway of least resistance and can be redirected through a series of less permeable layers embedded in the soil profile. Each of these types of preferential flow is caused by different physical mechanisms. Often, several mechanisms act simultaneously, which leads to a broad variety of flow patterns.

The objectives of this paper are to describe the mechanisms and processes that lead to uniform and preferential flow in the vadose zone, and to elucidate the differences in the types of flow patterns observed. The different types of preferential flow are discussed in terms of three different conceptual and physical models for water flow which are frequently used in vadose zone hydrology and lead to the recognition of three spatial scales. We present illustrative examples and provide the basis for conceptual models to describe preferential flow phenomena.

CONCEPTUAL AND PHYSICAL MODELS

The physical principles that govern flow and capillary processes in the vadose zone are well understood and many excellent text books are available that deal with this topic at both introductory (Campbell, 1985; Hanks and Ashcroft, 1986; Hillel, 1998; Jury et al., 1991; Koorevaar et al., 1983; Marshall and Holmes, 1979) and advanced levels (Bear, 1972; Childs, 1969; Corey, 1990; Dullien,

1992; Kirkham and Powers, 1972). Flows of incompressible Newtonian fluids such as water are mathematically described by the Navier-Stokes equations, which are nonlinear, second-order, partial differential equations. These equations are related to a conceptual model for the *pore scale* that is based on the concept of a fluid continuum filling the void space. This approach is valid if the size of the pore diameters is larger than the mean free path of the water molecules. Although the continuum condition is readily met in most flow conditions in the vadose zone, the intricacy of the Navier-Stokes equations allows only a few exact mathematical solutions (Currie, 1993). One of these is the Hagen-Poiseuille equation that describes laminar flow through circular tubes (flow through a pore), between two parallel plates (i.e., flow through a fracture), and over a plate (i.e., film flow).

For example, the water flux q_{fr} (m/s) through a saturated parallel, smooth-walled fracture with aperture opening b (m) under laminar flow conditions is

$$q_{fr} = -K_{fr} \frac{dH}{dz} \tag{5.1}$$

(Bear et al., 1993; Corey, 1990; Snow, 1969; Streeter et al., 1998), where H is total hydraulic head (m), z is vertical distance (m), and K_{fr} is the hydraulic conductivity in the fracture (m/s) defined as:

$$K_{fr} = \frac{\rho g}{\mu} \frac{b^2}{12} \tag{5.2}$$

where ρ is fluid density (kg/m^3), g is the acceleration due to gravity (m/s^2), and μ is the dynamic viscosity (kg/s $\cdot$ m).

The geometrical complexity of porous materials makes it very cumbersome to treat water flow by referring only to the fluid continuum filling the pore space. A solution for this problem is found by moving to a larger spatial scale, which we name in this paper the *Darcian scale*. Instead of trying to exactly describe pore geometries and corresponding boundary conditions, the actual multiphase porous medium is replaced by a fictitious representative volume consisting of many pores and solids over which an average is performed. This changes the conceptual model from one based on a fluid continuum at the *pore scale* to one based on the concept of a representative volume at a larger spatial scale.

At the *Darcian scale,* water movement through a one-dimensional, unsaturated, vertical soil column is mathematically expressed by Darcy-Buckingham's equation:

$$q = -K(h) \frac{dH}{dz} \tag{5.3}$$

where q is the water flux (m/s), $K(h)$ the unsaturated hydraulic conductivity (m/s), and H the total hydraulic head:

$$H = h + z \tag{5.4}$$

in which h is the (negative) water pressure (m) and z the elevation head or height above a reference level (m).

Equations 5.1 and 5.3 have the same functional form; that is, the flux is proportional to the total hydraulic gradient. However, the proportionality factors in the two flux laws are fundamentally different. The hydraulic conductivity in the fracture K_{fr} applies to a single aperture whereas the hydraulic conductivity $K(h)$ is defined over a representative volume of the porous medium. As a result $K(h)$ is much more complex than K_{fr}, which is, for a given fluid, completely defined by a single soil or rock parameter: the aperture width (see Equation 5.2).

Only empirical formulations of the hydraulic conductivity $K(h)$ exist. Several mathematical functions have been proposed to represent measured data. The complexity of the unsaturated hydraulic conductivity is apparent in the function proposed by Van Genuchten (1980):

$$K(\theta) = K_s \left[\frac{\theta - \theta_r}{\theta_s - \theta_r} \right]^{\lambda} \left[1 - \left[1 - \left(\frac{\theta - \theta_r}{\theta_s - \theta_r} \right)^{1/m} \right]^m \right]^2 \tag{5.5}$$

where K_s is the saturated hydraulic conductivity (m/s); θ is volumetric soil water content (m³/m³); θ_s is the saturated water content (m³/m³), often taken equal to the soil porosity; θ_r is residual water content (m³/m³); and the parameters n, m, and λ are empirical constants. The relationship between θ and h, the water retention characteristic, is:

$$\frac{\theta - \theta_r}{\theta_s - \theta_r} = \frac{1}{[1 + (\alpha h)^n]^m} \tag{5.6}$$

where α (m⁻¹), n, and m are parameters that determine the water retention curve shape. As it is often assumed that $m = 1 - 1/n$, the number of parameters needed to describe the hydraulic conductivity as a function of soil water pressure h totals six: K_s, θ_s, θ_r, n, λ, and α. The parameter values can be determined from measured θ-h and K-h data pairs using nonlinear curve fitting programs available in statistical software packages and spreadsheets, or by optimization software such as the package RETC developed by Van Genuchten et al. (1991).

At the *areal scale*, application of the Darcy-Buckingham equation is no longer practical since it would require long and expensive field campaigns to characterize and quantify the spatial variability of the vadose zone at the *Darcian scale*. One approach for the evaluation of water movement at such a big scale is to employ areal mass balance or soil moisture budgeting models (Hendrickx and Walker, 1997). For example, the groundwater recharge over a large area can be assessed by an areal water balance equation:

$$q_r = P + R - ET - \Delta W \qquad (5.7)$$

where q_r is the groundwater recharge (m/month), P is precipitation (m/month), R is the net runoff/runon (m/month), ET is actual evapotranspiration (m/month), and ΔW is the change in soil moisture storage in the vadose zone (m/month).

Table 5-1 follows previous work by Wagenet (1998), Wagenet et al. (1994), and Wheatcraft and Cushman (1991) to summarize the principal characteristics of the conceptual and physical models discussed above. One immediate observation of great practical significance is the fact that conceptual models at different scales result in different physical models and mathematical equations. Moreover, each of these equations requires a completely different set of input parameters. The complexity of input parameters for physical models increases with the spatial scale. Flow in a fracture requires a measurement of its width; unsaturated flow through a soil profile requires measurements or indirect determination of the six soil parameters K_s, θ_s, θ_r, n, λ, and α; evaluation of a regional water balance in the vadose zone requires long-term monitoring of soil water contents, meteorological variables, and groundwater levels. This leads to the observation that the timeframe of a study often will increase if the spatial scale of its conceptual model becomes larger. Although the measurement of fracture widths at depth is no simple matter, it takes less time than the many years of monitoring key environmental parameters needed to assess the water balance of a watershed. Moreover, while measurements of fracture widths and the Van Genuchten soil parameters will yield estimates of their true values within relatively small confidence limits that can be used in a deterministic manner, the nature of environmental parameters often gives studies at areal scales a stochastic character. The results of such studies frequently have to rely more on statistical interpolations of field measurements than on well determined causal and physical relationships. For given weather conditions in a specific soil profile, the changes of soil water fluxes with depth and time can be predicted quite well once the Van Genuchten soil parameters have been determined. However, for the same weather conditions, determination of regional groundwater recharge using Equation 5.7 will become a stochastic exercise using statistical techniques for the interpolation and averaging of soil physical and meteorological measurements. For this reason, areal-scale methods can only yield reliable results if the averaging process does not create havoc with the true flow mechanisms.

Our heuristic approach for the discussion of the three different conceptual models and their respective scales suggests an increasing spatial dimension from pore, to Darcian, to areal scale. Although such an increase is a common feature in many vadose zones, there are also hydrological observations that demonstrate at least some overlap between the spatial dimensions of the three distinctive scales. For example, a Darcian approach can be applied to soil volumes as small as a few cubic millimeters, while the Navier-Stokes equations in principle can be used to describe water flow through pores with diameters of centimeters. A Darcian

TABLE 5-1 Scales, Conceptual Models, Critical Parameters, and Measurements Relevant to Flow Mechanisms in the Vadose Zone

Spatial Scale	Domain	Conceptual Model	Physical Model	Critical Parameters	Smallest Temporal Measurements	Scale
Pore	Macropores, Fractures	Fluid Continuum	Hagen-Poisseuille Equation 5.1	Fracture Width	Thin sections, NMR	Minutes Days
Darcian	Laboratory, Soil Profiles	Representative Volume	Darcy-Buckingham Equation 5.3	Hydraulic Properties	TDR, Neutron Attenuation, Tensiometers	Hours Months
Areal	Field, Local Depression, Landscape Element	Mass Balance	Mass Balance Equation 5.7	Weather, Soil Moisture	Meteorological Station, TDR, Neutron Attenuation, Remote Sensing, Groundwater Level	Days Years

approach may be quite appropriate for a large uniform landscape element covering squares of kilometers but fail on a meter scale in a heterogeneous environment. An areal water balance approach can be applied to volumes as small as a flower pot as well as to areas covering an entire continent. For this reason we have chosen names for the spatial scales that reflect their link with a particular physical model rather than a spatial dimension.

FLOW MECHANISMS AT DIFFERENT SCALES

Water flow in the vadose zone occurs at different spatial and temporal scales under a wide variety of conditions. This makes it problematic to classify vadose zone flow mechanisms in a consistent manner. Another complicating factor is that processes at the pore scale determine processes at larger scales. For these reasons we have selected a practical classification criterion based upon the three conceptual models discussed in the previous section which lead to three typical spatial scales often encountered in vadose zone studies. The *pore scale* deals with water flow processes described by Hagen-Poiseuille's equations, the *Darcian scale* with processes considered to take place within fictitious representative volumes and described by Darcy's equation, and the *areal scale* with processes affected by major landscape elements such as local depressions, faults, and discontinuous layers in the vadose zone. Sometimes processes at the areal scale can be described quantitatively using Darcy's law in numerical models such as HYDRUS2D, while in more complicated situations only a qualitative approach is feasible. Figures 5-1 and 5-2 illustrate the different flow mechanisms and their relation to the spatial scale. On the pore scale, we observe several types of macropore flow and at the Darcian scale, stable wetting as well as funnel and unstable flow (Figure 5-1). On the areal scale, we observe preferential flow due to localized recharge caused by topographic depressions, pipe flow, and funnel flow (Figure 5-2).

Pore Scale

The lucidity of the Hagen-Poiseuille Equation 5.1 makes the pore scale very attractive for the investigation of water flow through soils and rocks since the only material parameter needed for its application is the pore size. Unfortunately, its application is only practical in materials with a relatively simple pore geometry, such as capillary tube models (Bear, 1972; Scheidegger, 1974), network models (Dullien, 1992; Luxmoore and Ferrand, 1992), fractures (Bear, 1993; Rasmussen, 1987; Schrauf and Evans, 1986), and macropores.

There is abundant evidence that many soils are susceptible to rapid water flow through macropores. Macropores are often defined in terms of a specific radius (Luxmoore, 1981; Beven and German, 1982); however, no accepted general definition of a macropore exists. The definition of a specific radius is rather

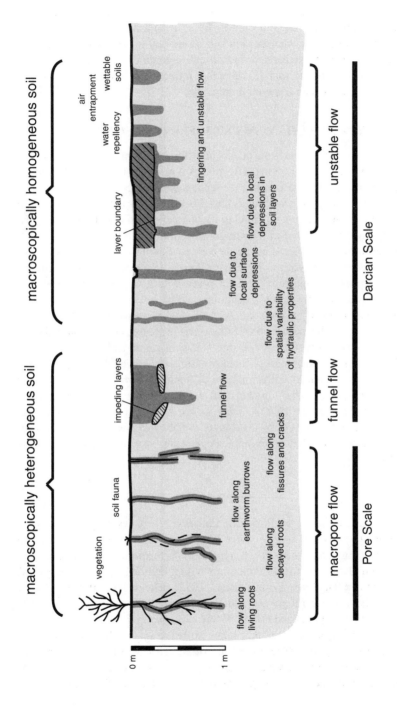

FIGURE 5-1 Schematic showing different preferential flow mechanisms observed at pore and Darcian scales.

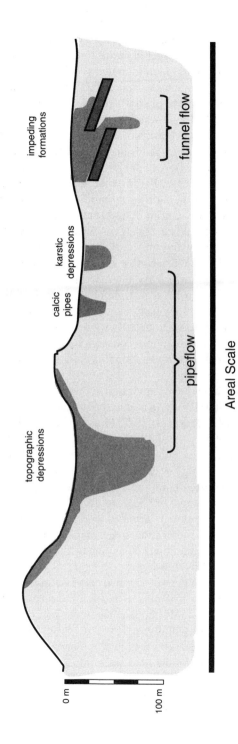

FIGURE 5-2 Schematic showing different preferential flow mechanisms observed at the areal scale.

arbitrary, and for operational purposes in this paper we consider a macropore to be a pore or fracture considerably larger in radius than the bulk of the pores and fractures in the porous soil or rock. In unsaturated porous media, macropores do not necessarily conduct large amounts of water. Two conditions must be met in order for macropores to contribute to rapid water flow. First, the macropore must be partially or completely filled with water, and, second, the pore needs to continuously extend over a significant portion of the porous medium. Obviously, the definition of a significant distance depends on the system of interest, and therefore the relevance of macropore flow might be very different depending on the spatial and temporal scale of interest. For instance, one common type of macropore flow in soils is flow through earthworm burrows or root channels. Earthworm burrows are typically 1-3 mm in diameter and can extend up to 6 m in vertical length. As such, water and chemicals can bypass the topsoil by traveling in earthworm burrows, and pollutants and pesticides can potentially contaminate shallow groundwater resources. When earthworm burrows end at a certain depth, water and dissolved chemicals will leave the macropore and flow through the porous matrix, thus slowing down vertical migration considerably. It is rather unlikely that continuous vertical flow paths extend over several dozens of meters in unsaturated soil, and therefore macropore flow will stop at a certain depth. The bypass of the topsoil, however, has serious consequences for many contaminants because sorption and degradation processes are usually strongest in the topsoil and cease with increasing soil depth.

Several dye tracing studies have demonstrated that macropore flow is rather common in many agricultural soils, particularly in fine-textured soils with a pronounced soil structure (Bouma et al., 1977; Flury et al., 1994; Germann, 1990; Mohanty et al., 1998; Petersen et al., 1997; Stamm et al., 1998). Out of 14 different soils investigated, Flury et al. (1994) found a majority of soils susceptible to macropore flow. Typical flow patterns observed are shown in Figure 5-3. Depending on whether the macropores are more planar or cylindrical in shape, the flow patterns appear more areal or linear, respectively.

The process of macropore flow is depicted in Figure 5-4. When the overall water input from precipitation or irrigation, $q^*(t)$, exceeds infiltration capacity of the soil, $i(t)$, a horizontal overland flow, $o(t)$, is generated that causes a water flux into the macropores, $q(0, t)$. This flux causes water content inside the macropore, $w(z, t)$, to increase. A fraction of the water, r, that occupies a macropore at a given depth will be absorbed by the soil matrix through the macropore walls. The remainder will percolate downwards into the macropore, $q(z, t)$. The interplay of precipitation or irrigation rates with dynamics of the infiltration rate over time add to the macropore flow mechanism complexity. For example, the infiltration rate of a soil depends not only on the time since infiltration started but also on antecedent water content (Philip, 1969). When the infiltration rate, $i(t)$, decreases with time and with increasing antecedent soil water content, the opportunity for overland flow, $o(t)$, and macropore flow, $q(0, t)$ increases.

159

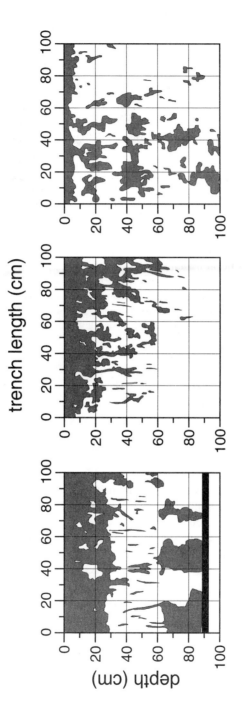

FIGURE 5-3 Typical macropore flow patterns observed in structured, fine-textured soils. Patterns depict vertical cross sections of soil profiles after sprinkling irrigation of 40 mm dye solution. The horizontal bar in the left graph indicates the maximum excavation depth. After Flury, M., H. Flühler, W. A. Jury, and J. Leuenberger, 1994. Susceptibility of soils to preferential flow of water: A field study. Water Resources Research 30: 1945-1954. Copyright by American Geophysical Union.

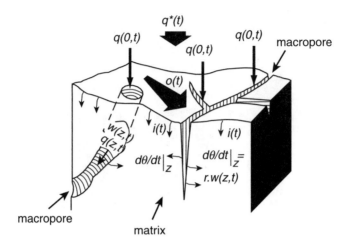

FIGURE 5-4 Schematic presentation of fluxes occurring during infiltration into a macro-porous soil: $q^*(t)$ represents overall water input (precipitation, irrigation); $i(t)$ represents infiltration into the topsoil matrix; $o(t)$ represents short duration/short distance overland flow when $q^*(t) > i(t)$; $q(0, t)$ represents volume flux density into the soil macropores ($q(0, t) > 0$ when $o(t) > 0$); $q(z, t)$ and $w(z, t)$ represent volume flux density and volumetric soil moisture, respectively, in the macropore system; $r.w(z, t)$ represents water sorbance from macropores into the soil matrix. *Note*: All volume flux densities are per cross-sectional area for the entire soil. From Beven, K., and P. Germann, 1982. Macropores and water flow in soils. Water Resources Research 18: 1311-1325. Copyright by American Geophysical Union.

Macropore flow is usually initiated when a portion of the soil matrix reaches saturation or at least is close to saturation in order to exceed the water entry potential of the macropore. This phenomenon can occur during an infiltration event, for instance, when the water front reaches textural boundaries. An example of such preferential flow initiation at textural boundaries is shown in Plate 4. In this experiment, conducted by Forrer et al. (1999), water was infiltrated under two different rates, 5 and 25 mm/day. A dye tracer was applied as a line source to the soil surface to visualize the flow patterns. Under the high infiltration rate, the water front started to pond at the layer boundary between plow horizon and subsoil at about 20 cm depth, and water tended to enter macropores. As a consequence, the tracer pulse split into several pulses, and channeled through a smaller cross section of the subsoil. In contrast, under the low infiltration rate, no ponding occurred at the layer boundary, and water and tracer movement followed rather uniform pathways. Horizontally and vertically averaged concentration distributions, also shown in Plate 4, depict multimodal characteristics under the high infiltration rates typical for macropore flow. Much smoother and unimodal distributions were observed under the low infiltration rate.

Another example is presented by Germann (1986), who analyzed drainage responses to storms observed during a 7-year period in the Coshocton monolith lysimeters (Northeast Experimental Watershed, USDA-ARS, Coshocton, Ohio). Rains of only 10 mm/day caused a drainage response at 2.4 m depth on the same day as precipitation when volumetric water content in the upper 1 m of the undisturbed profile exceeded a threshold value of 0.3 m^3/m^3, whereas at soil water contents below this threshold value, storms greater than 50 mm/day were found not to cause any drainage flow.

Many experimental studies have confirmed the dependency of macropore flow on infiltration rates and water contents of the soil (Flury et al., 1994). The occurrence of macropore flow has always been reported in conjunction with pronounced soil structure, such as earthworm burrows, root channels, or fractures. Recent numerical analyses, however, have shown that preferential flow can also occur in macroscopically homogeneous soils with spatially variable hydraulic properties, but no pronounced macropore structure.

There is still considerable uncertainty about the physics of water flow in macropores in soils. Experiments indicate that macropores might not need to be completely filled with water in order to be conductive for water flow. Dye tracing studies by Bouma et al. (1977) and Bouma and Dekker (1978) indicate that water flows through only a small portion of planar macropore voids. Li and Ghodrati (1994) observed preferential flow of NO_3 in laboratory columns at water fluxes well below the flux expected for completely water-filled macropores. This might indicate the water could flow as film flow at macropore walls.

Darcian Scale

Experimental evidence (Darcy, 1856) and theoretical considerations (Bear 1972) have proven that Darcy's empirical Equation 5.3 is valid for the exploration of flow conditions in the vadose zone using the conceptual model of a *representative volume* at the *Darcian scale*. Indeed, under many conditions this concept yields very satisfactory results, although sometimes creative adaptations are required. The most straightforward application is found under **stable flow** conditions characterized by horizontal wetting fronts parallel to the soil surface. For example, Figure 5-5 shows a stable wetting front in a loam soil along the Rio Salado near Socorro observed in August, 1995, a few days after heavy rainfall. Wierenga et al. (1991) have observed stable wetting fronts during long-term infiltration experiments in desert profiles near Las Cruces. In addition, many infiltration studies under controlled conditions in laboratory columns and lysimeters indicate that stable flow is a common phenomenon.

A special case of *stable flow* takes place during *flow through stony soils*. During the last 60 years, theoretical studies and experimental investigations of

FIGURE 5-5 Stable wetting front in a loam soil along the Rio Salado near Socorro observed in August, 1995, after a few days with heavy rainfall. The wetting was exposed after the river bank collapsed shortly before the picture was taken by Hendrickx..

water flow through the vadose zone have focused on agricultural soils, while our knowledge of stony soils remained limited. However, stony environments are widespread in river bed and mountain-front hydrogeological provinces that are characterized by recharge infiltration through stony layers (Issar and Passchier, 1990). Channeling of water flow through stony soils has been quantified by Hendrickx et al. (1991), who evaluated travel times through a 100-m-deep, stony vadose zone in Baluchistan (Pakistan). They found that such stony zones shorten recharge travel times considerably. For example, yearly net infiltration by an artificial recharge of 40 cm would take 35 years to reach the water table at a depth of 100 m in a sand loam without stones, but only 17 years in the same soil with 60 percent stones. Likewise, Buchter et al. (1995) observed highly irregular water flow through a stony soil monolith of 77 percent (per weight) gravel. Only a small fraction of the cross-sectional area appeared to be active in water flow.

One of the least understood preferential pathway flow mechanisms is that of *unstable flow*. A stable wetting front is a horizontal front that moves downwards without breaking into fingers. The behavior of such a front can be simulated by one-dimensional computer models. Unstable wetting fronts start out as horizontal wetting fronts that, under certain conditions, break into fingers or preferential flow paths as the front moves downwards, much like rain running off a sheet of glass and breaking into streams. These fingers facilitate recharge flow and transport of contaminants to the groundwater at velocities many times those calculated if a stable horizontal front is assumed. For example, in a bromide tracer experiment, Hendrickx et al. (1993) measured in the field that after 5 weeks with 120 mm precipitation, bromide concentrations in the groundwater are 6 to 13 times higher under unstable wetting fronts than under stable wetting fronts.

The occurrence of what we now recognize as unstable wetting fronts has been reported in the literature since the beginning of this century (Deecke, 1906). Similar observations were later reported by Gripp (1961) on the island of Amrum in Germany, by Gees and Lyall (1969) on Cape Sable Island in Nova Scotia (Canada), and in The Netherlands by Schuddebeurs (1957), Mooij (1957), Lopes de Leao (1988), and Raats (1984). The conditions under which unstable wetting fronts form in the field are not yet fully understood, because systematic investigations of the phenomena in field soils have been rare. However, theoretical work (Du et al., 2001; Glass et al., 1989a, 1989c, 1991; Hillel and Baker, 1988; Parlange and Hill, 1976; Philip, 1975a, 1975b; Raats, 1973; Tabuchi, 1961) supported by laboratory experiments (Baker and Hillel, 1990; Diment and Watson, 1985; Glass et al., 1989b, 1990; Hill and Parlange, 1972; Selker et al., 1989, 1992; Tamai et al., 1987; White et al., 1977; Yao and Hendrickx, 1996, 2001) indicate that fingered flow may occur under the following conditions: (1) infiltration of ponded water with compression of air ahead of the wetting front; (2) water-repellent soils; (3) continuous nonponding infiltration; and (4) in layered soil profiles where coarse-textured soil layers are overlain by less permeable layers. We will present examples of these scenarios.

Air Entrapment

Compression of air below a wetting front has been shown to cause instability of the front. For example, Wang et al. (1998) observed in a series of laboratory experiments that air entrapment can cause an otherwise stable wetting front to break up into fingers.

Water-Repellent Soils

Dry soils readily absorb water because of a strong attraction between the mineral particles and water. However, the affinity of soils for water can be reduced by hydrophobic organic materials that are either mixed with the soil particles or form a coating around them. Such soils are called hydrophobic or water-repellent, and are found in many parts of the world under a variety of climatic conditions (De Bano, 1981; Jamison, 1969; Jaramillo et al., 2000; Mallik and Rahman, 1985; McGhie and Posner, 1980; Miyamoto et al., 1977; Richardson, 1984; Rietveld, 1978). When dry, water movement is severely limited and, consequently, precipitation will not infiltrate uniformly. Indeed, field evidence for irregular wetting patterns and considerable soil water content variation have been reported in water-repellent field soils by Bond (1964), De Bano (1969a, 1969b), Dekker and Jungerius (1990), Emerson and Bond (1963), Letey et al. (1975), and Meeuwig (1971).

Field studies (Dekker and Ritsema, 1994a, 1994b; Hendrickx et al., 1993; Ritsema et al., 1993; Ritsema and Dekker, 1994) and results from lysimeter experiments (Hendrickx and Dekker, 1991) demonstrate that the irregular and incomplete wetting pattern in water-repellent soils can be predicted and explained by unstable wetting front theory. Figure 5-6 demonstrates the extreme variability in volumetric water content for the top layer of a water-repellent soil. In flat areas, dry water-repellent topsoils will accelerate natural recharge since the infiltrated water is traveling through preferential flow paths toward the aquifer (Van Dam et al., 1990). However, on slopes, water repellency may enhance runoff and reduce recharge.

On the basis of numerous field observations and measurements of soil water content and bromide content, Ritsema et al. (1993) derived a conceptual model for unstable flow (Figure 5-7). Although their study deals with unstable flow in water-repellent soils, it appears valid wherever unstable flow takes place. They recognize not only preferential flow paths with patches of dry soil between them, but also a distribution layer and a divergence layer. The distribution layer receives the precipitation and feeds by lateral flow the preferential flow paths below it. The divergence layer is located underneath the preferential flow paths and laterally distributes the water and solute fluxes. Therefore, it is to some extent counteracting the rapid transport through the preferential flow paths.

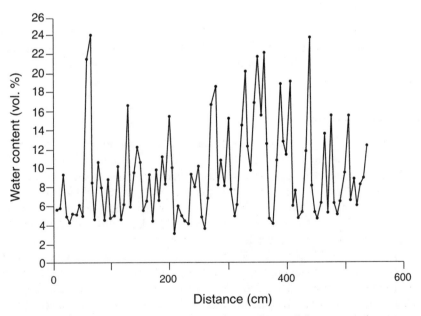

FIGURE 5-6 Variation in volumetric water content along a 5.5-m transect in a water-repellent soil at 0.08 m depth. From Hendrickx and Dekker (1991).

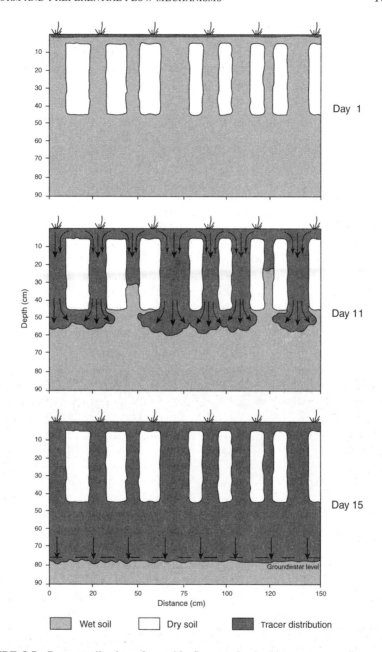

FIGURE 5-7 Conceptualization of unstable flow mechanism in a water-repellent soil. From Ritsema, C. J., L. W. Dekker, J. M. H. Hendrickx, and W. Hamminga, 1993. Preferential flow mechanism in a water-repellent sandy soil. Water Resources Research 29: 2183-2193. Copyright by American Geophysical Union.

Nonponding Infiltration

The infiltration theories of Green and Ampt (1911) and Philip (1969) predict sharp, stable wetting fronts during infiltration in dry, homogeneous wettable soils. These theories were mostly verified in laboratory columns with diameters of a few centimeters, but also in field soils after periods with precipitation (Hendrickx and Yao 1996; Figure 5-5 of this chapter). However, many observers (Deecke, 1906; Raats, 1984) noted field phenomena that appear to contradict the universal validity of these infiltration theories. A lysimeter experiment was thus conducted (Hendrickx and Dekker, 1991) to demonstrate that preferential flow paths observed in wettable dune sands are not caused by local runoff, soil cracks, or macropores. A nonweighing lysimeter (height 1.2 m, width 1.4 × 1.2 m) was carefully filled with wettable dune sand and, successively, exposed to 403 mm natural precipitation during a period of four months. The lysimeter was then sampled layer by layer. The wetted soil volume decreased from 100 percent at the surface to 22 percent at 30 cm depth. Between depths 40-80 cm only 11 percent of the soil had been wetted, a clear indication for preferential flow paths. Because the lysimeter surface was perfectly horizontal and careful packing had excluded the existence of macropores, the observed preferential flow paths are without doubt a result of unstable wetting (Figure 5-8).

FIGURE 5-8 Unstable wetting patterns at 0.20 m depth in wettable sand after 403 mm precipitation; dark spots are wet. From Hendrickx and Dekker (1991).

Layered Soil Profiles

Another common field situation with a potential for fingered flow is found where a less permeable layer overlies a more permeable layer. The occurrence of unstable wetting under these conditions has been demonstrated by laboratory experiments (Glass et al., 1990; Hill and Parlange, 1972; White et al., 1977) and field observations (Starr et al., 1978, 1986). Starr et al. (1978) also conducted a controlled experiment in an undisturbed soil column (diameter 1.8 m, height 3.6 m) with a sandy loam (0.0-0.6 m) overlying a gravelly coarse sand (0.6-3.0 m). They observed uniform flow through the top layer and fingered flow in the bottom layer, where twelve distinct fingers occupied only 5 percent of the total cross-sectional area at 1.0 m depth.

Spatial Variability of Hydraulic Properties

A different type of preferential flow may be caused by spatial variability of hydraulic properties. Roth (1995) simulated water flow in a variably saturated Miller-similar medium, where local heterogeneities were represented by a spatially correlated random field of scaling factors. Under steady-state water flow, distinct flow patterns developed, with regions of high flux and regions of low flux. The flow patterns observed depended on the magnitude of the steady water flux, and consequently on the water content of the porous medium. Interestingly, two states of hydraulic structures were identified, separated by a critical point: high-flux regions developed under high flow rates corresponded to low-flux regions under low flow rates, and vice versa. Qualitative comparisons showed that these types of simulations can represent many experimental observations reported from large-scale field experiments (Roth and Hammel, 1996).

Funnel Flow

A special case of preferential flow at the Darcian scale is caused by redirection of water at soil textural boundaries, a phenomenon commonly called funnel flow. Gardner and Hsieh (1959) demonstrated impressively how coarse sand or aggregate layers embedded in a finer textured soil can funnel water flow through a small fraction of the soil profile. Figure 5-9 shows a sequence of an infiltration event where unsaturated water flow is redirected at an inclined sand lens embedded in a finer texture soil material. A field experiment, conducted by Kung (1990), showed that such embedded coarse sand layers can funnel water flow through 1 percent of the soil matrix.

Flow Through Small Displacement Faults in Sand

There is evidence that faults in sand are relatively common features in tectonically active extensional regions like the Rio Grande rift. Because fault-zone

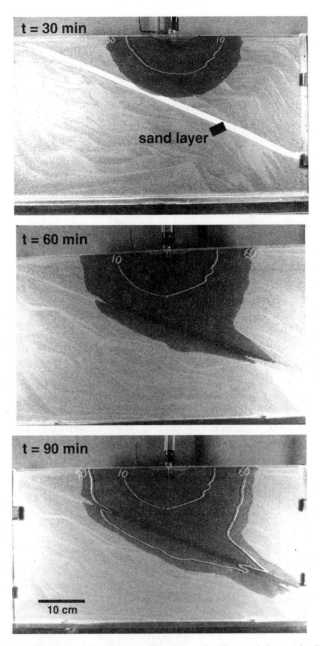

FIGURE 5-9 Infiltration of water into a fine textured soil containing an inclined coarse sand layer. The sequential pictures show the infiltration event after 30, 60, and 90 minutes. (W.H. Gardner, unpublished photographs, Department of Crop and Soil Sciences, Washington State University.)

deformation typically decreases the pore size, it decreases the saturated hydraulic conductivity. However, under unsaturated conditions the actual unsaturated hydraulic conductivity is likely to be greater than that of the adjoining undisturbed sand layers in regions with low soil water contents. It is hypothesized that fault zones can become preferred flow paths under the arid conditions of the southwestern United States (John Wilson, Laurel Goodwin, and Peter Mozley, written communication, 1996; Sigda, 1997). Figure 5-10 shows preferential wetting of conjugate faults in a sand near Socorro, New Mexico, after a period of precipitation.

Areal Scale

Just as the complexity of the pore geometry makes it all but impossible to use the Hagen-Poiseuille equation to describe flow through a real porous medium, the complexity of some landscape elements often makes it all but impossible to use a Darcian approach. We will illustrate this with four typical situations.

Karstic Vadose Zone

Gunn (1983) investigated mechanisms by which flow is concentrated and transmitted to the underlying aquifer for a karst area in the Waitomo district of

FIGURE 5-10 Displacement faults in sand near Socorro, New Mexico. The wet spots coincide with the faults. From J.M. Herrin (September 1997).

New Zealand. Mean annual precipitation and potential evapotranspiration in the study area are 2,370 mm and 775 mm, respectively. Gunn found that closed depressions (solution dolines, sinkholes, cockpits) act as funnels and collect near surface water through three concentrating mechanisms: (1) overland flow, defined as any water flowing along the ground surface; (2) throughflow, defined as any water flowing laterally within the soil; and (3) subcutaneous flow, defined as water flowing laterally through the upper, weathered layer of limestone.

Three vertical flow mechanisms were recognized for water transmission through the vadose zone: (1) film flow (named shaft flow by Gunn), defined as water flowing underground as films on the walls of vertical shafts; (2) fracture or macropore flow (named vadose flow by Gunn), defined as vertically moving water that flows for a major part of its course in enlarged joints and fractures; (3) vadose seepage, defined as vertically moving water that percolates through small, tight joints and fissures or as intergranular flow. Gunn (1983) combined these six flow mechanisms into a conceptual vadose zone flow model (Figure 5-11) and qualitatively validated it by determining water travel times for each flow mechanism using measurements of water temperature and calcium and magnesium ion

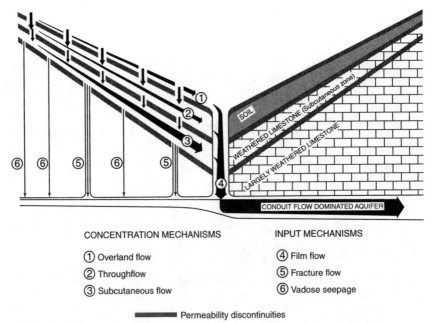

CONCENTRATION MECHANISMS   INPUT MECHANISMS

① Overland flow   ④ Film flow
② Throughflow   ⑤ Fracture flow
③ Subcutaneous flow   ⑥ Vadose seepage

▬▬▬▬ Permeability discontinuities

FIGURE 5-11 Idealized cross section of a closed depression showing flow concentration and mechanisms (arrows proportional to water flow in Waitomo depressions). After Journal of Hydrology, 61, Gunn, J., Point-recharge of limestone aquifers: A model from New Zealand karst, pp. 19-29, 1983, with permission from Elsevier Science.

concentrations. This validation revealed the principal characteristics for each model component.

Overland flow is quantitatively insignificant in the Waitomo district, while subcutaneous flow appears to supply more water to vertical shafts than does throughflow. Travel times for subcutaneous and throughflow vary between 0 and 14 weeks with a mean of approximately 6 weeks. The concentration of near surface water makes film flow more important than fracture flow, while vadose seepage contributes less than 5 percent to aquifer recharge. Two different categories of fracture flow were identified: flows through open joints and fissures, and flows through soil-filled fissures. The first responded directly to rainfall with a mean travel time of less than a week, whereas the second had a mean travel time of 8 weeks. The travel times for vadose seepage varied between 0 and 19 weeks, indicating that part of the water flows rapidly through small fractures as macropore flow while the remainder flows slowly through the porous material as capillary flow. Although Gunn (1983) cautions that in other climate-soil-regolith regimes the relative importance of these flow mechanisms may be different, his conceptual model and field observations offer considerable insight to the dynamics of flow processes through fractured rocks. Striking is the fact that capillary flow accounts for less than 5 percent of recharge volume.

Simultaneous Occurrence of Fast and Slow Flow

The simultaneous occurrence of capillary and macropore flow within the same soil mass without the presence of clearly defined macropores appears to be quite typical. Under these conditions *macropore flow* cannot be detected from visual observations in the field, but is inferred from analysis of solute profiles or drainage responses at the groundwater table.

Johnston (1987) used solute profiles in combination with groundwater table measurements to demonstrate macropore flow in a deep clayey regolith in southwest Western Australia. The regolith showed marked heterogeneity over horizontal and vertical distances of only a few meters, which resulted in a complex water movement pattern. Thirteen vertical concentration profiles of natural chloride were used to estimate recharge rates through the 16-m-deep unsaturated vadose zone. Although over most of the 700 m^2 experimental area recharge rates varied from 2.2-7.2 mm/year, a small portion of the site had rates between 50 and 100 mm/year. As a result of this preferred flow, a groundwater mound was observed in piezometer 1356 below the localized recharge area within 12-14 h of intense rainstorms and dissipated over a period of 2-4 days (Figure 5-12).

Scanlon (1992) demonstrated the existence of preferred pathway flow through fissured sediments in the Chihuahuan desert of Texas, with annual precipitation of 280 mm and approximate potential evapotranspiration of 3,000 mm. The term *fissure* refers to the alignment of discontinuous surface collapse structures, or gulleys where the underlying extensional feature is filled with sediment.

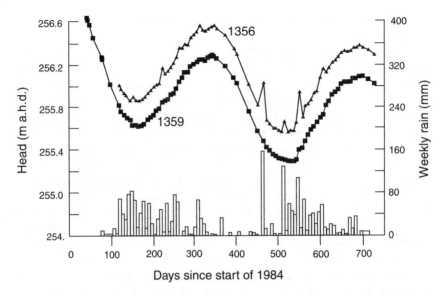

FIGURE 5-12 Variation in weekly precipitation and potentiometric head in piezometers 1359, located where capillary flow dominates, and 1356, located where macropore flow occurs. After Journal of Hydrology, 94, Johnston, C. D., Preferred water flow and localized recharge in a variable regolith, pp. 129-142, 1987, with permission from Elsevier Science.

A somewhat unexpected result from Scanlon's study is that water fluxes calculated from vertical chloride profiles in the fissured sediments ranged from 1-8 mm/year and were as much as 350 times higher than those calculated for adjacent ephemeral stream sediments. This difference could not be attributed to differences in soil texture since fissures and ephemeral streams have similar textures varying from muddy, sandy gravel to loamy sand and clay loam. However, open cavities and fractures in the fissured sediments were filled with loose sediments to give lower bulk densities and looser soil structure, so that fast macropore flows are promoted rather than slow capillary flows. Much of the precipitation occurring as high-intensity convective summer storms infiltrates to depths of only 0.1-0.3 m in the ephemeral stream beds and is readily lost by evapotranspiration, whereas deep infiltration in the fissured sediments prevents such losses and results in a much larger recharge rate.

Localized Recharge

In comparison to direct recharge, localized and indirect recharge are often considered at least as significant, if not the most important sources, of natural recharge in arid and semiarid lands (Gee and Hillel, 1988; Lerner et al., 1990; Stephens, 1994; Wood and Sanford, 1995). Localized recharge implies horizontal movement of surface and/or near-surface water and occurs in weathered bare hardrock or limestone terrain, topographical depressions, minor wadis or arroyos, and in mountain front systems. To account for localized recharge it is necessary to measure or estimate local runoff and runon volumes so that these can be included in the water balance. Such estimates and measurements are often complicated by subsurface components of runoff and runon flow (Anderson and Burt, 1990) and by the fact that they frequently occur on a scale too detailed to map for engineering studies (Lerner et al., 1990).

A classical example of localized recharge occurs in the numerous depressions dotting the Great Plains of North America. These features can measure tens to thousands of meters across, are often occupied by wetlands or lakes, and are referred to as "potholes," "sloughs," or "playas." Meyboom (1966) was one of the first investigators to quantify localized recharge during a one-year study of a till plain pothole with a watershed contributing area of 0.8 ha in south-central Saskatchewan (Canada). His pothole has a bottom diameter of 40 m and the height of its surrounding rim varies between 3 and 8 m. It was determined that the pothole or local depression, with 15 percent of the total surface area, contributed 70 percent of the recharge. Other investigators have also reported studies that confirm the large contribution of localized recharge described by Meyboom (1966); examples are Freeze and Banner (1970), Miller et al. (1985), and Winter (1986). Using chemical techniques, Wood and Sanford (1995) estimated that approximately half (4-5 mm/year) the annual recharge (9-10 mm/year) to the Ogallala Aquifer on the southern High Plains in the United States occurs through playa floors that cover only 6 percent of the area.

The effects of topographic, soil, and climatic conditions on the magnitude of depression-focused recharge for specific sites is difficult to measure in the field. Therefore, Nieber et al. (1993) and Boers (1994) have developed mathematical models. Nieber et al. (1993) assumed for simplicity that their catchment was circular in form and contained a circular-shaped depression with a single drainage outlet. Boers (1994) developed a similar procedure for the design of rainwater harvesting catchments in arid and semiarid zones that can also be used for the assessment of localized recharge. His method is based on actual evapotranspiration predictions using a numerical soil water balance model, while the runoff component is predicted by a runoff model (Boers et al., 1986). The *microcatchment* in Figure 5-13 illustrates how the components of rainwater harvesting interact. A *microcatchment* consists of a *runoff area* with a maximum flow distance of 100 m and an adjacent *basin area* (the depression) with a tree, bush, or row crop. The objective of rainwater harvesting is to induce runoff, collect and store the water in the basin area, and conserve it in the root zone for consumptive use by the vegetation. The components shown in Figure 5-13 yield the following water balance on an annual basis:

$$D = P + R - E_i - E_w - E_{act} - T_{act} - \Delta W \qquad (5.8)$$

where D is deep percolation or recharge, P is precipitation, R is runoff calculated over the basin area where it is collected, E_i is evaporation of water intercepted by the vegetation, E_w is open water evaporation, E_{act} is the evaporation from bare soil, T_{act} is actual transpiration by the vegetation, and ΔW is the increase in soil water storage in the root zone. Table 5-2 presents the annual water balance components at Sadoré (Niger) for one Neem tree in a basin of 8 m^2 with a soil profile comprising 3 m fine sand above 2 m laterite gravel. If the basin receives no runon water, no recharge takes place during a dry, average, or even a wet year. In a dry year, runoff areas of 20 and 40 m^2 do not generate any recharge, although they do increase actual transpiration and, thus, the growth rate of the tree. In an average year a modest increase in runoff area from 20-40 m^2 produces a 22-fold increase in recharge from 5-113 mm, whereas in a wet year a similar increase in area ratio produces a 4-fold increase in recharge from 38-185 mm. These and other model studies by Tosomeen (1991) demonstrate the great sensitivity of localized recharge to rather small changes in topography, soil type, and climate.

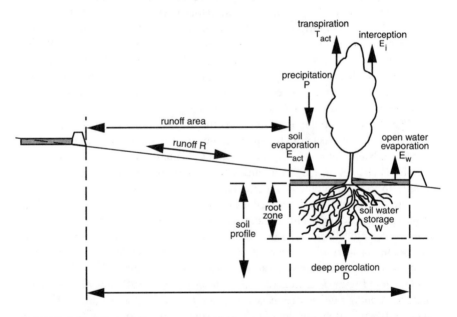

FIGURE 5-13 Microcatchment consisting of runoff area and basin area with tree. Rainfall induces runoff, which collects in the basin area where the water infiltrates, is stored, and is available for root water uptake and transpiration. In the basin area, losses occur by interception, soil evaporation, and deep percolation below the root zone. After Boers (1994).

TABLE 5-2   Annual Precipitation (P), Predicted Runoff (R), Actual Transpiration (T_{act}), and Recharge (D) (mm) for One Neem Tree in a 8 m^2 Basin at Sadoré, Niger, for Precipitation Only and for Precipitation and Runoff from 20 m^2 and 40 m^2 Runoff Areas in Three Precipitation Years (Boers, 1994)

	Rainfall Only	Rainfall and Runoff	
Average Year		Runoff area 20 m^2	Runoff area 40 m^2
P	545	545	545
R	0	232	465
T_{act}	409	633	755
D	0	5	113
Dry Year		Runoff area 20 m^2	Runoff area 40 m^2
P	258	258	258
R	0	78	155
T_{act}	138	205	277
D	0	0	0
Wet Year		Runoff area 20 m^2	Runoff area 40 m^2
P	673	673	673
R	0	285	571
T_{act}	481	720	849
D	0	38	185

Pipes Through Calcic Horizons

The La Mesa surface in southwestern New Mexico has developed on Rio Grande deposits mainly consisting of sands and gravels. The surface was abandoned by the Rio Grande River during the middle Pleistocene and since that time a calcic soil with an indurated calcic horizon has formed. A characteristic of such an indurated calcic horizon is the presence of dissolution pipes that develop through the horizon to the underlying sediments. Most of the upper soil profile has been eroded, exposing the calcic horizon. However, in the last few hundred years the exposed calcic horizon has been covered with sand dunes with a loamy sand texture. The sand has taken the form of coppice dunes that develop underneath the shrub vegetation. Consequently, the pipes in the carbonate horizon are mostly buried underneath 0.5-1.5 m of loamy sand and are very difficult to identify from the surface. Rodríguez-Marín and Hendrickx observed in October 1998 on the La Mesa surface pipe densities from one per 14-38 m^2 along two transects of 5.6 and 3.2 km in a 2.2-m-deep trench dug for a gas pipeline (Figure 5-14). Since the pipes cover approximately 15-18 percent of the total surface, the ratio between catchment area (around the pipe) and through flow area (inside the pipe) varies from about 7 to 5 on the La Mesa surface. It is anticipated that such

FIGURE 5-14 A pipe through an indurated calcic horizon on the La Mesa surface in New Mexico (Rodríguez-Marín and Hendrickx, personal communication, 1999).

localized pipes serve as preferential flow conduits for water flow and may increase groundwater recharge by an order of magnitude.

Since 1998, Harrison and Hendrickx have observed many pipes in highway cuts near Hatch and Albuquerque and along escarpments with exposed calcic horizons near Socorro and El Paso. These field observations along a 400-km-long stretch from El Paso to Albuquerque indicate that pipes through indurated calcic horizons are widespread in New Mexico and West Texas.

MEASUREMENT OF PREFERENTIAL FLOW PHENOMENA

Preferential flow phenomena are difficult to measure. Often, preferential flow is inferred from unexpectedly early water and chemical breakthrough or from unexpectedly deep water and chemical migration in the vadose zone. Bimodal or multimodal chemical concentration profiles are often interpreted as a result of preferential flow. Such measurements, even when preferential flow inference is very plausible, only give indirect evidence of the causal flow mechanisms. More direct evidence can be obtained by staining techniques with dyes or water, which have been extensively used to visualize preferential flow at the pore scale, such as macropore and fracture flow (Bouma et al., 1977), and preferential flow at the Darcian scale, such as unstable (Glass et al., 1988; Hendrickx et al., 1988) or funnel flow (Gardner and Hsieh 1959; Kung, 1990). Probably the most accurate quantitative measure of macropore flow is obtained through intercepting macropores with tubing to collect the throughflow (Edwards et al., 1989).

Macropore and fracture flow are often experimentally quantified by using tension infiltrometers. When infiltrating water is under negative pressure, flow through macropores and fractures can be more or less controlled according to the Laplace equation. The fraction of macropores of different sizes can thereby be measured by using a series of infiltration tests with different tensions (Watson and Luxmoore, 1986). More recently, tomographic techniques and nuclear magnetic imaging have been applied to visualize complicated pore structures and associated flow phenomena (Heijs et al., 1996; van As and van Dusschoten, 1997). Such techniques, and improvements thereof in the future, will help elucidate the preferential flow mechanisms at the pore scale and at those Darcian scales that cover centimeters and decimeters.

At larger spatial scales covering entire fields or landscape elements, flow phenomena are indirectly assessed by measuring soil moisture and soil structure through remote sensing, ground-penetrating radar, neutron-probe moisture measurements, and other geophysical techniques that allow large-scale mapping of pedological and geological features. Deep coring tests analyzed for chloride distributions often provide insight into flow mechanisms. The most convincing evidence for preferential flow at large spatial scales comes from visual observations, such as the pipes through calcic horizons in New Mexico (Figure 5-14) or the potholes in the Great Plains (Meyboom, 1966). Such structures are testimony to long-term persistent preferential flow processes that lead to the formation of observable landscape features.

FORMULATION OF CONCEPTUAL MODELS

The formulation of criteria for determining whether a conceptual model is an adequate characterization for a specific vadose zone is extremely complicated since there are so many factors that affect flow mechanisms. Nevertheless, the case studies presented in this review have a number of general characteristics in common that can serve as a guideline for the development of conceptual models.

All three spatial scales of preferential flow discussed have in common that an initially spatially uniform flux is disturbed and water flow is confined to a smaller cross-sectional area of the vadose zone. Even though the mechanisms that lead to the formation of preferential flow patterns are very different, the phenomenological appearances and the environmental consequences are often very similar. Figure 5-15 depicts a schematic view of different flow regimes that can be distinguished during a preferential flow process: (1) lateral distribution flow in the attractor zone where preferential flow is initiated; (2) downward preferential flow in the transmission zone where water moves along preferential flow pathways and thus bypasses a considerable portion of the porous media matrix; and (3) lateral and downward dispersive flow in the dispersion zone where preferential flow pathways are interrupted and water flow becomes uniform again.

The attractor zone can vary considerably in thickness (Figure 5-3), and may even be the soil surface itself when preferential flow is initiated by runon into localized surface depressions. Figures 5-3, 5-4, 5-5, 5-7, and 5-14 clearly show an attractor zone located close to the soil surface. The pipe in Figure 5-14 receives water by lateral collection over the calcic horizon from the surrounding areas. Therefore, it is concluded that an attractor zone at or below the soil surface is a definitive feature of preferential flow at whatever scale it takes place.

The dispersion zone can be recognized in Figures 5-7 and 5-11 but is missing in the other figures. It is often not included in graphical presentations of preferential flow when it occurs so deep in the vadose zone that water rarely even reaches this zone (Figures 5-3, 5-4, 5-14). Figure 5-10 shows that the dispersion zone for one layer with preferential flow can be the attractor zone for another deeper layer. Indeed, in a deep vadose zone, finding a sequence of attractor and dispersion zones is expected.

Stable flow can be considered as a flow mechanism that did not yet reach the transmission zone (Figure 5-5). This is the case, for example, where the amount of precipitation is not sufficient to wet the attractor zone to a depth that allows unstable wetting to occur (Hendrickx and Yao, 1996). Apparently stable flow can be dealt with as a special case in the general mechanism of preferential flow.

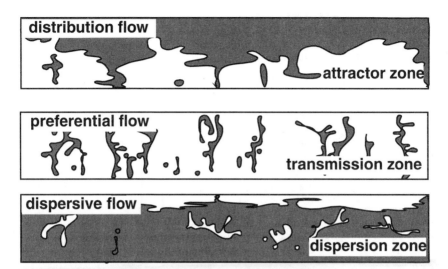

FIGURE 5-15 Schematic view of flow processes in the vadose zone. After Geoderma, 70, Flühler, H., W. Durner, and M. Flury, Lateral solute mixing processes: A key for understanding field-scale transport of water and solutes. Pp. 165-183, 1996, with permission from Elsevier Science.

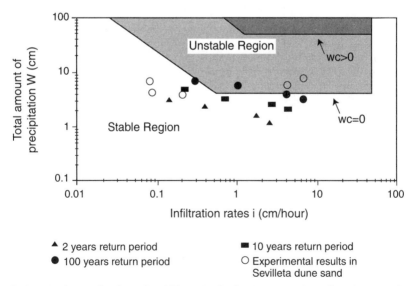

FIGURE 5-16 Application of stability criteria for determination of preferential flow caused by unstable wetting fronts. After Geoderma, 70, Hendrickx, J. M. H. and T. Yao, Prediction of wetting front stability in dry field soils using soil and precipitation data, pp. 265-280, 1996, with permission from Elsevier Science.

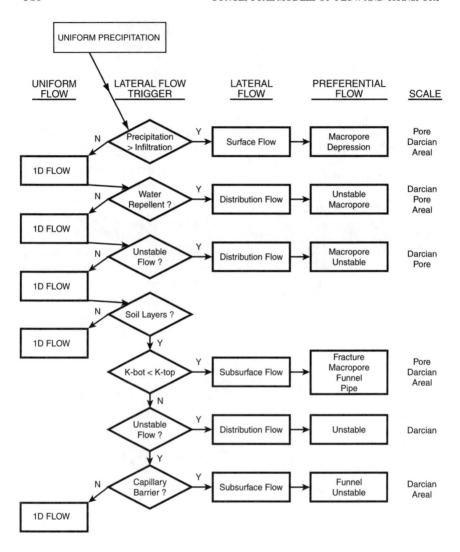

FIGURE 5-17 Flow diagram for evaluation of vadose conditions that will lead to preferential flow.

All three spatial scales of preferential flow discussed also have in common that preferential flow can start only after a certain minimum amount of precipitation has fallen at an intensity sufficiently high to cause the attractor zone to feed the transmission zone. Precipitation at a low intensity will infiltrate into the soil surface and result in a stable flow mechanism under a wide variety of conditions, from the fracture at the pore scale to the depression at the areal scale. However, a small amount of rain at a sufficiently high intensity may immediately trigger preferential flow at the pore scale as well as the areal scale. We need also to consider the effect of antecedent soil water content that affects the amount of water that can infiltrate into the soil at the surface or at the walls of a macropore. An example of how precipitation amount, precipitation intensity, and antecedent soil water content affect the occurrence of unstable flow in a wettable sand soil is given in Figure 5-16.

In Figure 5-17 we present a flow diagram for the evaluation of vadose zone conditions that will cause preferential flow at different scales as a result of a *lateral flow trigger*. Since precipitation is spatially uniform, it will result in uniform one-dimensional flow through the vadose zone unless a *lateral flow trigger* is present. We recognize different lateral flow triggers, such as a precipitation rate that exceeds the infiltration rate, vertical changes in unsaturated hydraulic conductivities where the lower layer has a lower conductivity, water repellency, and other mechanisms that cause unstable flow. Although the principles of the flow diagram are straightforward and will result in a first assessment of the propensity for preferential flow, no evaluation is complete without considering in which manner antecedent soil water content affects the flow processes in the attractor and transmission zones as a function of precipitation amount and intensity. This is the most difficult question but also the most crucial, since the threshold values for the minimum amount of precipitation and the intensity determine the occurrence of preferential flow at any scale.

REFERENCES

Anderson, M. G., and T. P. Burt, 1990. Subsurface runoff. Chapter 11 in M. G. Anderson and T. P. Burt (eds.). Process Studies in Hillslope Hydrology.
Baker, R. S., and D. Hillel, 1972. Laboratory tests of a theory of fingering during infiltration into layered soils. Soil Sci. Soc. Am. J. 54: 20-30.
Bear, J., 1972. Dynamics of Fluids in Porous Media. New York: Dover Publications, Inc., 764 pp.
Bear, J., 1993. Modeling flow and contaminant transport in fractured rocks. In: J. Bear, C. Tsang, and G. De Marsily, eds. Flow and Contaminant Transport in Fractured Rock. San Diego, Calif.: Academic Press, Inc., pp. 1-37.
Bear, J., T. Chin-Fu, and G. De Marsily (eds.), 1993. Flow and Contaminant Transport in Fractured Rock. New York: Academic Press, 560 pp.
Beven, K., and P. Germann, 1982. Macropores and water flow in soils. Water Resources Research 18: 1311-1325.

Boers, Th. M., 1994. Rainwater harvesting in arid and semi-arid zones. Wageningen, The Netherlands: International Land Reclamation and Improvement Institute (ILRI), 132 pp.

Boers, Th. M., M. De Graaf, R. A. Feddes, and J. Ben-Asher, 1986. A linear regression model combined with a soil water balance model to design micro-catchments for water harvesting in arid zones. Agricultural Water Management 11: 187-206.

Bond, R. D., 1964. The influence of the microflora on the physical properties of soils. II. Field studies on water repellent sands. Australian Journal of Soil Research 2: 123-131.

Bouma, J., and L. W. Dekker, 1978. A case study on infiltration into dry clay soil. I. Morphological observations. Geoderma 20: 27-40.

Bouma, J., A. Jongerius, O. Boersma, A. Jager, and D. Schoonderbeek, 1977. The function of different types of macropores during saturated flow through four swelling soil horizons. Soil Sci. Soc. Am. J. 41: 945-950.

Buchter, B., C. Hinz, M. Flury, and H. Fluhler, 1995. Heterogeneous flow and solute transport in an unsaturated stony soil monolith. Soil Sci. Soc. Am. J. 59: 14-21.

Campbell, G. S., 1985. Soil Physics with Basic, Transport Models for Soil-Plant Systems. New York: Elsevier, 150 pp.

Childs, E. C., 1969. The Physics of Soil Water Phenomena. New York: John Wiley and Sons.

Corey, A. T., 1990. Mechanics of Immiscible Fluids in Porous Media. Littleton, Colo.: Water Resources Publications, 255 pp.

Currie, I. G., 1993. Fundamental Mechanics of Fluids, 2nd ed. New York: McGraw-Hill, Inc.

Darcy, H. 1856. Les Fontaines Publique de la Ville de Dijon. Paris: Dalmont.

DeBano, L. F., 1969a. Water movement in water-repellent soils. In: Proc. Symp. on Water-repellent Soils, Riverside, California, 61-89.

DeBano, L. F., 1969b. Water-repellent soils: A worldwide concern in management of soil and vegetation. Agric. Sci. Rev. 7: 11-18.

DeBano, L. F., 1981. Water-repellent soils: A state of the art. Gen. Tech. Rep. PS-W-46, Pacific Southwest Forest and Range Experiment Station, 21 pp.

Deecke, W., 1906. Einige Beobachtungen am Sandstrande. Centralbl. für Mineral. Geol. und Paläont., 721-727.

Dekker, L. W., and P. D. Jungerius, 1990. Water repellency in the dunes with special reference to The Netherlands. Catena Suppl. 18: 173-183.

Dekker, L. W., and C. J. Ritsema, 1994a. How water moves in a water repellent sandy soil. 1. Potential and actual water repellency. Water Resources Research 30: 2507-2517.

Dekker, L. W., and C. J. Ritsema, 1994b. Fingered flow: The creator of sand columns in dune and beach sands. Earth Surface Processes and Landforms 19: 153-164.

Diment, G. A., and K. K. Watson, 1985. Stability analysis of water movement in unsaturated porous materials. 3. Experimental studies. Water Resources Research 21: 979-984.

Du, X. H., T. Yao, W. D. Stone, and J. M. H. Hendrickx. 2001. Stability analysis of the unsaturated water flow equation, 1. Mathematical derivation. Water Resources Research, in press.

Dullien, F. A. L., 1992. Porous media: Fluid transport and pore structure. New York: Academic Press, 574 pp.

Edwards, W. M., M. J. Shipitalo, L. B. Owens, and L. D. Norton, 1989. Water and nitrate movement in earthworm burrows within long-term no-till cornfields. J. Soil Water Conserv. 44: 240-243.

Emerson, W. W., and R. D. Bond, 1963. The rate of water entry into dry sand and calculation of the advancing contact angle. Australian Journal of Soil Research 1: 9-16.

Flühler, H., W. Durner, and M. Flury, 1996. Lateral solute mixing processes: A key for understanding field-scale transport of water and solutes. Geoderma 70: 165-183.

Flury, M., H. Flühler, W. A. Jury, and J. Leuenberger, 1994. Susceptibility of soils to preferential flow of water: A field study. Water Resources Research 30: 1945-1954.

Forrer, I., R. Kasteel, M. Flury, and H. Flühler, 1999. Longitudinal and lateral dispersion in an unsaturated field soil. Water Resources Research 35(10): 3049-3060.

Freeze, R. A., and J. Banner, 1970. The mechanism of natural groundwater recharge and discharge. 2. Laboratory column experiments and field measurements. Water Resources Research 6: 138-155.

Gardner, W. H., and J. C. Hsieh, 1959. Water movement in soils. Dept. of Crop and Soil Sciences, Washington State University, [Video], Pullman, Wash.

Gee, G. W., and D. Hillel, 1988. Groundwater recharge in arid regions: Review and critique of estimation methods. Hydrological Processes 2: 255-266.

Gees, R. A., and A. K. Lyall, 1969. Erosional sand columns in dune sand, Cape Sable Island, Nova Scotia, Canada. Canadian Journal of Earth Sciences 6: 344-347.

Germann, P. F., 1986. Rapid drainage response to precipitation. Hydrol. Processes 1: 3-13.

Germann, P. F., 1990. Macropores and hydrologic hillslope processes. In: M. G. Anderson and T. P. Burt, eds. Process Studies in Hillslope Hydrology, 327-363.

Glass, R. J., T. S. Steenhuis, and J.-Y. Parlange, 1988. Wetting front instability as a rapid and far-reaching hydrologic process in the vadose zone. J. Contam. Hydroll. 3: 207-226.

Glass, R. J., T. S. Steenhuis, and J.-Y. Parlange, 1989a. Wetting front instability. 1. Theoretical discussion and dimensional analysis. Water Resources Research 25: 1187-1194.

Glass, R. J., T. S. Steenhuis, and J.-Y. Parlange, 1989b. Wetting front instability. 2. Experimental determination of relationships between system parameters and two-dimensional unstable flow field behavior in initially dry porous media. Water Resources Research 25: 1195-1207.

Glass, R. J., T. S. Steenhuis, and J.-Y. Parlange, 1989c. Mechanism for finger persistence in homogeneous, unsaturated, porous media: Theory and verification. Soil Sci. 148: 60-70.

Glass, R. J., J. King, S. Cann, N. Bailey, J.-Y. Parlange, and T. S. Steenhuis, 1990. Wetting front instability in unsaturated porous media: A three-dimensional study. Transp. Porous Media 5: 247-268.

Glass, R. J., J.-Y. Parlange, and T. S. Steenhuis, 1991. Immiscible displacement in porous media: Stability analysis of three-dimensional, axisymmetric disturbances with application to gravity-driven wetting front instability. Water Resources Research 27: 1947-1956.

Green, W. H., and G. A. Ampt, 1911. Studies on soil physics. I. The flow of water and air through soils. J. Agric. Sci. 4: 1-24.

Gripp, K., 1961. Über Werden und Vergehen von Barchanene an der Nordsee-Küste Schleswig-Holsteins. Zeitsch. für Geomorphologi, Neue Folge 5: 24-36.

Gunn, J., 1983. Point-recharge of limestone aquifers: A model from New Zealand karst. J. Hydrol. 61: 19-29.

Hanks, R. J., and G. L. Ashcroft, 1986. Applied Soil Physics: Advanced Series in Agricultural Sciences 8. New York: Springer-Verlag, 159 pp.

Heijs, A. W., C. J. Ritsema, and L. W. Dekker, 1996. Three-dimensional visualization of preferential flow patterns in two soils. Geoderma 70: 101-116.

Hendrickx, J. M. H., L. W. Dekker, M. H. Bannink, and H. C. van Ommen, 1988. Significance of soil survey for agrohydrological studies. Agricultural Water Management 14: 195-208.

Hendrickx, J. M. H., and L. W. Dekker, 1991. Experimental evidence of unstable wetting fronts in non-layered soils. In: Proc. Natl. Symp. on Preferential Flow. Chicago, Ill. 16-17, Dec. 1991, St. Joseph, Mich.: Am. Soc. Agric. Eng., 22-31.

Hendrickx, J. M. H., S. Khan, M. H. Bannink, D. Birch, and C. Kidd, 1991. Numerical analysis of groundwater recharge through stony soils using limited data. J. Hydrol. 127: 173-192.

Hendrickx, J. M. H., L. W. Dekker, and O. H. Boersma, 1993. Unstable wetting fronts in water repellent field soils. Journal of Environmental Quality 22: 109-118.

Hendrickx, J. M. H., and T. Yao, 1996. Prediction of wetting front stability in dry field soils using soil and precipitation data. Geoderma 70: 265-280.

Hendrickx, J. M. H., and G. Walker, 1997. Recharge from precipitation. Chapter 2, In: I. Simmers, ed. Recharge of Phreatic Aquifers in (Semi)-Arid Areas. Rotterdam, The Netherlands: Balkema.

Hill, D. E., and J.-Y. Parlange, 1972. Wetting front instability in layered soils. Soil Sci. Soc. Am. J. 36: 697-702.

Hillel, D., 1998. Environmental Soil Physics. San Diego: Academic Press.

Hillel, D., and R. S. Baker, 1988. A descriptive theory of fingering during infiltration into layered soils. Soil Sci. 146: 51-56.

Issar, A., and R. Passchier, 1990. Regional hydrogeological concepts. In: D. N. Lerner et al., eds. Groundwater Recharge. International Contributions to Hydrogeology, Vol. 8, Hannover, Germany: Int. Assoc. Hydrogeologists, Verlag Heinz Heise, 21-98.

Jamison, V. C., 1969. Wetting resistance under citrus trees in Florida. In: Proc. Symp. on Water Repellent Soils, Riverside, California, 9-15.

Jaramillo, D. F., L. W. Dekker, C. J. Ritsema, and J. M. H. Hendrickx. 2000. Occurrence of soil water repellency in arid and humid climates. Journal of Hydrology 231/232: 105-114.

Johnston, C. D., 1987. Preferred water flow and localized recharge in a variable regolith. J. Hydrol. 94: 129-142.

Jury, W. A., W. R. Gardner, and W. H. Gardner, 1991. Soil Physics, New York: John Wiley and Sons, 328 pp.

Kirkham, D., and W. L. Powers, 1972. Advanced Soil Physics. New York: John Wiley and Sons, 534 pp.

Koorevaar, P., G. Menelik, and C. Dirksen, 1983. Elements of soil physics. New York: Elsevier, 230 pp.

Kung, K.-J. S., 1990. Preferential flow in a sandy vadose zone. 1. Field observation. Geoderma 46: 51-58.

Lawes, J. B., J. H. Gilbert, and R. Warington, 1881. On the amount and composition of the rain and drainage-waters collected at Rothamsted, Part I and II. J. Royal Agric. Soc. of England, London 17: 241-279.

Lerner, D. N., A. S. Issar, and I. Simmers, 1990. Groundwater Recharge. A Guide to Understanding and Estimating Natural Recharge. International Contributions to Hydrogeology, Vol. 8. Hannover: Internat. Assoc. of Hydrogeologists, Heise.

Letey, J., J. F. Osborn, and N. Valoras, 1975. Soil water repellency and the use of nonionic surfactants. Calif. Water Res. Center (Contribution 154) 85 pp.

Li, Y., and M. Ghodrati, 1994. Preferential transport of nitrate through soil columns containing root channels. Soil Sci. Soc. Am. J. 94: 653-659.

Lopes De Leao, L. R., 1988. Hé. Grondboor en Hamer 3/4: 111-112.

Luxmoore, R. J., 1981. Micro-, meso-, and macroporosity of soil. Soil Sci. Soc. Am. J. 45: 671-672.

Luxmoore, R. J., and L. A. Ferrand, 1992. Water flow and solute transport in soils: Modeling and applications. In: D. Russo and G. Dagan, eds. New York: Springer Verlag, pp. 45-60.

Mallik, A. U., and A. A. Rahman, 1985. Soil water repellency in regularly burned Calluna heathlands: Comparison of three measuring techniques. Journal of Environmental Management 20: 207-218.

Marshall, T. J., and J. W. Holmes, 1979. Soil Physics. Cambridge University Press, 345 pp.

McGhie, D. A., and A. M. Posner, 1980. Water repellency of a heavy-textured Western Australian surface soil. Aust. J. Soil Res. 18: 309-323.

Meeuwig, R. O., 1971. Infiltration and Water Repellency in Granitic Soils. U.S. Dept. of Agriculture Forest Service, Research Paper INT-111, Ogden, Utah, 20 pp.

Meyboom, P., 1966. Unsteady groundwater flow near a willow ring in hummocky moraine. J. Hydrol. 4: 38-62.

Miller, J. J., D. F. Acton, and R. J. St. Arnaud, 1985. The effect of groundwater on soil formation in a morainal landscape in Saskatchewan. Canadian Journal of Soil Science 65: 293-307.

Miyamoto, S., A. Bristol, and W. I. Gould, 1977. Wettability of coal-mine spoils in Northwestern New Mexico. Soil Sci. 123: 258-263.

Mohanty, B. P., R. S. Bowman, J. M. H. Hendrickx, J. Simunek, and M. Th. van Genuchten, 1998. Preferential transport of nitrate to a tile drain in an intermittent-flood-irrigated field: Model development and experimental evaluation. Water Resources Research 34: 1061-1076.

Mooij, J., 1957. Aeolian destruction forms on a sand beach. Grondboor en Hamer 1: 14-18.

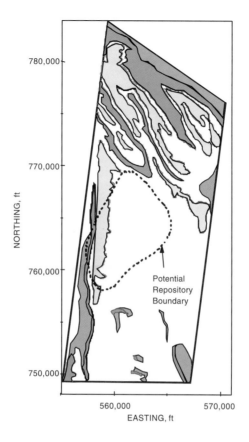

PLATE 1 Spatially distributed infiltration in bedrock units. From Flint and Flint (1994, Figure 4 and Table 3). Copyright 1994 by the American Nuclear Society, La Grange Park, Illinois.

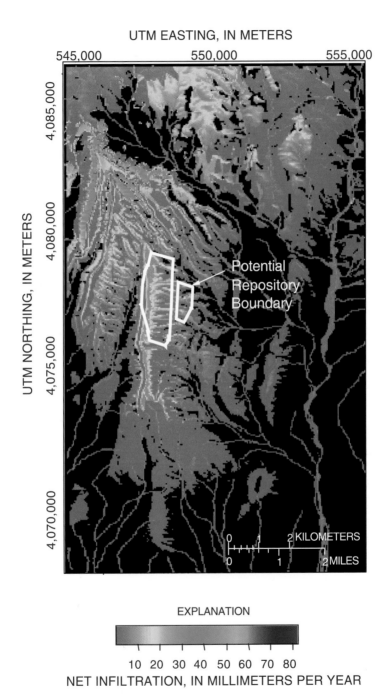

UTM EASTING, IN METERS

545,000   550,000   555,000

Potential
Repository
Boundary

0   1   2 KILOMETERS

0   1   2 MILES

EXPLANATION

10 20 30 40 50 60 70 80

NET INFILTRATION, IN MILLIMETERS PER YEAR

PLATE 2 Spatial distribution of net infiltration. From Flint et al. (1996a, Figure 39).

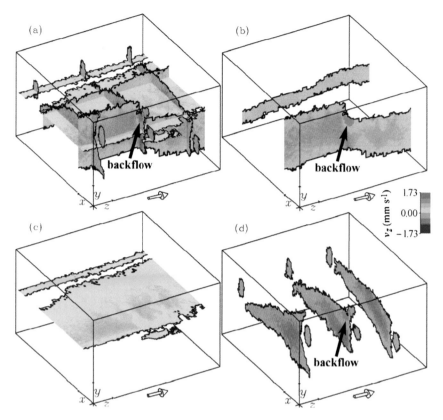

PLATE 3 NMRI flow profiles in fractures showing 2D slices through a 3D image of the principal (z) velocity component of water flowing inside the rock fracture. Fracture sample dimensions are ~36 mm by ~18 mm. (a) Two yz-slices with $x = 4.7$ and 15.2 mm, and an xz-slice with $y = 4.0$ mm, parallel to the principal flow direction (indicated by the round arrow); and three xy-slices with $z = 2.8$, 12.7, and 22.5 mm, perpendicular to the principal flow direction; (b) two yz-slices with $x = 4.7$ and 17.3 mm, parallel to the principal flow direction; (c) an xz-slice with $y = 4.0$ mm, parallel to the principal flow direction; (d) three xy-slices with $z = 2.8$, 12.7, and 21.1 mm, perpendicular to the principal flow direction. Darker shades represent regions of higher velocity. The vertical scale is highly exaggerated, since the image scale is in units of measured volume elements. Control tubes are clearly visible on the two sides of the fracture sample. After Dijk et al. (1999).

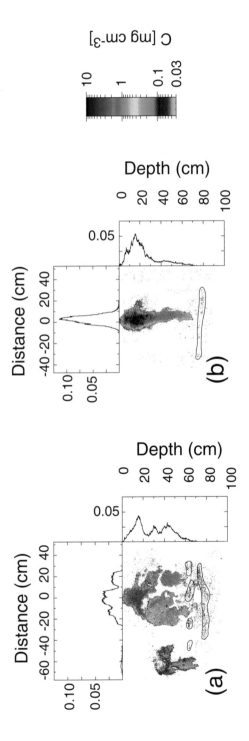

PLATE 4 Flow patterns observed in a loamy soil, showing (a) high infiltration rate of 25 mm/day, and (b) low infiltration rate of 5 mm/day (after Forrer, I., R. Kasteel, M. Flury, and H. Flühler, 1999. Longitudinal and lateral dispersion in an unsaturated field soil. Water Resources Research 35(10): 3049-3060. Copyright by American Geophysical Union). The graphs represent two-dimensional tracer distributions and the corresponding one-dimensional averages observed on a vertical soil profile.

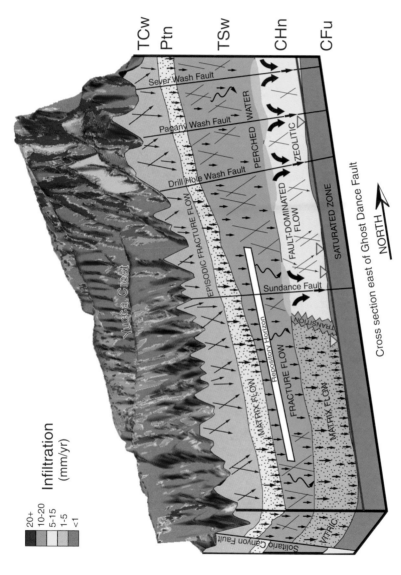

Infiltration (mm/yr)

20+
10-20
5-15
1-5
<1

Yucca Crest

TCw
Ptn
TSw
CHn
CFu

Sever Wash Fault
Pagany Wash Fault
Drill Hole Wash Fault
Sundance Fault
Solitario Canyon Fault

EPISODIC FRACTURE FLOW
MATRIX FLOW
Repository Horizon
FRACTURE FLOW
MATRIX FLOW
VITRIC
TRANSITION
PERCHED WATER
ZEOLITIC
FAULT-DOMINATED FLOW
SATURATED ZONE

Cross section east of Ghost Dance Fault

NORTH

PLATE 5 A simplified schematic diagram showing conceptualized water flow through Yucca Mountain.

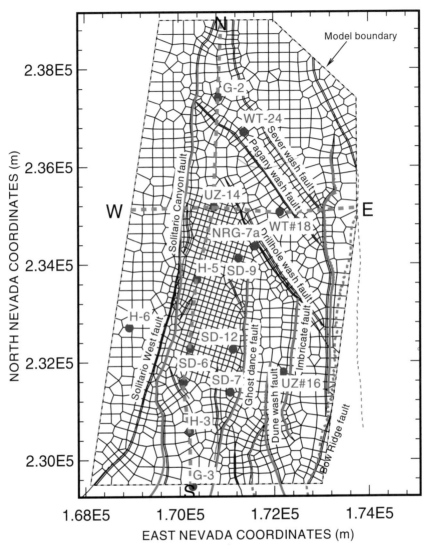

PLATE 6 A plan view of the three-dimensional UZ site-scale model domain, showing grid and locations of boreholes (including perched water boreholes) used for model calibration.

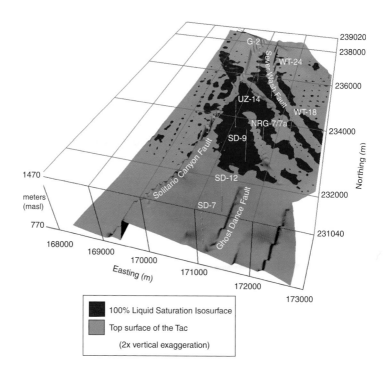

PLATE 7 Three-dimensional perched-water bodies near the base of the TSw, using the permeability barrier conceptual model.

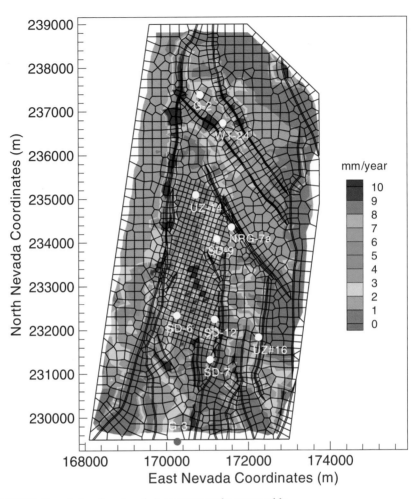

PLATE 8 Percolation flux (mm/yr) contours at the water table.

Nieber, J. L., C. A. S. Tosomeen, and B. N. Wilson, 1993. A stochastic-mechanistic model of depression-focused recharge. In: Y. Eckstein and A. Zaporozec, eds. Hydrogeologic Inventories and Monitoring and Groundwater Modeling. Proc. of Industrial and Agricultural Impacts on the Hydrologic Environment, 207-229.

Parlange, J.-Y., and D. E. Hill, 1976. Theoretical analysis of wetting front instability in soils. Soil Sci. 122: 236-239.

Petersen, C. T., S. Hansen, and H. E. Jensen, 1997. Tillage-induced horizontal periodicity of preferential flow in the root zone. Soil Sci. Soc. Am. J. 61: 586-594.

Philip, J. R., 1969. Theory of infiltration. Advances in Hydroscience 5: 215-296.

Philip, J. R., 1975a. Stability analysis of infiltration. Soil Sci. Soc. Am. Proc. 39: 1042-1049.

Philip, J. R., 1975b. The growth of disturbances in unstable infiltration flows. Soil Sci. Soc. Am. Proc. 39: 1049-1053.

Raats, P. A. C., 1973. Unstable wetting fronts in uniform and nonuniform soils, Soil Sci. Soc. Am. Proc. 37: 681-685.

Raats, P. A. C., 1984. Tracing parcels of water and solutes in unsaturated zones. In: B. Yaron, G. Dagan, and J. Goldshid, eds. Pollutants in Porous Media: The Unsaturated Zone Between Soil Surface and Ground Water. Berlin: Springer, 4-16.

Rasmussen, T. C., 1987. Computer simulation model of steady fluid flow and solute transport through three-dimensional networks of variably saturated, discrete fractures. In D. D. Evans and T. J. Nicholson, Flow and Transport Through Unsaturated Fractured Rock. Geophysical Monograph 42, American Geophysical Union, Washington D.C., pp. 107-114.

Richardson, J. L., 1984. Field observation and measurement of water repellency for soil surveyors. Soil Survey Horiz. 25: 32-36.

Rietveld, J. J., 1978. Soil Nonwettability and Its Relevance as a Contributing Factor to Surface Runoff on Sandy Dune Soils in Mali. Report of project Production primaire au Sahel, Agric. Univ., Wageningen, The Netherlands, 179 pp.

Ritsema, C. J., L. W. Dekker, J. M. H. Hendrickx, and W. Hamminga, 1993. Preferential flow mechanism in a water-repellent sandy soil. Water Resources Research 29: 2183-2193.

Ritsema, C. J., and L. W. Dekker, 1994. How water moves in a water-repellent sandy soil. 2. Dynamics of fingered flow. Water Resources Research 30: 2517-2531.

Roth, K., 1995. Steady state flow in an unsaturated, two-dimensional, macroscopically homogeneous, Miller-similar medium. Water Resources Research 31: 2127-2140.

Roth, K., and K. Hammel, 1996. Transport of conservative chemical through an unsaturated two-dimensional Miller-similar medium with steady state flow. Water Resources Research 32: 1653-1663.

Scanlon, B. R., 1992. Moisture and solute flux along preferred pathways characterized by fissured sediments in desert soils. J. of Contaminant Hydrol. 10: 19-46.

Scheidegger, A. E., 1974. The physics of flow through porous media. Toronto: University of Toronto Press.

Schrauf, T. W., and D. D. Evans, 1986. Laboratory studies of gas flow through a single natural fracture. Water Resources Research 19: 1253-1265.

Schuddebeurs, A. P, 1957. Cone sand and sandstone. Grondboor en Hamer. 2. 21-25.

Selker, J. S., T. S. Steenhuis, and J.-Y. Parlange, 1989. Preferential flow in homogeneous sandy soils without layering. Paper No. 89-2543, Am. Soc. Agric. Eng., Winter Meeting, New Orleans, La., 22 pp.

Selker, J. S., T. S. Steenhuis, and J.-Y. Parlange, 1992. Wetting front instability in homogeneous sandy soils under continuous infiltration. Soil Sci. Soc. Am. J. 56: 1346-1350.

Sigda, J. M., 1997. Effects of small-displacement faults on the permeability distribution of poorly consolidated Santa Fe Group sands, Rio Grande Rift New Mexico. M.S. thesis, New Mexico Tech, Socorro.

Snow, D. T., 1969. Anisotropic permeability of fractured media. Water Resources Research 5: 1273-1289.

Stamm, C., H. Flühler, R. Gächter, J. Leuenberger, and H. Wunderli, 1998. Preferential transport of phosphorus in drained grassland soils. J. Environ. Qual. 27: 515-522.

Starr, J. L., H. C. DeRoo, C. R. Frink, and J.-Y. Parlange, 1978. Leaching characteristics of a layered field soil. Soil Sci. Soc. Am. J. 42: 386-391.

Starr, J. L., J.-Y. Parlange, and C. R. Frink, 1986. Water and chloride movement through a layered field soil. Soil Sci. Soc. Am. J. 50: 1384-1390.

Stephens, D. B., 1994. A perspective on diffuse natural recharge mechanisms in areas of low precipitation. Soil Sci. Soc. Am. J. 58: 40-48.

Streeter, V. L., E. B. Wylie, and K. W. Bedford, 1988. Fluid Mechanics, 9th ed. Boston: McGraw-Hill.

Tabuchi, T., 1961. Infiltration and ensuing percolation in columns of layered glass particles packed in laboratory. Nogyo dobuku kenkyu, Bessatsu. Trans. Agr. Eng. Soc., Japan, 1, 13-19 (in Japanese, with a summary in English).

Tamai, N., T. Asaeda, and C. G. Jeevaraj, 1987. Fingering in two-dimensional, homogeneous, unsaturated porous media. Soil Sci. 144: 107-112.

Tosomeen, C. A. S., 1991. Modeling the effects of depression focusing on groundwater recharge. M.S. thesis, Dept. of Agricultural Engineering, University of Minnesota.

Van As, H., and D. van Dusschoten, 1997. NMR methods for imaging of transport processes in micro-porous systems. Geoderma 80: 389-403.

Van Dam, J. C., J. M. H. Hendrickx, H. C. van Ommen, M. H. Bannink, M. Th. Van Genuchten, and L. W. Dekker, 1990. Simulation of water and solute transport through a water-repellent sand soil, J. Hydrol. 120: 139-159.

Van Genuchten, M. Th., 1980. A closed-form equation for predicting the hydraulic conductivity of unsaturated soils. Soil Sci. Soc. Am. J. 44: 892-898.

Van Genuchten, M. Th., F. J. Leij, and S. R. Yates, 1991. The RETC code for quantifying the hydraulic functions of unsaturated soils. EPA/600/2-91/065. R. S. Kerr Environ. Res. Lab., U.S. Environmental Protection Agency, Ada, Okla., 93 pp.

Wagenet, R. J., 1998. Scale issues in agroecological research chains. Nutrient Cycling in Agroecosystems 50: 23-34.

Wagenet, R. J., J. Bouma, and J. L. Hutson, 1994. Modelling water and chemical fluxes as driving forces in pedogenesis. In R. B. Bryant, R. W. Arnold, and M. R. Hoosbeek, eds. Quantitative Modelling of Soil Forming Processes. SSSA Special Publ. 39, ASA/CSSA/SSSA, Madison, Wis., pp. 17-35.

Wang, Z., J. Feyen, M. Th. Van Genuchten, and D. R. Nielsen, 1998. Air entrapment effects on infiltration rate and flow instability. Water Resources Research 34: 213-222.

Watson, K., and R. J. Luxmoore, 1986. Estimating macroporosity in a watershed by use of a tension infiltrometer. Soil Sci. Soc. Am. J. 50: 578-582.

Wheatcraft, S. W., and J. H. Cushman, 1991. Hierarchical approaches to transport in heterogeneous porous media. In: U.S. National Report to International Union of Geodesy and Geophysics, Rev. Geophysics (supplement), pp. 263-269, American Geophysical Union. Washington, D.C.

White, I., P. M. Colombera, and J. R. Philip, 1977. Experimental studies of wetting front instability induced by gradual change of pressure gradient and by heterogeneous porous media. Soil Sci. Soc. Am. J. 41: 483-489.

Wierenga, P. J., R. G. Hills, and D. B. Hudson, 1991. The Las Cruces trench site: Characterization, experimental results, and one-dimensional flow predictions. Water Resources Research 27: 2695-2705.

Winter, T. C., 1986. Effect of ground-water recharge on configuration of the water table beneath sand dunes and on seepage in lakes in the sandhills of Nebraska, USA. J. Hydrol. 86: 221-237.

Wood, W. W., and W. E. Sanford, 1995. Chemical and isotopic methods for quantifying ground-water recharge in a regional, semiarid environment. Ground Water 33: 458-468.

Yao, T., and J. M. H. Hendrickx, 1996. Stability of wetting fronts in homogeneous soils under low infiltration rates. Soil Science Society of America Journal 60: 20-28.

Yao, T., and J. M. H. Hendrickx, 2001. Stability analysis of the unsaturated water flow equation: 2. Experimental Verification. Water Resources Research, in press.

6

Modeling Macropore Flow in Soils:
Field Validation and Use
for Management Purposes

Nicholas Jarvis[1] and Martin Larsson[1]

ABSTRACT

This paper reviews the development and application of models to predict the impact of preferential flow in the unsaturated zone on nonpoint-source pollution of groundwater and surface waters by agrochemicals, focusing mainly on the effects of macropores in structured soils. A brief review of the various modeling approaches that have been adopted is first presented. We then discuss the use of macropore flow models for management and regulatory purposes, focusing especially on data requirements and procedures for field validation and parameter estimation for predictive applications. Results of field-scale tests of one widely used macropore flow model (MACRO) are presented as a case study to illustrate the consequences of macropore flow for nitrate and pesticide leaching to shallow groundwater.

It is concluded that regulatory use of macropore flow models is necessary, but such use requires transparency and uniformity of approach to ensure that simulations are free of subjective bias. Regulatory applications of macropore flow models should be based either on precalibrated scenarios or on objective automatic parameter estimation routines (pedotransfer functions). Recommendations and procedures currently under development for use of the MACRO model in E.U. harmonized pesticide registration are also described.

[1] Department of Soil Sciences, Swedish University of Agricultural Sciences, Uppsala, Sweden

INTRODUCTION

Nonequilibrium or preferential flow of water enhances the risk of leaching of surface-applied contaminants to receiving water bodies, since much of the buffering capacity of the biologically and chemically reactive topsoil may be bypassed. In this way, chemicals can quickly reach subsoil layers where degradation and sorption processes are generally less effective. The term "preferential flow" implies that recharge is concentrated to a small fraction of the total pore space where vertical flow rates are much faster than rates of lateral equilibration with slowly moving resident soil water (Flühler et al., 1996). For example, in structured soils, nonequilibrium flow in soil macropores (e.g., shrinkage cracks, worm channels, old root holes) may dominate the soil hydrology, especially in fine-textured soils, where they operate as high-conductivity flow pathways bypassing the impermeable soil matrix (Beven and Germann, 1982). Preferential flow also occurs in structureless sandy soils in the form of unstable wetting fronts or fingering (Hillel, 1987), although this aspect of nonequilibrium flow behavior is not discussed further in this paper.

This paper reviews the development and application of models to predict the impact of macropore flow in the soil unsaturated zone on nonpoint-source pollution of groundwater and surface waters, focusing especially on agrochemicals. A brief review of the various modeling approaches that have been adopted is first presented. The dual-porosity model MACRO (Jarvis, 1994a; Jarvis and Larsson, 1998) is briefly described, and results of field tests of the model are presented that illustrate the consequences of macropore flow for nitrate and pesticide leaching. Concepts of model validation and calibration are discussed, and the data requirements for field validation of macropore flow models are reviewed. Current trends and recommendations for use of these models in pesticide registration procedures are also described.

MACROPORE FLOW: PHYSICAL PRINCIPLES

Macropores are large, continuous, structural pores that constitute preferred flow pathways for infiltrating water in most soils. At the macroscopic scale of measurement, this is reflected in large increases in unsaturated hydraulic conductivity across a small soil water pressure head range close to saturation (Clothier and Smettem, 1990; Jarvis and Messing, 1995). At the pore scale, macropore flow is generated when the water pressure locally increases to near saturation at some point on the interface with the surrounding soil matrix, such that the water-entry pressure of the pore is exceeded. Macropore flow can be sustained if the vertical flux rates in the macropore are large in relation to the lateral infiltration losses into the matrix due to the prevailing capillary pressure gradient. This is most likely to be the case in clay soils with an impermeable matrix. These lateral losses can be further restricted by relatively impermeable interfaces between macropores and the bulk soil, including cutans on aggregate surfaces and organic linings in biotic pores (Thoma et al., 1992). Thus, macropore flow can sometimes be significant even in lighter-textured soils of large matrix hydraulic conductivity.

The impact of macropore flow on solute transport depends strongly on the nature of the solute ion under consideration, particularly the size of the molecule as it affects lateral spreading between pore domains by diffusion, its sorption characteristics, and the nature of source/sink terms (i.e., biological transformations) affecting the transport process. For example, the effects of macropore flow on leaching are opposite for solutes that are foreign to the soil (e.g., pesticides) compared to indigenous solutes produced within the soil (e.g., nitrate mineralization from organic matter, without fertilizer applications). In the former case, macropore flow may strongly increase leaching losses, while in the latter case, leaching should be decreased by macropore flow (Jarvis, 1998).

CURRENT APPROACHES TO MODELING MACROPORE FLOW IN SOILS

In this section, some existing models accounting for macropore flow are described. This review is not intended to be exhaustive, but rather to give some idea of the different types of models available, their advantages and limitations, and their potential as management tools. Most existing models take a macroscopic, continuum approach by lumping individual preferential flow pathways in the soil into two or more pore domains within a one-dimensional numerical scheme. A large number of models of this type have been developed in recent years. They vary with regard to the number of flow domains considered (two-domain or multidomain models), the degree of simplification and empiricism involved in process descriptions (e.g., functional vs. mechanistic models), and also the manner in which exchange between flow domains is represented. In the following sections, some of these existing models are described.

Functional Models

In recent years, a number of simple empirical (or functional) models have been developed that can account for preferential flow (Addiscott, 1977; Corwin et al., 1991; Hall, 1993; Nicholls and Hall, 1995). Barraclough (1989a) described a simple three-region model based on a capacity ("tipping bucket") approach to describe water flow, the mobile-immobile water concept to characterize solute transport in two domains (see next section), and a simple instantaneous bypass routine to account for macropore flow. The model was apparently implemented only for nonreactive solute transport, although it was compared to nitrate leaching data during three winter seasons (Barraclough, 1989b).

Some functional models may be rather limited in their ability to simulate preferential flow due to simplifications in the treatment of soil water flow. For example, these models usually assume that water outflow from each soil layer is zero until filled to field capacity. Thus, rapid and deep-penetrating bypass flow in dry, macroporous soils cannot be simulated. Difficulties may also be caused by the fact that the normal time step in these models is one day, whereas an appropriate time scale to characterize preferential flow and transport processes would be of the order of hours, or even minutes.

Dual-Porosity/Single Permeability Models

Analytical models have been developed based on the two-region (mobile-immobile water) convection-dispersion equation (CDE) for both nonreactive and sorbing and degrading solutes (van Genuchten and Wierenga, 1976; van Genuchten and Wagenet, 1989; Gamerdinger et al., 1990). These analytical solutions of the CDE require idealized initial and boundary conditions and assume steady-state water flow. Thus, although such approaches may provide some theoretical insights into solute transport processes occurring in structured soils, they are not appropriate to field situations characterized by time-varying soil water content due to evaporation, root water uptake, and intermittent inputs of rainfall. For this reason, their use as management tools is rather limited.

Numerical solutions of the mobile-immobile water concept have also been implemented. However, even when the mobile-immobile water CDE is solved numerically, it is often either coupled to Richards' equation or steady water flow is assumed (Lafolie and Hayot, 1993; Zurmühl and Durner, 1996). Such an approach only provides at best a partial description of nonequilibrium flow and transport phenomena, since it cannot account for rapid movement of infiltrating water in macropores to depth in dry soil. For transient conditions, one unresolved question is how best to define the fractional mobile water content in relation to the time-varying total water content. Zürmuhl and Durner (1996) discussed this problem and recommended defining a variable ratio between the mobile water content and the total water content based on the shape of the hydraulic conductivity function, such that the conductivity at the boundary between mobile and immobile water represented a constant fraction of the conductivity at the current water content. Van Dam et al. (1990) presented a simple solution of the mobile-immobile water CDE, coupled to Richards' equation, which may be appropriate to water-repellent soils where a part of the soil remains unwetted during infiltration.

Jarvis (1989) and Armstrong et al. (1995) described a model particularly suited to cracking clay soils. In this model (CRACK), vertical water flow and solute transport occur in a crack system that is defined, for each layer in the soil profile, by a fixed spacing and a porosity that varies due to swelling and shrinkage. Vertical transport is neglected in the soil aggregates, which act as sinks for water and sources/sinks for solute. In the case of solute, the aggregates are divided into numerical slices and lateral diffusion and exchange is modeled explicitly using Fick's law. In the latest version of the model, subroutines have been implemented to describe nitrate transformations and pesticide sorption and degradation (Armstrong et al., 1995).

Dual-Porosity/Dual-Permeability Models

In these models, two pore domains are each characterized by a porosity, a water pressure (and water content), and solute concentration. In contrast to the mobile-immobile water approach, vertical water flow and solute transport are calculated for both domains, with mass exchange between domains treated as source/sink terms in the one-dimensional (vertical) model structure. This mass exchange between domains is calculated using approximate first-order equations, based either on an effective diffusion pathlength related to the macroscale soil geometry (aggregate size) or empirical mass transfer coefficients.

Gerke and van Genuchten (1993a) described a model of water flow and nonreactive solute transport based on the van Genuchten/Mualem model for the soil hydraulic functions (van Genuchten, 1980; Mualem, 1976), applied to a bimodal pore size distribution (Othmer et al., 1991; Durner, 1992). Richards' equation and the CDE are used to calculate water flow and solute transport, respectively, in both pore domains, while mass exchange of water and solutes is calculated using approximate first-order equations accounting for both convective and diffusive transfer (Gerke and van Genuchten, 1993b). Due to the geometry of the flow system, the rate of mass exchange between the domains is inversely proportional to the square of an effective diffusion pathlength (van Genuchten, 1985; Youngs and Leeds-Harrison, 1990).

Jarvis (1994a) and Jarvis and Larsson (1998) described a dual-porosity model (MACRO) in which Richards' equation and the convection-dispersion equation are used to model soil water flow and solute transport in soil micropores, while a simplified capacitance-type approach is used to calculate water and solute movement in macropores. This description of gravity-driven water flow in macropores can be considered as the numerical equivalent of the analytical kinematic wave model described by Germann (1985). Later in this paper, some case studies are presented of field applications of MACRO. Therefore, this particular model is now described in more detail.

In MACRO, a simple cut-and-join or two-line method is used to define the hydraulic functions (Smettem and Kirkby, 1990), rather than the additive superimposition of two pore regions employed, for example, by Gerke and van Genuchten (1993a). The "boundary" between the two pore domains is defined by the air-entry pressure head in the Brooks and Corey (1964) equation, with an equivalent water content and hydraulic conductivity defining the saturated state of the micropores (Figure 6-1). The boundary condition at the soil surface is an important part of any flow and transport model. In MACRO, a flux boundary

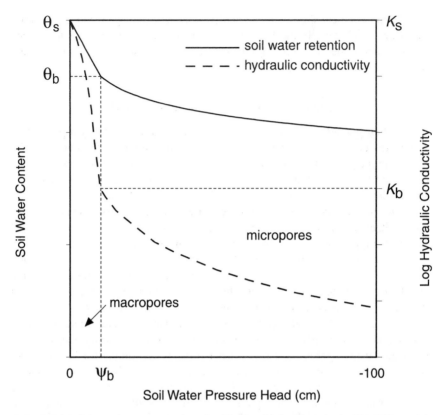

FIGURE 6-1 Schematic representation of soil hydraulic functions in the MACRO model: K_s is the saturated hydraulic conductivity, θ_s is the saturated water content, ψ_b is the pressure head defining the boundary between macropores and micropores, θ_b is the equivalent water content, and K_b is the hydraulic conductivity at ψ_b. From Larsson and Jarvis (1999b).

condition is used to partition incoming rainfall between the two pore regions depending on the infiltration capacity of the micropores. With respect to solute transport, the water infiltrating in macropores is characterized by a concentration c_{ma}^{*} calculated assuming complete mixing with solute stored in the solution phase of a shallow "mixing" depth z_d:

$$c^{*}_{ma} = \frac{\left(\left(\dfrac{z_d}{\Delta z}\right)Q_s\right) + (R\,c_r)}{R + \left(z_d\left(\theta + \left((1-f)\gamma_b k_d\right)\right)\right)},\qquad (6.1)$$

where Q_s is the amount of solute stored in the top layer of the soil, Δz is the layer thickness, R is the net rainfall, c_r is the concentration in the rain, θ and γ are the water content and bulk density in the top layer, and f is the fraction of sorption sites in the macropore region.

Water and solute exchange rates between the flow domains (M_w and M_s, respectively) are calculated as a function of an effective diffusion pathlength d, using approximate physically based first-order expressions (Booltink et al., 1993; Gerke and van Genuchten, 1993b). For solute, both diffusive and convective mass transfer are considered:

$$M_w = \left(\frac{3D_w\gamma_w}{d^2}\right)(\theta_b - \theta_{mi}) \tag{6.2}$$

and

$$M_s = \left(\frac{3D_e\theta_{mi}}{d^2}\right)(c_{ma} - c_{mi}) + M_w c' \tag{6.3}$$

The subscripts *ma* and *mi* refer to macropores and micropores, θ_b is the saturated water content of the micropores (Figure 6-1), γ_w is a scaling factor introduced to match approximate and exact solutions to the diffusion problem (van Genuchten, 1985), c' is the solute concentration in either macropores or micropores depending on the direction of convective transfer, D_e is an effective diffusion coefficient, and D_w is an effective water diffusivity given by:

$$D_w = \left(\frac{D_{\theta b} + D_{\theta mi}}{2}\right)S_{ma} \tag{6.4}$$

where $D_{\theta b}$ and $D_{\theta mi}$ are the water diffusivities at θ_b and θ_{mi}, respectively, and S_{ma} is the degree of saturation in the macropores, which is introduced to account for the effects of incomplete wetted contact area between the two pore domains.

A full water balance is considered in MACRO, including precipitation, evapotranspiration, and root water uptake; deep seepage; and lateral fluxes to tile drains in saturated soil. Solute transport and transformation processes in the model include convective-dispersive transport, Freundlich equilibrium sorption with sorption sites partitioned between the two pore domains, microbial degradation according to first-order kinetics, plant uptake, and canopy interception/washoff.

Multiporosity/Multipermeability Models

Steenhuis et al. (1990) described a simple multidomain model of solute movement based on a piecewise linear approximation to the hydraulic conductivity function and the assumption of a unit hydraulic gradient. Solute exchange between pore classes is calculated by mixing a fixed fraction of the water and solute in each pore class in a common pool before being redistributed. If the mixing fraction is set to zero, the multidomain model reduces to a parallel stream-tube model, whereas if it is set to unity (complete lateral mixing), it becomes equivalent to the CDE.

Hutson and Wagenet (1995) described a multiregion model of water flow and solute transport (TRANSMIT) based on the existing single-domain model LEACHM (Wagenet and Hutson, 1987). TRANSMIT can be used to investigate the impacts of preferential flow occurring at the porescale (e.g., macropore flow) by simulating multiple pore domains, and also physical nonequilibrium induced by soil heterogeneity through modeling multiple, interacting soil columns. A wide range of solutes can be dealt with, including pesticides and nitrate. Convective and diffusive mass exchange between all combinations of columns and/or pore domains is calculated using empirical expressions.

Gwo et al. (1995) described a multiporosity model of water flow and solute transport (MURF/MURT) based on the overlapping pore continua concept (Gerke and van Genuchten, 1993a). Richards' equation and the CDE are used to model transport in each pore domain, while water exchange is regulated by empirical exchange coefficients in simple first-order expressions. Solute exchange is driven by both convective transfer and first-order diffusive exchange.

Multiporosity models can, of course, default to the dual-porosity case when required, and there may be good reasons for doing so. As the number of pore classes in the model increases, the description of mass exchange processes necessarily becomes more uncertain and less mechanistic. Thus, although multidomain models may provide valuable insights into the nature of solute transport in heterogeneous soils, and also allow much greater flexibility in matching the observed behavior of solute transport in soil compared to two-domain models (Gwo et al., 1995), they may require extensive calibration and cannot easily be used predictively for management purposes (Hutson and Wagenet, 1995).

Two-Dimensional Models

Nieber and Misra (1995) described a two-dimensional, dual-porosity, finite element model of water flow and nonreactive solute transport in tile-drained soil. The mass exchange terms between the domains were taken from Gerke and van Genuchten (1993a, 1993b). A sensitivity analysis for the model was presented showing how chemical mass flux arriving at the drains was strongly regulated by the strength of interaction between the domains.

APPLICATION OF MACROPORE FLOW MODELS

Model Validation

Regulatory authorities and other management users must have confidence in the output of macropore flow models if these are to be used as tools to aid decision-making. This implies that the model must be considered sufficiently valid or validated in the minds of the users for the intended purpose. It is often claimed, from a logical and philosophical point of view, that models cannot be validated because they are only ever approximations to reality, and because the truth of theory cannot be proven (Oreskes et al., 1994). However, this widely held belief is based on a rather narrow view of what constitutes validation. A pragmatic alternative definition can be summarized by the truism that all models are wrong, but some are useful (see reviews by Rykiel, 1996, and Steefel and van Cappellen, 1998). It is helpful here to distinguish between conceptual validity, which involves subjective judgement, and model validation, which should be an objective procedure. Conceptual validity implies that the model concepts, simplifications, and assumptions are considered justifiable and reasonable for the stated use of the model, based on a consensus of understanding in the scientific community. Of course, it is always possible to criticize some aspects of the theoretical or conceptual basis of any model, since all models are to a greater or lesser extent simplifications of reality. Thus, operational model validation is a necessary procedure that identifies the limits and conditions of conceptual validity. Thus, operational validation is the process of determining whether the model meets specified performance criteria, defined as the level of accuracy required for a given purpose and context of use. In an interesting review of the concept and practice of model validation, Rykiel (1996) stated that "validation is not a procedure for testing scientific theory or for certifying the 'truth' of current scientific understanding. Validation means that a model is acceptable for its intended use because it meets specified performance requirements." Seen in this light, operational validation is a rigorous quantitative procedure similar to hypothesis testing in statistics. It is also the process by which model concepts gain credibility for a given use or, after repeated testing, are ultimately rejected in favor of what are perceived to be improved descriptions. Thus, conceptual validity can only be achieved by extensive and repeated operational validation.

The importance of context of use can be demonstrated by a simple example. Richards' equation and the CDE may be conceptually valid for predicting water flow and solute transport in homogeneous porous media, but would repeatedly fail operational validation tests for predicting leaching to groundwater in naturally heterogeneous soils, and lose credibility for this purpose.

Apart from difficulties with semantics, one reason for the continuing confusion over the meaning and purpose of model validation is that there are no widely accepted, standard, objective performance criteria for testing models (Rykiel, 1996). In statistics, a probability level of 0.05 is often used as a test of significance. This is an arbitrary standard, but widely accepted. Similar standard measures are required for operational model validation. In this respect, it is important that representatives of the user-community (regulatory authorities, industry) take an active role in specifying performance criteria, simply because the level of accuracy required depends on the use of the model and the target quantity being predicted (i.e., concentrations, loads, average values, maximums). The confusion that arises when performance criteria for model validation are not clearly expressed a priori can be demonstrated by the current situation in the European Union with respect to harmonized pesticide registration procedures. E.U. directives state that models "validated at the community (E.U.) level" are to be used to predict environmental concentrations in soil, groundwater, and surface waters. However, validation is not defined, nor are any criteria specified for assessing validity. This has led to a situation where a large number of different models are being used by regulators and industry, with considerable confusion (and costs) resulting from the very wide range of different predictions that are possible.

Model Calibration

The success of the operational validation procedure depends critically on distinguishing between conceptual errors and parameter errors (Loague and Green, 1991). As noted above, conceptual errors result from incorrect or undue simplification of process descriptions in the model and also neglect of significant processes (Russell et al., 1994). Clearly, some degree of conceptual error is inevitable, since models necessarily represent simplifications of reality. However, in principle, these should be minimized when mechanistic process descriptions are used (Wauchope, 1992) and in detailed models that include as many relevant processes as possible. Parameter error is the use of inappropriate parameter values. These errors arise either because the required data are not available or because the measurements are themselves subject to error and/or uncertainty due to spatial and temporal variability, or can be interpreted ambiguously, or for some other reason do not adequately reflect the prevailing field conditions. These errors may be potentially serious for detailed simulation models and for those parameters for which the relevant model outcome (e.g., leaching) is especially sensitive. Calibration is the process of minimizing parameter error in order to determine the extent of conceptual error, and is normally the first step in operational validation. Calibration is needed for all but the simplest models because direct measurements of all input parameters are rarely if ever available, either because of lack of time or money, or because the parameter is too difficult, too variable, or in many cases, impossible to measure. Bearing in mind the need for calibration, it is vital that field data utilized for operational validation meet certain quality criteria.

Data Requirements for Model Validation

The data required for model validation can be divided into two groups: measurements of model input parameters, and corroboratory validation measurements of the state of the system to be modeled. As a general rule, the modeler must be constrained as far as possible by both sets of measurements, such that the degree of freedom during model calibration is minimized. Ideally, for macropore flow and transport models, the parameterization should be based on actual measurements of soil water retention and unsaturated hydraulic conductivity, including the critical pressure head range close to saturation. In this way, parameters defining the hydraulic properties of the macropore region can be determined by curve-fitting the model functions to measurements made across an appropriate range of pressure heads (Jarvis et al., 1999).

The modeler must also be constrained by detailed measurements of the state of the system to be modeled (i.e., validation data). In this respect, it is especially critical that, as far as possible, (1) complete measured mass balances are available for both water and the solute of interest, and (2) both resident concentrations/ amounts and solute fluxes are measured. This is because resident and flux concentrations will differ considerably if macropore flow is a significant process. Also, coring methods alone cannot reliably detect preferential movement of strongly sorbing and degrading solutes, since only small (though environmentally significant) amounts are transported. This is because the analytical detection limits for resident concentrations obtained by core samples are relatively high, and the spatial variability of the solute distribution is often large in relation to the sampling intensity. In contrast, flux measurements give a measure of integrated transport for larger areas. Thus, ideally, measurements should be made of changes of both water and solute storage in the soil unsaturated zone, together with gains and losses from the system at the soil surface and at some depth in the soil, preferably below the maximum rooting depth. Storage measurements should be made with a high spatial resolution, both vertically and laterally, while flux measurements at the boundaries should be made with a high temporal resolution to capture the rapid and dynamic nature of macropore flow processes. The requirement for flux measurements means that either lysimeter or field-scale experiments on monitored tile-drainage systems are preferred. As an example of the latter, the nonreactive tracer experiment and model application reported by Larsson and Jarvis (1999a) can serve as a useful example to illustrate the range of field data required to properly validate macropore flow models such as MACRO. The measurements utilized by Larsson and Jarvis (1999a) are listed in Table 6-1, together with the spatial and temporal resolution achieved in each case. The results of this application of the MACRO model are discussed in the following section.

Case Studies with the MACRO Model

The MACRO model is probably the most widely used and tested model dealing with macropore flow in the unsaturated zone. Applications of the model to field and lysimeter experiments concerning nonreactive tracer movement, salt leaching from salinized soils, and pesticide leaching have been reported by, among others, Saxena et al. (1994), Jarvis et al. (1994), Andreu et al. (1994, 1996), Jarvis (1995), Jarvis et al. (1995), Gottesbüren et al. (1995), Bergström

TABLE 6-1 Field Measurements Utilized by Larsson and Jarvis (1999a)

Component	Measurement	Resolution
Surface boundary	Precipitation	Hourly
	Meteorological variables	Daily averages
	Solute application	30 collector trays to estimate mean and spatial variability of bromide spray application
Storage changes	Water and bromide	8 to 12 replicate cores, on 6 occasions following application during a 1-year period samples taken at 10-cm-depth intervals to 90-cm depth
Fluxes	Drainflow at 1-m depth	Continuous
	Bromide concentrations in drainflow	Every 1.5 mm of drainflow
Bottom boundary	Bromide concentrations in groundwater at 2-m depth	Weekly

(1996), Besien et al. (1997), Bourgault de Coudray et al. (1997), Beulke et al. (1998), Villholth and Jensen (1998), Jørgensen et al. (1998), and Jarvis et al. (2000). One of the most comprehensive field tests of the MACRO model to date was reported by Larsson and Jarvis (1999a, 1999b), who compared model predictions with measurements of bromide, herbicide, and nitrate leaching to tile-drains in a structured silty clay soil in Sweden. Bromide and the weakly sorbing herbicide bentazone were applied together to bare soil in autumn 1994, and the subsequent leaching from a 0.4-ha plot monitored during a one-year period (Larsson and Jarvis, 1999a). Model predictions of nitrate leaching were compared with measurements made during an eight-year period on an adjacent plot at the same site (Larsson and Jarvis, 1999b). These experiments and model applications are now discussed in more detail.

Site Hydrology and Nonreactive Transport

A reasonable description of the site hydrology is a prerequisite for accurate predictions of solute transport (Armstrong et al., 1996). Therefore, as a first step in the model application, Larsson and Jarvis (1999a) compared model predictions and measurements of drain discharge, concentrating especially on estimating soil macroporosity $(\theta_s - \theta_b)$ and saturated hydraulic conductivity by matching the timing and shape of individual flow hydrographs (i.e., time to peak flows and recessions). These parameters were estimated by calibration even though some direct measurements were available. This is because the drain response predicted by the model is highly sensitive to these parameters, which in turn are normally characterized by large spatial and temporal variability. Model predictions were checked against water content profiles measured on six occasions during a one-year period, and were found to be satisfactory. In the second stage of model calibration, estimates of sensitive parameters regulating the impact of macropore flow on preferential solute transport, particularly the mixing depth, z_d (Equation 6.1), and the effective diffusion pathlength, d (Equations 6.2 and 6.3), were obtained by comparing model predictions and measurements of the initial bromide breakthrough to the drains occurring 26 days after application. Figure 6-2 shows that

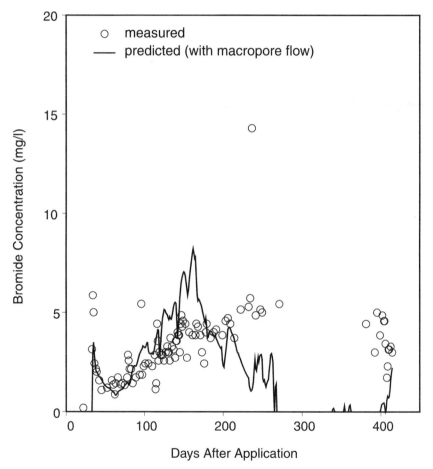

FIGURE 6-2 A comparison of bromide concentrations measured in tile drainage outflow at Lanna with those simulated by the MACRO model. Redrawn from Journal of Hydrology, 215, Larsson, M. H., and N. J. Jarvis, Evaluation of a dual-porosity model to predict field-scale solute transport in macroporous soil, pp. 153-171, 1999, with permission from Elsevier Science.

the model was able to capture reasonably well the initial breakthrough of bromide due to macropore flow, and also the subsequent slower convective-dispersive movement of bromide through the matrix, assuming d values of 150, 100, and 300 mm in topsoil and upper and lower subsoil horizons, respectively. The initial rapid breakthrough to tile drains was accompanied by significant bromide concentrations measured in groundwater at 2-m depth (10 μg l^{-1} measured 42 days after application). The model accurately matched the timing of this response in the groundwater, but the concentrations were somewhat underestimated (3 μg l^{-1}). The model satisfactorily matched the observed distributions of bromide in the soil profile throughout the one-year period following application (see Figure 6-3 for one example), but only if a rather large pore volume for anion exclusion was assumed for this clayey textured soil (23 percent). It is interesting to note that, even without macropore flow, the model could have been calibrated to achieve an excellent fit to the bromide profiles (for example, by changing the anion exclusion volume), because the proportion of the dose moving deep into the soil by macropore flow was quite small (about 10 percent) and the variability in the measured profiles was quite large. However, it would not be possible to match the pattern of flux concentrations observed in tile-drain outflow without considering macropore flow. This emphasizes that validation of macropore flow models requires measurements of both resident and flux concentrations, and that without flux measurements, there is a clear risk that misleading positive conclusions can be drawn concerning the validity of models that do not account for preferential movement.

Pesticide Leaching

Following this calibration, the model was validated by comparing model predictions and measurements of bentazone concentrations in tile drain outflow (Figure 6-4) and resident concentrations in the soil profile (Figure 6-5), with no changes to the parameterization of soil hydraulic functions. Figure 6-4 shows that the very weakly sorbed herbicide bentazone (k_{oc} = 5 cm^3 g^{-1}) behaved rather similarly to bromide, so that a more complete validation of the model for the Lanna site ought to include a more strongly sorbed solute. Nevertheless, the dual-porosity approach does seem to accurately reflect field-scale solute transport in this well-structured soil. Model simulations were also run without macropore flow by setting the effective diffusion pathlength to 1 mm (nominal zero), thereby ensuring complete equilibrium between the pore domains. A comparison of these simulation results enables the impact of macropore flow to be quantified. Figures 6-4 and 6-5 show that in the Lanna soil, macropore flow results in the rapid movement of a small pulse of applied herbicide, deep in the soil, but that this by-pass flow also effectively reduces the rate of convective movement of the bulk of the compound in the matrix. This retardation is especially significant because the longer transit time of the compound in the topsoil results in much larger degrada-

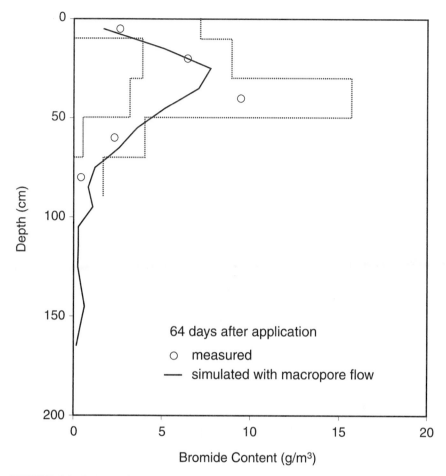

FIGURE 6-3 A comparison of bromide contents measured by soil coring at Lanna with those simulated by the MACRO model. Redrawn from Journal of Hydrology, 215, Larsson, M. H., and N. J. Jarvis, Evaluation of a dual-porosity model to predict field-scale solute transport in macroporous soil, pp. 153-171, 1999, with permission from Elsevier Science. Dotted lines indicate one standard deviation of the mean.

tion losses, since biodegradation in the Lanna subsoil is negligible (Bergström et al., 1994). Thus, the net effect of macropore flow was a reduction of leaching of the weakly sorbed compound bentazone, by about 50 percent. However, it should be noted that significant quantities of bentazone remained in the soil at the end of the experiment, so that the net reduction in leaching due to macropore flow probably would have been smaller, had the measurements continued to the point when all the compound had been either degraded or leached.

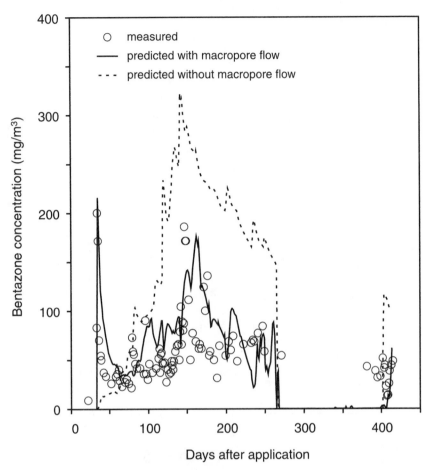

FIGURE 6-4 A comparison of bentazone concentrations measured in tile drainage outflow at Lanna with those simulated by the MACRO model both with and without considering macropore flow. Redrawn from Journal of Hydrology, 215, Larsson, M. H., and N.
J. Jarvis, Evaluation of a dual-porosity model to predict field-scale solute transport in
macroporous soil, pp. 153-171, 1999, with permission from Elsevier Science.

    Larsson (1999) utilized the calibrated Lanna parameter set as input to purely
predictive scenario simulations investigating the interactions between pesticide
compound properties and macropore flow effects on leaching. It was demonstrated that for mobile leachable compounds, macropore flow at Lanna results in
small decreases in leaching, or has no effect at all, while for intermediate leach-
ers, macropore flow may increase leaching by up to three to four orders of

magnitude. Conversely, for very strongly sorbed compounds (i.e., nonleachers), macropore flow again had apparently little effect on leaching. However, it should be borne in mind that these predictions take no account of colloid-facilitated transport of pesticide, a process that may be important for strongly sorbed compounds in macroporous soils (Brown et al., 1995). Larsson (1999) showed that to a good approximation, the total leaching loss in the presence of macropore flow

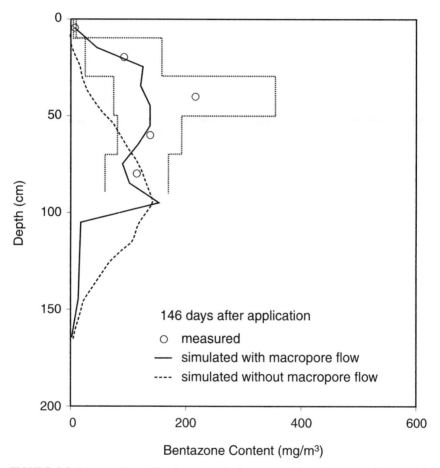

FIGURE 6-5 A comparison of bentazone contents measured by soil coring at Lanna with those simulated by the MACRO model both with and without considering macropore flow. Redrawn from Journal of Hydrology, 215, Larsson, M. H., and N. J. Jarvis, Evaluation of a dual-porosity model to predict field-scale solute transport in macroporous soil, pp. 153-171, 1999, with permission from Elsevier Science. Dotted lines indicate one standard deviation of the mean.

could be expressed as a simple linear function of the loss predicted without macropore flow, at least for compounds where leaching was more than 0.001 percent of the applied dose (see Figure 6-6). The slope of the best-fit line in Figure 6-6 reflects the tendency of macropore flow to reduce the significance of compound sorption and degradation properties in determining leaching losses. One practical consequence of this result is that reductions in applied dose would appear to be a relatively more effective means of reducing pesticide leaching in the presence of macropore flow (Larsson, 1999).

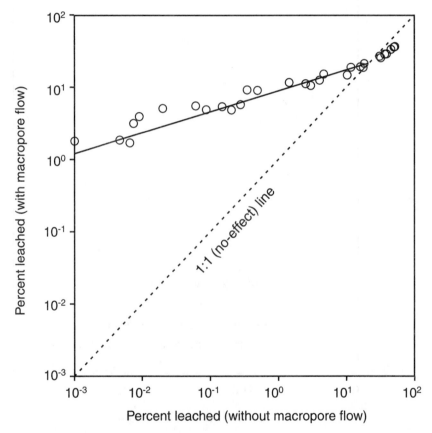

FIGURE 6-6 Leaching at Lanna (expressed as a fraction of the applied dose) for a range of hypothetical autumn-applied pesticides simulated with and without macropore flow. The one-to-one line indicates no effect of macropore flow on leaching, while the slope of the best-fit line represents the extent to which the effect of compound properties (sorption, degradation) on leaching is reduced by macropore flow. Redrawn from Larsson (1999).

The results of the scenario simulations presented by Larsson (1999) and Jarvis (1994b) can also be used to illustrate the importance of year-to-year variability in the rainfall pattern following surface application in determining leaching losses of agrochemicals in the presence of macropore flow. Figure 6-7 shows that for spring applications, annual leaching losses of bentazone predicted during a 10-year period were strongly correlated to the total rainfall in the three- to four-week period following application (Larsson, unpublished data). For less mobile pesticide compounds, the overwhelming proportion of the total leaching load during a 10-year period might be expected to occur during 1 year with heavy rainfall following spring applications (Jarvis, 1994b).

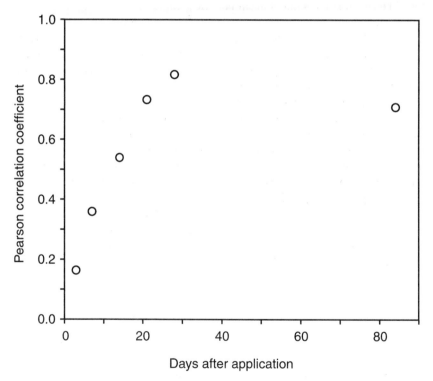

FIGURE 6-7 The relationship between predicted annual leaching loads of the spring-applied herbicide bentazone and precipitation amounts following application. The strength of the relationship is expressed in terms of the Pearson rank correlation coefficient calculated for a 10-year period.

Nitrate Leaching

Based on the calibrated parameter set derived from the model application to the bromide tracer experiment described above, Larsson and Jarvis (1999b) quantified macropore flow effects on nitrate leaching at the Lanna site by coupling MACRO to the nitrogen turnover model SOILN (Johnsson et al., 1987). The combined modeling system was applied to an eight-year field experiment carried out on an adjacent plot at Lanna, with a rotation of spring cereals and rape receiving normal rates (about 100 kg N ha^{-1} yr^{-1}) of commercial fertilizer. The calibration procedure focused especially on the crop nitrogen uptake function in SOILN, in order to achieve a close match to measured nitrogen in harvested grain, which is the largest component of the mass balance of nitrogen. A comparison of model simulations (run with and without macropore flow) with measurements of mineral nitrogen in the soil profile and nitrate concentrations in tile drainage showed that (1) the model could only satisfactorily predict all measured components of the nitrogen mass balance when macropore flow was accounted for; (2) the simulation without macropore flow could be made to match measured nitrogen contents in both the soil profile and harvested grain, but nitrate concentrations in drain flow were then generally overestimated; (3) in one year, significant increases in nitrate leaching due to macropore flow were observed during a short period following spring fertilizer application; and (4) nevertheless, during the eight-year period studied, macropore flow reduced nitrate leaching at Lanna overall by about 28 percent, but with large variations occurring from year to year (reductions varied from 3 percent to 45 percent in individual years). These results clearly show that assessments of the impact of macropore flow on nitrate leaching based on short-term experiments may be misleading, especially if the experimental period immediately follows fertilizer application. For the particular soil/climate/management scenario characterized by the Lanna experiment, and for the eight-year period in question, Larsson and Jarvis (1999b) concluded that increases in leaching due to macropore flow generated by occasional periods of heavy rain following fertilizer applications in spring were clearly outweighed by reductions in leaching losses occurring during the winter periods when rainfall of low nitrogen concentrations bypasses the resident soil water with larger nitrogen concentrations.

Regulatory Use of Macropore Flow Models: Current Trends

The regulatory use of macropore flow models requires care to eliminate user bias and strict application of predefined protocols to ensure that the model is used in an appropriate manner. A transparent and uniform approach to using macropore flow models for regulatory purposes can be achieved in two possible ways. The first is the use of precalibrated scenarios, where parameter values describing soil properties, including those describing the macropore region, are fixed a priori from previous model applications to experimental data obtained at representative sites. In this approach, the only model parameters the user is free to vary are those related to the chemical compound in question, for example, in the case of pesticides, the degradation and sorption properties and the use/application pattern. Such an approach is currently being developed for E.U. harmonized pesticide registration procedures by the FOCUS advisory group (Adriaanse et al., 1997), whereby a limited number of representative E.U. soil scenarios, for which the MACRO model has been previously calibrated, will form the basis of exposure assessments for pesticide movement to surface waters via subsurface drainage systems. Similarly, the Danish Environmental Protection Agency (DEPA) has developed two fixed national scenarios representing the main soil types in Denmark (sandy and loamy moraines). These are currently used together with the MACRO model to make pesticide exposure assessments for regulatory purposes.

Precalibrated scenarios offer a reliable means of predicting leaching with macropore flow models, but for some management applications, this approach may be too restrictive. For example, geographically distributed predictions of likely pollutant impacts on groundwater and surface waters are often required at regional and national scales. Here, macropore flow models must be used entirely predictively, with no possibility to calibrate difficult parameters. Objective and reliable estimation procedures (pedotransfer functions) are then needed to automatically derive model parameter values from widely available soils data. This approach is only likely to be possible for the simplest case of dual-porosity models, since the number of model parameters multiplies disproportionately as the number of domains increases (e.g., Gwo et al., 1995). Jarvis et al. (1997) described one such decision-support tool, or expert system (MACRO_DB, Figure 6-8), based on MACRO, in which the model is automatically parameterized from linked soils, cropping, weather, and pesticide properties databases through a combination of simple rules (based on expert judgement) and pedotransfer functions, while a geohydrological classification scheme (Boorman et al., 1991) is used to assess likely routes of pesticide contamination (surface water vs. groundwater). For example, in MACRO_DB, the diffusion pathlength is estimated from field observations of not only the size, but also the strength of development of the aggregate structure. The strength of structural development reflects the presence of cutans and organic linings at macropore/matrix interfaces, which may restrict mass exchange processes between pore domains (e.g., Thoma et al., 1992). The boundary hydraulic conductivity K_b (Figure 6-1) is one of the most sensitive hydraulic parameters in MACRO because it partitions water flow between macropores and micropores. In MACRO_DB, K_b is estimated from soil water retention curve parameters (Laliberte et al., 1968). Beulke et al. (1998) tested the predictive ability of the MACRO_DB system against a number of field and lysimeter experiments on pesticide leaching and concluded that the strength of macropore flow was significantly underestimated compared both to the measurements and to predictions made by an expert user of the stand-alone MACRO

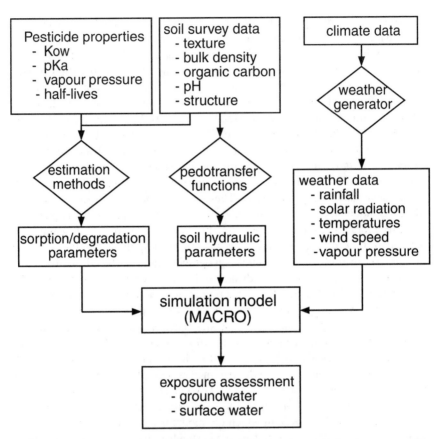

FIGURE 6-8 Schematic diagram illustrating MACRO_DB, a decision-support tool for making pesticide exposure assessments for surface waters and groundwater. From Environmental Modelling and Software, 12, Jarvis, N. J., J. M. Hollis, P. H. Nicholls, T. Mayr, and S. P. Evans, MACRO_DB: A decision-support tool to assess the fate and mobility of pesticides in soils, pp. 251-265, 1997, with permission from Elsevier Science.

model. One reason for this may be the difficulty in estimating one critical parameter for reactive solutes, namely, the partitioning of sorption sites between the pore domains. The simple approximation currently used in MACRO_DB is to distribute the sorption sites in proportion to the relative volumes of the pore domains. In many cases, macropore interfaces and linings should be more reactive than the surface area of the bulk soil (e.g., Mallawatantri et al., 1996), but this may be more than compensated for by the much smaller surface area per soil volume (Luxmoore et al., 1990).

CONCLUDING REMARKS

Although macropore flow models are now well established as research tools, there are still few examples of management applications in the literature, and only in recent years are they starting to be used for regulatory purposes. Despite the extra difficulties of parameterizing these models, it is important that they be used for regulation, simply because the consequences of preferential flow and transport for contaminant transport are often dramatic, and because models that do not treat these processes are limited in their applicability. Thus, future regulatory use of macropore flow models is necessary, but requires strict protocols to ensure that simulations are free of subjective user bias. Regulatory applications of macropore flow models should therefore be based either on precalibrated scenarios or on objective parameter estimation routines (pedotransfer functions). In this respect, it is vital that experience in model calibration and parameterization gained in research applications is effectively made available to management users, through further development and refinement of user-friendly decision-support tools that link data sources to state-of-the-art models and parameter estimation routines. Hopefully, as research experience accumulates, macropore flow models will be increasingly used by decision-makers in risk assessments and evaluation of mitigation strategies to minimize pollution impacts on water resources.

REFERENCES

Addiscott, T.M., 1977. A simple computer model for leaching in structured soils. Journal of Soil Science 28: 554-563.

Adriaanse, P., R. Allen, V. Gouy, J. Hollis, J. Hosang, N. Jarvis, T. Jarvis, M. Klein, R. Layton, J. Linders, H. Schäfer, L. Smeets, and D. Yon, 1997. Surface water models and EU registration of plant protection products. Report 6476-VI-96 (EU Commission), Regulatory Modelling Group, FOCUS, 218 pp.

Andreu, L., F. Moreno, N. J. Jarvis, and G. Vachaud, 1994. Application of the model MACRO to water movement and salt leaching in drained and irrigated marsh soils, Marismas, Spain. Agricultural Water Management 25: 71-88.

Andreu, L., N. J. Jarvis, F. Moreno, and G. Vachaud, 1996. Simulating the impact of irrigation management on the water and salt balance in drained marsh soils (Marismas, Spain). Soil Use and Management 12: 109-116.

Armstrong, A. C., A. M. Matthews, A. M. Portwood, and N. J. Jarvis, 1995. CRACK-NP: A model to predict the movement of water and solutes from cracking clay soils. ADAS Report, Land Research Centre, Gleadthorpe, Mansfield, Notts., U.K., 63 pp.

Armstrong, A. C., A. M. Portwood, P. B. Leeds-Harrison, G. L. Harris, and J. A. Catt, 1996. The validation of pesticide leaching models. Pesticide Science 48: 47-55.

Barraclough, D., 1989a. A useable mechanistic model of nitrate leaching. I. The model. Journal of Soil Science 40: 543-554.

Barraclough, D., 1989b. A useable mechanistic model of nitrate leaching. II. Application. Journal of Soil Science 40: 555-562.

Bergström, L., N. J. Jarvis, and J. Stenström, 1994. Pesticide leaching data to validate simulation models for registration purposes. Journal of Environmental Science and Health A29(6): 1073-1104.

Bergström, L., 1996. Model predictions and field measurements of Chlorsulfuron leaching under non-steady-state flow conditions. Pesticide Science 48: 37-45.

Beulke, S., C. D. Brown, and I. Dubus, 1998. Evaluation of the use of preferential flow models to predict the movement of pesticides to water sources under UK conditions. Final Report on MAFF project PL0516 SSLRC, Cranfield University, Silsoe, U.K.

Besien, T. J., N. J. Jarvis, and R. J. Williams, 1997. Simulation of water movement and isoproturon behaviour in a heavy clay soil using the MACRO model. Hydrology and Earth System Sciences 1: 835-844.

Beven, K., and P. Germann, 1982. Macropores and water flow in soils. Water Resources Research 18: 1311-1325.

Booltink, H.W.G., R. Hatano, and J. Bouma, 1993. Measurement and simulation of bypass flow in a structured clay soil: A physico-morphological approach. Journal of Hydrology 148: 149-168.

Boorman, D. B., J. M. Hollis, and A. Lilly, 1991. The production of the hydrology of soil types (HOST) data set. In: British Hydrological Society, 3rd National Hydrology Symposium, Southampton, U.K., pp. 6.7-6.13.

Bourgault de Coudray, P. L., D. R. Williamson, and W. D. Scott, 1997. Prediction of chloride leaching from a non-irrigated, de-watered saline soil using the MACRO model. Hydrology and Earth System Sciences 1: 845-851.

Brooks, R. H., and A. T. Corey, 1964. Hydraulic properties of porous media. Hydrology Paper no. 3, Colorado State University, Ft. Collins, Colo., 27 p.

Brown, C. D., R. A. Hodgkinson, D. A. Rose, J. K. Syers, and S. J. Wilcockson, 1995. Movement of pesticides to surface waters from a heavy clay soil. Pesticide Science 43: 131-140.

Clothier, B. E., and K. R. J. Smettem, 1990. Combining laboratory and field measurements to define the hydraulic properties of soil. Soil Science Society of America Journal 54: 299-304.

Corwin, D. L., B. L. Waggoner, and J. D. Rhoades, 1991. A functional model of solute transport that accounts for bypass. Journal of Environmental Quality 20: 647-658.

Durner, W., 1992. Predicting the unsaturated hydraulic conductivity using multi-porosity water retention curves. In: Proc. of the International Workshop, Indirect Methods for Estimating the Hydraulic Properties of Unsaturated Soils (M. Th. van Genuchten, F. Leij, and L. Lund, eds.), University of California, Riverside, pp. 185-201.

Flühler, H., W. Durner, and M. Flury, 1996. Lateral solute mixing processes: A key for understanding field-scale transport of water and solutes. Geoderma 70: 165-183.

Gamerdinger, A. P., R. J. Wagenet, and M. Th. van Genuchten, 1990. Application of two-site/two-region models for studying simultaneous nonequilibrium transport and degradation of pesticides. Soil Science Society of America Journal 54: 957-963.

Gerke, H. H., and M. Th. van Genuchten, 1993a. A dual-porosity model for simulating the preferential movement of water and solutes in structured porous media. Water Resources Research 29: 305-319.

Gerke, H. H., and M. Th. van Genuchten, 1993b. Evaluation of a first-order water transfer term for variably saturated dual-porosity flow models. Water Resources Research 29:1225-1238.

Germann, P., 1985. Kinematic wave approach to infiltration and drainage into and from soil macropores. Trans. ASAE 28: 745-749.

Gottesbüren, B., W. Mittelstaedt, and F. Führ, 1995. Comparison of different models to simulate the leaching behaviour of quinmerac predictively. In: Proceedings of the BCPC Symposium Pesticide Movement to Water (A. Walker, R. Allen, S. W. Bailey, A. M. Blair, C. D. Brown, P. Günther, C. R. Leake, and P. H. Nicholls, eds.), Warwick, U.K., pp. 155-160.

Gwo, J. P., P. M. Jardine, G. V. Wilson, and G. T. Yeh, 1995. A multiple-pore-region concept to modeling mass transfer in subsurface media. Journal of Hydrology 164: 217-237.

Hall, D. G. M., 1993. An amended functional leaching model applicable to structured soils. Journal of Soil Science 44: 579-588.

Hillel, D., 1987. Unstable flow in layered soils: A review. Hydrological Processes 1: 143-147.

Hutson, J. L., and R. J. Wagenet, 1995. A multiregion model describing water flow and solute transport in heterogeneous soils. Soil Science Society America Journal 59: 743-751.

Jarvis, N. J., 1989. CRACK: A model of water and solute movement in cracking clay soils. Technical description and user notes. Report 159, Division of Agricultural Hydrotechnics, Department of Soil Sciences, Swedish University of Agricultural Sciences, Uppsala, 38 pp.

Jarvis, N. J., 1994a. The MACRO model (Version 3.1): Technical description and sample simulations. Reports and Dissertations, 19, Department of Soil Sciences, Swedish University of Agricultural Sciences, Uppsala, 51 pp.

Jarvis, N. J., 1994b. The implications of preferential flow for the use of simulation models in the registration process. In: Proceedings of the 5th International Workshop Environmental Behaviour of Pesticides and Regulatory Aspects (COST) (A. Copin, G. Houins, L. Pussemier, J. F. Salembier, eds.), Brussels, April, pp. 464-469.

Jarvis, N. J., M. Stähli, L. Bergström, and H. Johnsson, 1994. Simulation of dichlorprop and bentazon leaching in soils of contrasting texture using the MACRO model. Journal of Environmental Science and Health A29(6): 1255-1277.

Jarvis, N. J., 1995. Simulation of soil water dynamics and herbicide persistence in a silt loam soil using the MACRO model. Ecological Modeling 81: 97-109.

Jarvis, N. J., and I. Messing, 1995. Near-saturated hydraulic conductivity in soils of contrasting texture as measured by tension infiltrometers. Soil Science Society of America Journal 59: 27-34.

Jarvis, N. J., M. Larsson, P. Fogg, and A. D. Carter, 1995. Validation of the dual-porosity model MACRO for assessing pesticide fate and mobility in soil. In: Proc. BCPC Symposium Pesticide Movement to Water (A. Walker, R. Allen, S. W. Bailey, A. M. Blair, C. D. Brown, P. Günther, C. R. Leake, and P.H. Nicholls, eds.), Warwick, U.K., pp. 161-170.

Jarvis, N. J., J. M. Hollis, P. H. Nicholls, T. Mayr, and S. P. Evans, 1997. MACRO_DB: A decision-support tool to assess the fate and mobility of pesticides in soils. Environmental Modelling and Software 12: 251-265.

Jarvis, N. J., 1998. Modeling the impact of preferential flow on nonpoint source pollution. In: Physical Non-Equilibrium in Soil: Modeling and Application (H. M. Selim and L. Ma, eds.), pp. 195-221. Ann Arbor Press.

Jarvis, N. J., and M. H. Larsson, 1998. The MACRO model (Version 4.1): Technical description. http://130.238.110.134:80/bgf/Macrohtm/macro.htm. Department of Soil Sciences, Swedish University of Agricultural Sciences, Uppsala.

Jarvis, N. J., I. Messing, M. H. Larsson, and L. Zavattaro, 1999. Measurement and prediction of near-saturated hydraulic conductivity for use in dual-porosity models. In: Characterization and Measurement of the Hydraulic Properties of Unsaturated Porous Media (M. Th. van Genuchten, F. Leij, and L. Wu, eds.). University of California, Riverside, pp. 839-850.

Jarvis, N. J., C. D. Brown, and E. Granitza, 2000. Sources of error in model predictions of pesticide leaching: A case study using the MACRO model. Agricultural Water Management 44: 247-262.

Johnsson, H., L. Bergström, P.-E. Jansson, and K. Paustian, 1987. Simulated nitrogen dynamics and losses in a layered agricultural soil. Agriculture, Ecosystems and Environment 18: 333-356.

Jørgensen, P. R., T. Schrøder, G. Felding, A. Helweg, N.-H. Spliid, M. Thorsen, J.-C. Refsgaard, and O.-H. Jacobsen, 1998. Validation and development of pesticide leaching models. Pesticides Research no. 47, Danish Environmental Protection Agency, Copenhagen, 150 pp.

Lafolie, F., and Ch. Hayot, 1993. One-dimensional solute transport modelling in aggregated porous media. Part 1. Model description and numerical solution. Journal of Hydrology 143: 63-83.

Laliberte, G. E., R. H. Brooks, and A. T. Corey, 1968. Permeability calculated from desaturation data. J. Irrig. and Drainage Div., Proc. ASCE, 94: 57-69.

Larsson, M. H., and N. J. Jarvis, 1999a. Evaluation of a dual-porosity model to predict field-scale solute transport in macroporous soil. Journal of Hydrology 215: 153-171.

Larsson, M. H., and N. J. Jarvis, 1999b. A dual-porosity model to quantify macropore flow effects on nitrate leaching. Journal of Environmental Quality 28: 1298-1307.

Larsson, M. H., 1999. Quantifying macropore flow effects on nitrate and pesticide leaching in a structured clay soil: Field experiments and modelling with the MACRO and SOILN models. Acta Universitatis Agriculturae Sueciae, Agraria 164, SLU, Uppsala, Sweden, 34 pp.

Loague, K. M., and R. E. Green, 1991. Statistical and graphical methods for evaluating solute transport models: Overview and application. Journal of Contaminant Hydrology 7: 51-73.

Luxmoore, R. J., P. M. Jardine, G. V. Wilson, J. R. Jones, and L. W. Zelazny, 1990. Physical and chemical controls of preferred path flow through a forested hillslope. Geoderma 46: 139-154.

Mallawatantri, A. P., B. G. McConkey, and D. J. Mulla, 1996. Characterization of pesticide sorption and degradation in macropore linings and soil horizons of Thatuna silt loam. Journal of Environmental Quality 25: 227-235.

Mualem, Y., 1976. A new model for predicting the hydraulic conductivity of unsaturated porous media. Water Resources Research 12: 513-522.

Nicholls, P. H., and D. G. M. Hall, 1995. Use of the pesticide leaching model (PLM) to simulate pesticide movement through macroporous soils. In Proc. BCPC Symposium Pesticide Movement to Water (A. Walker, R. Allen, S. W. Bailey, A. M. Blair, C. D. Brown, P. Günther, C. R. Leake, and P. H. Nicholls, eds.), Warwick, U.K., pp. 187-192.

Nieber, J. L., and D. Misra, 1995. Modeling flow and transport in heterogeneous, dual-porosity drained soils. Irrigation and Drainage Systems 9: 217-237.

Oreskes, N., K. Schrader-Frechette, and K. Belitz, 1994. Verification, validation and confirmation of numerical models in the earth sciences. Science 263: 641-646.

Othmer, H., B. Diekkrüger, and M. Kutilek, 1991. Bimodal porosity and unsaturated hydraulic conductivity. Soil Science 152: 139-150.

Russell, M. H., R. J. Layton, and P. M. Tillotson, 1994. The use of pesticide leaching models in a regulatory setting: an industrial perspective. Journal of Environmental Science and Health A29: 1105-1116.

Rykiel, E. J., 1996. Testing ecological models: The meaning of validation. Ecological Modelling 90: 229-244.

Saxena, R., N. J. Jarvis, and L. Bergström, 1994. Interpreting non-steady state tracer breakthrough experiments in sand and clay soils using a dual-porosity model. Journal of Hydrology 162: 279-298.

Smettem, K. R. J., and C. Kirkby, 1990. Measuring the hydraulic properties of a stable aggregated soil. Journal of Hydrology 117: 1-13.

Steefel, C. I., and P. van Cappellen, 1998. Reactive transport modeling of natural systems. Journal of Hydrology 209: 1-7.

Steenhuis, T. S., J.-Y. Parlange, and M. S. Andreini, 1990. A numerical model for preferential solute movement in structured soils. Geoderma 46: 193-208.

Thoma, S. G., D. P. Gallegos, and D. M. Smith, 1992. Impact of fracture coatings on fracture/matrix flow interactions in unsaturated, porous media. Water Resources Research 28: 1357-1367.

van Dam, J.C., J. M. H. Hendrickx, H. C. van Ommen, M. H. Bannink, M. Th. van Genuchten, and L. W. Dekker, 1990. Water and solute movement in a coarse-textured water-repellent field soil. Journal of Hydrology 120: 359-379.

van Genuchten, M. Th., and P. J. Wierenga, 1976. Mass transfer in sorbing porous media. I. Analytical solutions. Soil Science Society of America Journal 40: 473-480.

van Genuchten, M. Th., 1980. A closed-form equation for predicting the hydraulic conductivity of unsaturated soils. Soil Science Society of America Journal 44: 892-898.

van Genuchten, M. Th., 1985. A general approach for modeling solute transport in structured soils. In: Proceedings of the 17th International Congress IAH, Hydrogeology of Rocks of Low Permeability. Memoires IAH, 17, pp. 513-526.

van Genuchten, M. Th., and R. J. Wagenet, 1989. Two-site/two-region models for pesticide transport and degradation: Theoretical development and analytical solutions. Soil Science Society of America Journal 53: 1303-1310.

Villholth, K. G., and K. H. Jensen, 1998. Flow and transport processes in a macroporous subsurface-drained glacial till soil. II. Model analysis. Journal of Hydrology 207: 121-135.

Wagenet, R. J., and J. L. Hutson, 1987. LEACHM: Leaching estimation and chemistry model. A process-based model of water and solute movement, transformations, plant uptake and chemical reactions in the unsaturated zone. Continuum 2, Water Resources Institute, Cornell University, Ithaca, N.Y.

Wauchope, R. D., 1992. Environmental risk assessment of pesticides: Improving simulation model credibility. Weed Technology 6: 753-759.

Youngs, E. G., and P. B. Leeds-Harrison, 1990. Aspects of transport processes in aggregated soils. Journal of Soil Science 41: 665-675.

Zurmühl, T., and W. Durner, 1996. Modeling transient water and solute transport in a biporous soil. Water Resources Research 32: 819-829.

7

Free-Surface Films

Maria Ines Dragila[1] and Stephen W. Wheatcraft[2]

ABSTRACT

A new transport mechanism is proposed to explain the anomalous fast transport observed at fractured vadose sites. Recent field observations have shown that fluid transport rates may be orders of magnitude greater than existing models can predict. This implies that a key transport mechanism is acting which is not being included in these models. It is imperative that the acting transport mechanism be identified because fractured vadose sites are being marked for potential toxic waste repositories.

A free-surface film may provide a mechanism by which fluids in the unsaturated zone are transported through air-filled fractures much more rapidly than plug flow or porous matrix transport. A rigorous mathematical model is developed to analyze the character of free surface films. The model shows free-surface films are chaotic. The chaotic character of the system is used to develop a hydrologic model for free-surface film transport in unsaturated fractures. This model is applied to the example of seepage from the matrix into a fracture and is shown to generate velocities of the order necessary to explain the fast transport rates observed in the field.

---

[1] Department of Geological Sciences, University of Nevada, Reno; presently at Department of Crop and Soil Science, Oregon State University, Corvallis
[2] Department of Geological Sciences, University of Nevada, Reno

INTRODUCTION

Recent field observations at fractured vadose study sites have determined that fluid transport rates through the earth's vadose zone may be orders of magnitude greater than existing models can predict. This implies that a key transport mechanism is acting which is not being included in these models. It is imperative that this transport mechanism be identified because fractured vadose sites are being identified for potential toxic waste repositories. Miscalculations regarding fluid transport could have a major impact on the future integrity of any repository.

A free-surface film may provide a mechanism by which fluids in the unsaturated zone are transported through air-filled fractures much more rapidly than plug flow or porous matrix transport. In fact, flow rates predicted by this particular mechanism are sufficient to explain recently observed behavior.

The vadose zone, also known as the unsaturated zone, is the region between the ground surface and water table. Porous media in this region are only partially saturated with fluid. Existing theory states that flow progresses only through filled and interconnected pores. Conceptual models for flow within the vadose zone are based on capillary theory, where the largest pore size to be saturated is determined by the local pressure head (matric potential). Natural extension to fractures of flow through porous media theory suggests that fluid will fill the smallest fracture apertures first, progressively filling larger apertures. In addition, the fracture will not conduct unless there is a connected fluid pathway. Flow through porous media, whether saturated or unsaturated, is a very slow process.

Models developed to date for flow through unsaturated fractures retain some of the basic maxims from porous media transport. An unsaturated fracture is one that sustains both an air phase and a fluid phase. Until recently it had been assumed that the fluid phase is confined to, and completely saturates, regions where the aperture is sufficiently small, as defined by the matric potential of the surrounding porous media. The larger aperture regions of the fracture are air-saturated. Dual-porosity models simply replace the fracture with a permeable medium of higher porosity than the surrounding matrix. Although the result is that flow is faster through the more permeable conduit, it is not sufficiently fast to explain the recently observed fast-flow phenomena. Cubic law models approximate the fracture aperture by a set of parallel plates, and the flow rate is calculated using what has come to be known as the cubic law, where flow rate is proportional to the cube of the fracture aperture (Figure 7-1) (Nitao et al., 1993). Further studies were performed to characterize the effect of aperture variability on tortuosity and the development of channels allowing for higher flow rates through regions of wider aperture (Tsang, 1984; Tsang and Tsang, 1987). Fractures sustaining this type of two-phase capillary saturation have also been shown to develop flow instabilities as the two saturated phases interact (Pruess and Tsang, 1990; Geller et al., 1996; Su et al., 1999). Instabilities have also been seen at the wetting front in fracture flow leading to fingering phenomena (Nicholl et

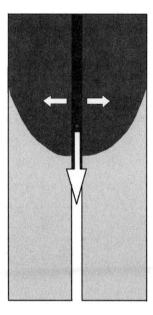

FIGURE 7-1 Parallel plate model for fracture flow with absorption into the matrix.

al., 1994). Although fingers are capable of generating faster flow than the remainder of the wetting front, they sustain capillary saturation and the flow rate is limited by the aperture geometry as defined by the cubic law.

Central to these models is the assumption that a continuous saturated path is required for fluid conduction. At high matrix saturations the entire fracture may fill and be capable of conducting fluid. However, at low capillary pressures, because pressure equilibrium is required between the fracture and matrix, the largest liquid-filled aperture in the fracture will be on the order of magnitude of the largest pore size that is liquid-filled in the surrounding porous medium. Fractures in arid regions, where the porous medium is at very low saturation and very low capillary pressures, are likely to be saturated only in sections where the aperture is very small, such as where two porous blocks touch. These saturated sections are most likely disconnected, in which case, a continuous fluid-filled conduit would not exist. These small sections of trapped water in the fracture are held against gravity by contact angle hysteresis. Using this scenario, and applying it to the arid conditions found in the Yucca Mountain area, Wang and Narasimhan (1985) hypothesized that flow in vadose zone fractures would be unlikely in arid regions. Their model is shown in Figure 7-2.

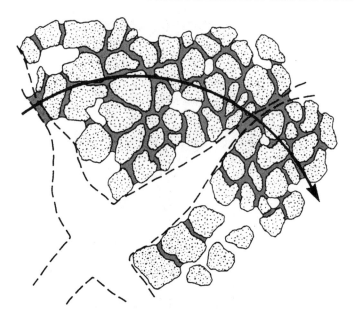

FIGURE 7-2 Model for unsaturated fracture showing flow only through capillary bridges between porous blocks. From Wang, J.S.Y., and T.N. Narasimhan, 1985. Hydrologic mechanism governing fluid flow in a partially saturated, fractured, porous medium. Water Resources Research 21(12): 1861-1874. Copyright from American Geophysical Union.

Because of this hypothesis, over the last two decades arid fractured vadose environments have been suggested as possible sites for placement of long-term nuclear waste repositories (Nitao and Buscheck, 1991). Since water is most likely the primary agent for movement of hazardous materials in the subsurface, the combination of an arid environment, a low permeability matrix, and fractures that possibly separate the matrix blocks and further delay transport, seemed ideal conditions for the isolation of toxic wastes. The sensitivity of the proposed use and the severe consequences of failure have significantly increased the importance of understanding the hydrologic behavior within an arid fractured vadose zone.

Contrary to the prediction of models mentioned above, radionuclides have been detected in arid vadose regions at much greater distances from the source than expected. At the Exploratory Studies Facility (ESF), Yucca Mountain, Nevada, a bomb-pulse chlorine-36 signature was found hundreds of feet below surface, even though the bomb-pulse signal is less than 50 years old (Fabryka-Martin et al., 1998). This anomalous rapid transport rate is occurring in an arid region where the vadose zone is highly unsaturated and fractured. Unfortunately, porous matrix theory, dual-porosity models, and plug flow have not been able to predict such rapid movement of fluid in a region that is highly unsaturated and disconnected from surface fractures that could transmit possible episodic flooding events.

Consequently, a new transport mechanism needs to be identified that is able to more accurately predict the large velocity flows observed within the fractured vadose zone. It is proposed that a free-surface film may be the mechanism by which fluids in unsaturated fractures are transported very rapidly through the vadose zone.

CONCEPTUAL MODEL FOR FREE-SURFACE FILM

The free-surface film model consists of a thin film of fluid flowing down the wall of an air-filled fracture in the vadose zone. The film is in contact with the porous matrix along one fracture wall and sustains an air phase between itself and the opposing fracture wall, as shown in Figure 7-3 (Dragila and Wheatcraft, 1997). The film flows under the influence of gravity, exclusive of capillary forces within the fracture. This mechanism can generate very rapid transport deep into the vadose zone. Transport velocities for a free-surface film can be up to four times that for an equivalently thick saturated plug (White, 1991). Furthermore,

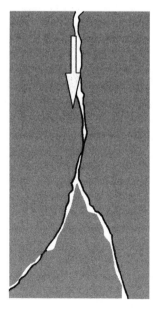

FIGURE 7-3 Model for flow in unsaturated fractures where fluid flows in the form of free surface films down the fracture walls while sustaining a continuous air phase.

velocities generated by wavy chaotic films may be two times higher still (Dragila and Wheatcraft, 1998).

The free-surface film model differs from the standard fracture flow models in that the fracture aperture does not directly control flow rate during transport by a free-surface film. Film flow can occur in small-aperture as well as large-aperture fractures. This characteristic eliminates the restrictions that constrain the standard fracture flow models.

Much experimental work has been done to study the behavior of flow in unsaturated fractures (Persoff and Pruess, 1985; Pruess and Tsang, 1990; Nicholl et al., 1994; Geller et al., 1996: Podgorney et al., 1998; Su et al., 1999). Most experimental and theoretical work on free-surface films is limited to the fields of mechanical and chemical engineering, where they have been studied in relation to many industrial operations, from wrinkle-free paint applications to solute transfer in chemical air strippers (Kapitza, 1965; Nakoryakov et al., 1977; Wasden and Dukler, 1992). The work by Tokunaga and Wan (1997) is the first reported experiment that specifically focuses on the characteristics of a free-surface film in a natural fracture.

The experiment presented by Tokunaga and Wan (1997) showed development of free-surface films on a fractured sample of Bishop Tuff. They established steady-state conditions in which flow delivered to the upper surface of their sample was quickly distributed to both the rock matrix and fracture surface while the matric potential was kept near saturation. Their experiment showed that these free-surface films can generate very high transport rates and velocity while sustaining subatmospheric pressures within the film (see section on Seepage Generated Film Flow below).

MATHEMATICAL MODEL

The mathematical model presented here was developed to understand the characteristics and transport capacity of film flow. Governing equations are mathematical expressions of fundamental laws that constrain the behavior of nature. In the case of fluid flow, these fundamental laws are conservation of mass (Equation 7.1), conservation of momentum (Equation 7.2), and conservation of energy (Equation 7.3):

$$\frac{\partial \rho}{\partial t} + \nabla \bullet (\rho \vec{u}) = 0, \tag{7.1}$$

$$\rho \frac{\partial \vec{u}}{\partial t} + \rho (\vec{u} \bullet \nabla)\vec{u} = \rho \vec{g} - \nabla p + \mu \nabla^2 \vec{u}, \tag{7.2}$$

$$\rho \frac{Dh}{Dt} = \frac{dp}{dt} + \nabla \bullet (k\nabla T) + \phi. \tag{7.3}$$

The symbols used represent the following parameters: ρ is the fluid density; μ is the fluid dynamic viscosity; u is the velocity vector; t is the time variable; g is the constant of gravitational acceleration; p is the pressure; h is the enthalpy; T is the temperature; k is the thermal constant, and ϕ is a dissipation function. The momentum equations, also known as the Navier-Stokes equations, are fundamental and rigorous. They are also nonlinear, complex, nonunique, and difficult to solve. In order to solve the problem of a film of fluid flowing down an air-filled fracture, we have assumed the following simplified model: the fracture surface is vertical, smooth, and impermeable; the air phase is quiescent; and there is no mass transfer into the vapor phase. A sketch of the model is provided in Figure 7-4. A steady-state solution can be obtained by integrating the one-dimensional momentum equation and applying the appropriate boundary conditions. Nevertheless, a solution to the film-flow problem poses a particular difficulty. The free surface makes the thickness of the film a dynamic variable, and thus cannot be used as a boundary condition. This lack of boundary condition can be resolved by assuming that at steady state the free surface is flat and the film thickness constant (Nusselt, 1916). As will be shown later, although this assumption is incorrect, it provides a

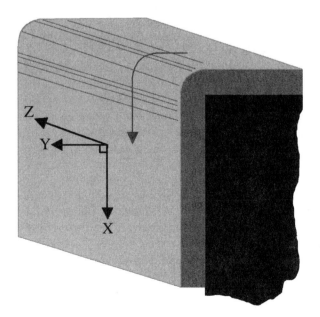

FIGURE 7-4 Geometry used for development of free-surface film mathematical model.

very useful solution. The steady-state solution to the velocity profile for a free-surface film flowing under the action of gravity along a vertical plane is

$$u_N(y) = \left(\frac{g}{\eta}\right)\left(h_N y - \frac{y^2}{2}\right). \tag{7.4}$$

The mean velocity across the thickness of the film at steady-state is

$$\bar{u}_N = \left(\frac{g}{3\eta}\right)h_N^2, \tag{7.5}$$

where h_N is the steady-state film thickness; u_N is the steady-state velocity, and η is the kinematic viscosity ($\eta = \mu/\rho$). The coordinate convention used throughout this paper is as follows: the x-coordinate points in the direction of flow, the y-coordinate points normal to the fracture wall, and the z-coordinate points parallel to the wall and normal to the direction of flow. The value for the steady-state film thickness, h_N, can be obtained using the concept of conservation of mass. Once the film has attained a steady state, the flow rate per unit width can be expressed as

$$Q' = h_N \bar{u}_N = \left(\frac{g}{3\eta}\right)h_N^3. \tag{7.6}$$

As this equation shows, film thickness is purely a function of the incoming, or source, flow rate. It is a dynamic variable.

To obtain the steady-state solution, we assumed that the film would sustain a flat free-surface film as it flows. However, these flat films are inherently un-stable. Instability is commonly seen for a wide range of fluids and has been the advantage as well as the plague of many industrial applications (Dukler and Bergelin, 1952; Yih, 1963; Pumir et al., 1983). Chemical engineers use the natu-rally developing wavy surface to increase the two-phase contact area for systems such as heat exchangers and chemical air strippers. Conversely, in the field of industrial paints, wavy surfaces appearing in the form of wrinkles in paint coat-ings are to be avoided. Experimental observations indicate that films develop large-amplitude waves that have either a two-hump or one-hump structure. These solitary waves appear to ride over a thin fluid substrate. They travel faster than the mean film speed, can have amplitudes two to five times the substrate thick-ness, and appear to trap a recirculating region of fluid (Figure 7-5) (Wasden and Dukler, 1992).

A more rigorous mathematical model for the film-flow problem is presented below. It was developed in an attempt to understand the character of the wavy structure (Dragila, 1999). The following simplifying assumptions were made: the system is assumed to be two-dimensional; the fracture surface is vertical, smooth, and impermeable; the air phase is quiescent and there is no mass transfer into the

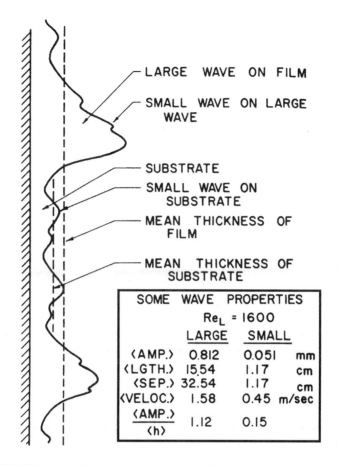

FIGURE 7-5 Schematic showing free-surface wave characteristics. From Chu and Dukler (1974).

vapor phase; the restorative force of surface tension due to the waviness of the surface is included in the fifth boundary condition; and the film is allowed to vary in thickness. The complete family of governing equations for this model follows. The x-momentum equation is

$$\rho \frac{\partial u}{\partial t} + \rho \left(u \frac{\partial}{\partial x} + v \frac{\partial}{\partial y} \right) u = -\frac{\partial P}{\partial x} + \rho g + \mu \left(\frac{\partial^2}{\partial x^2} + \frac{\partial^2}{\partial y^2} \right) u, \qquad (7.7)$$

the y-momentum equation is

$$\rho \frac{\partial u}{\partial t} + \rho \left(u \frac{\partial}{\partial x} + v \frac{\partial}{\partial y} \right) v = -\frac{\partial P}{\partial y} + \mu \left(\frac{\partial^2}{\partial x^2} + \frac{\partial^2}{\partial y^2} \right) v, \tag{7.8}$$

and the conservation of mass equation is

$$\frac{\partial u}{\partial x} + \frac{\partial y}{\partial y} = 0. \tag{7.9}$$

The first boundary condition is no slip at the wall,

$$\text{at } y = 0, \ u = 0; \tag{7.10}$$

The second boundary condition is no slip at the wall and no matrix absorption,

$$\text{at } y = 0, \ v = 0; \tag{7.11}$$

the third boundary condition is the kinematic condition at the free surface,

$$\text{at } y = h, \ v = h_t + u h_x, \tag{7.12}$$

where the subscript signifies differentiation with respect to that variable;
the fourth boundary condition is balance of shear stresses at the free surface,

$$\text{at } y = h, \ \left(\frac{\partial v}{\partial y} - \frac{\partial u}{\partial x} \right) \frac{2 h_x}{1 - h_x^2} + \frac{\partial u}{\partial y} + \frac{\partial v}{\partial x} = 0, \tag{7.13}$$

and the fifth boundary condition is balance of normal stresses at the free surface,
at $y = h$,

$$0 = [-p_0] - \left[-p - \mu \left(\frac{\partial u}{\partial y} + \frac{\partial v}{\partial x} \right) \sin 2\beta + 2\mu \left(\frac{\partial u}{\partial x} \sin^2 \beta + \frac{\partial v}{\partial y} \cos^2 \beta \right) \right] + \frac{\sigma h_{xx}}{(1 + h_x)^2} \tag{7.14}$$

To solve this family of eight nonlinear partial differential equations, the model was further simplified. A boundary layer technique was used, as developed by Prandtl (1904), that takes advantage of the peculiar geometry of the flow system, where the scale of space and motion in one coordinate dimension is orders of magnitude greater than in the other dimension. Furthermore, the definition of the stream function was used, which reduces the number of unknowns at the cost of increasing the order of differentiation. Lastly, a coordinate system transformation was made into one that moves with the wave generated by the perturbation. Yu et al. (1995) used a similar approach to characterize the waviness of thicker films used in chemical air stripping equipment. This approach has led to development of a family of ordinary differential equations in a coordinate system moving with the free-surface disturbance. The final set of equations is a family of six ordinary differential equations that can be used to study the evolution of perturbations with time:

$$
\begin{aligned}
\dot{x}_1 &= x_2 \\
\dot{x}_2 &= x_3 \\
\dot{x}_3 &= f_3(\bar{x}) \\
\dot{x}_4 &= x_5 \\
\dot{x}_5 &= f_5(\bar{x}) \\
\dot{x}_6 &= f_6(\bar{x}).
\end{aligned}
\tag{7.15}
$$

The components of the six-dimensional vector are:

$$
\bar{x} = (x_1, x_2, x_3, x_4, x_5, x_6) = (h, \dot{h}, \ddot{h}, a_2, \dot{a}_2, a_3)
\tag{7.16}
$$

where h is the film thickness and a_1 are the coefficients of the stream function expansion. The three functions in Equation 7.15, f_3, f_5, f_6, are defined as follows:

$$
f_3(\bar{x}) = \frac{1}{x_1^3 \, \mathrm{Re} \, We} \left(\begin{array}{l} \dfrac{3}{4} x_2 (Ce-1)^2 \left(\dfrac{Ce x_1}{Ce-1} - 3 \right) \mathrm{Re} - 120 Ce(x_1 - 1) + \\[2mm] 12 x_1^3 (2x_6 - 1) + 88 x_4 x_1^2 - 120 \end{array} \right)
\tag{7.17}
$$

$$
f_5(\bar{x}) = \frac{-1}{x_1^4} \left(\frac{1}{8} Ce \mathrm{Re} x_1^4 x_5 + 45 Ce(x_1 - 1) - 18 x_1^3 x_6 - 30 x_1^2 x_4 + 45 \right)
\tag{7.18}
$$

$$
f_6(\bar{x}) = \left(\frac{1}{270(Ce - 1 x_1^4 - 270 Ce x_1^5 + 18 x_1^7 x_6 + 81 x_1^6 x_4)} \right) [6930 We x_1^3 f_3(\bar{x})
$$

$$
+ \frac{1}{\mathrm{Re}} (-55440 x_1^2 x_4 + 83160 x_1^3) + x_2 [3015 Ce^2 (3 - x_1^2) + 1380 x_1^2 x_4 (1 - Ce)
$$

$$
+ x_1^4 (810 Ce x_6 - 558 x_4^2) + 540 x_1^3 \left(\left(\frac{46}{9} x_4 - x_6 \right) Ce + x_6 \right) - 19890 Ce
$$

$$
-45 x_1^6 x_6^2 - 324 x_1^5 x_4 x_6 + 9945] + 13880 x_1^4 x_5 Ce - 81 x_1^6 x_5 x_6 - 392 x_1^5 x_4 x_5]
$$

$$
\tag{7.19}
$$

The ODEs in Equations 7.15 are dimensionless and apply to films of any thickness within the limitations of the boundary conditions and assumptions used. At the two extremes are very thick films and very thin films. For very thick vertical films (greater than a millimeter), the Reynolds number is large and the flow dynamics are outside the bounds of Poiseuille-type flow and approach the domain of turbulent dynamics. Very thin films, less than a few microns, would require a different order of magnitude analysis in order to simplify the equations, and this would result in a different set of equations. All films within the valid thickness range are susceptible to shear stresses due to surface-tension gradients generated by surface active agents and temperature variability. This possible source of perturbation has not been included in the mathematical model, though it may be one source for generating unstable perturbations. The solution to the family of ODEs follows the development of a perturbation traveling at a given speed (celerity), Ce, and for a given Reynolds number, Re, and Weber number, We. Celerity is the wave speed normalized to the steady-state mean film speed, u_N. The family of ODEs (Equations 7.16) can be solved for the film thickness, h (or x_1), given a specific celerity and Reynolds number.

The solution to the family of ordinary differential equations above shows that the film-flow model has a distinctive chaotic behavior. Bifurcation of the solution appears with decreasing wave speed. Figure 7-6 shows phase diagrams for three celerity values indicating the solution at various stages of bifurcation and chaos for a sample film of Reynolds number 150 (dictating a film thickness of about 1/3 mm). For a celerity of 1.75 the behavior is sinusoidal. As the celerity is reduced below 1.748 the perturbation grows to a two-hump structure. Below a celerity of 1.73805 the behavior of the film thickness is chaotic. It exhibits a range of wave shapes including the characteristic single-hump solitary wave that has been experimentally observed. This analysis leads one to hypothesize that solitary waves that naturally form in free-surface films form as a result of a chaotic attractor natural to the system.

A very important result of this model is the link between the structure of the predicted chaotic disturbance and the shape of the naturally forming solitary wave. Since the attractor exists only for perturbations with celerity less than a specific value, we can hypothesize that the solitary waves forming on the free surface will travel at an approximate speed equal to the product of the transition celerity times the predicted steady-state film speed, u_N. Analysis of the chaotic transitions as a function of film thickness allows determination of a general equation for the speed of the solitary wave (Equation 7.23). Experimental observations indicate that the amount of liquid in these waves at any point in time can be as high as 50 percent. Due to their high velocity, these waves are responsible for transporting as much as 90 percent of the liquid. This large amount of liquid moving in lumps has major implications when building models for transport into and by these films (Telles and Dukler, 1970). Figure 7-7 shows a photograph of the solitary waves and the underlying substrate (Kapitza, 1965). Experimental observations by Nakoryakov et al. (1977) indicate that the velocity profile within the thin substrate is parabolic. This observation allows us to use the parabolic

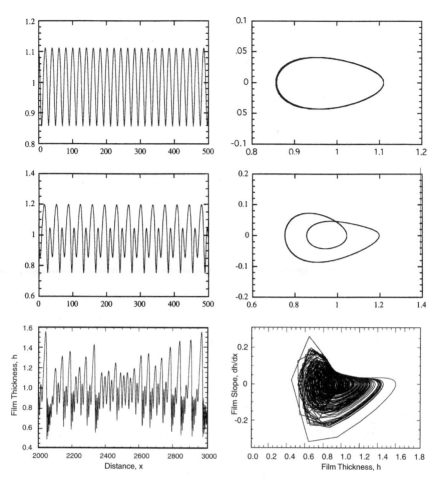

FIGURE 7-6 Phase diagrams for waves at various celerity values. From Yu (1995).

velocity profile solution derived from steady-state theory for the substrate. The development of solitary waves has implications for fracture flow related to transport rate, dispersion of contaminants, solute transfer between matrix and fluid, and sporadic flow behavior. Transport rate will be enhanced by the higher speed of the solitary waves. The existence of these waves has been attributed to increases of up to 500 percent in solute transfer between porous walls and fluid films (Striba and Hurt, 1955). Sporadic flow behavior may also be the result of the solitary waves, since contact with the opposite fracture wall during development of these waves could locally saturate the fracture and temporarily reduce the

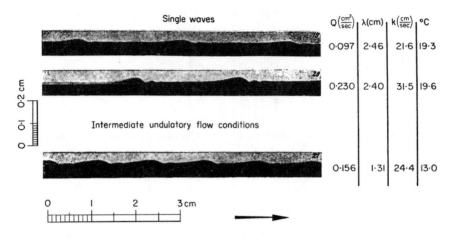

FIGURE 7-7 Photographs of fluid films on vertical glass rod; flow is from left to right. From Kapitza (1965).

flow rate through the network. These observations indicate that these waves may have important implications for hydrologic systems.

HYDROLOGIC MODEL

The mathematical model has permitted determination of key characteristics of the chaotic film that were used to develop a more simplified hydrologic model for fluid transport. The following simple expressions account for the dual-mode transport by a chaotic film:

$$\begin{aligned} Q_{substrate} &\approx 0.06\, Q_{source} \\ Q_{soliton} &\approx 0.94\, Q_{source} \end{aligned} \tag{7.20}$$

$$\bar{u}_{substrate} \approx 0.16\, u_N = 0.16\left(\frac{gQ_{source}^2}{3\eta} \right)^{1/3} \tag{7.21}$$

$$\bar{h}_{substrate} \approx 0.4\, h_N = 0.4\left(\frac{3\eta Q_{source}}{g} \right)^{1/3} \tag{7.22}$$

$$\bar{u}_{soliton} \approx 2.1\, \bar{u}_N = 2.1\left(\frac{gQ_{source}'^2}{3\eta} \right)^{1/3} \tag{7.23}$$

APPLICATIONS

The equations above can be used to analyze the impact that film development would have on transport through an unsaturated fracture. These are applied to the simple fracture model sketched in Figure 7-8. The model consists of a fracture with aperture variable in the z-direction, but no variability in the direction of flow, resembling a set of curtains. The fracture was allowed to sustain free-surface flow in the largest apertures and plug flow in the smallest saturated pathways. Figure 7-9 shows the transport distance after 1 second as predicted by the three models, free-surface films, saturated plug flow, and parallel plates. The mean aperture used for parallel plates was 0.31 mm. Film thickness was assumed to be that which "almost" saturates the conduit so as to properly compare it to the plug flow model. Both the plug flow and film flow models predict channeling of flow down the fracture; however, a more subdued behavior is predicted by the plug flow model.

One of the central interests in using these models is for determination of contaminant first arrivals. Equations 7.20-7.23 have been used in conjunction

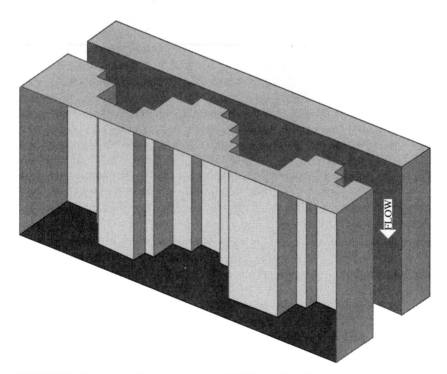

FIGURE 7-8 Fracture model; no aperture variability in flow direction.

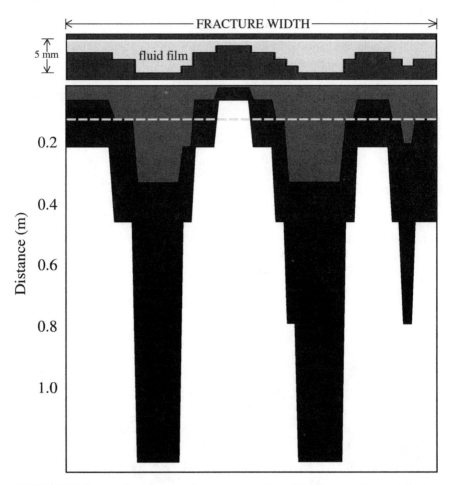

FIGURE 7-9 Top portion of figure shows cross-section of fracture aperture used for the calculations. Bottom portion of figure shows distance of contaminant first arrival for three models: free surface film model (dark grey), plug flow model (light grey), and parallel plate model (dashed line).

with the conservation of mass equation to generate the following predictive algorithm for determining the fastest velocities for a simple fracture model:

$$u_{max} = \frac{2.1g}{3\eta} \left(\frac{2\eta Q_{source}}{gWw_F} - \frac{2}{w_F} \sum_{i=1}^{n} w_i b_i^3 \right)^{2/3}, \qquad (7.24)$$

where Q_{source} is the source volumetric flow rate into the fracture; b the half aperture; W the total fracture width; w_i the percent of fracture width having

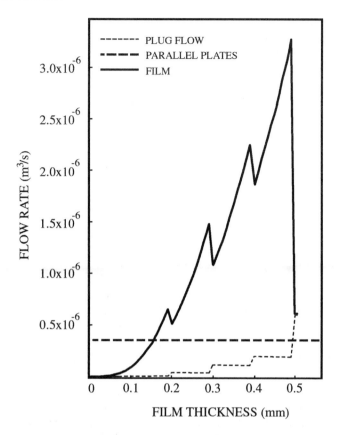

FIGURE 7-10 Flow capacity within a fracture for free-surface film model (solid line), plug model (dashed line), and parallel plate model (bold dashed line).

aperture $2b_i$; and w_F the percent of fracture width exhibiting free-surface film flow. Width weights are calculated from the discretization of the model aperture, total width of the fracture, and total flow rate. This algorithm allows prediction of first arrival times from fracture statistics and knowledge of the incoming flow rate. It should be noted that this algorithm represents idealized conditions for a simple fracture model. This first-order equation will solve for the fast end of the transport envelope. It does not take into account matrix absorption, path tortuosity, or chaotic resaturation from wave growth, all of which serve to slow the transport rate.

Free-surface films in a fracture can also generate sharp local changes in flow rate by two mechanisms; resaturation of the conduit and desaturation into a film.

A dry fracture that receives a small amount of source flow rate will carry flow in the form of a free-surface film. If flow rate is increased, smaller-aperture sections of the fracture will become saturated, reducing the local capacity for flow. If the flow rate is increased to the point where the free-surface film touches the opposing fracture wall, the conduit will completely saturate and the flow rate will be correspondingly reduced. This behavior is shown in Figure 7-10, which was generated by using the simple fracture model in Figure 7-8. For comparison, the graph also shows curves for the cubic law model and the equivalent permeability model. Note that as the fracture completely saturates, the flow rate drops to the plug flow value. Sudden saturation events can occur due to increases in flow rate, changes in fracture aperture, and growth of chaotic waves.

Sudden episodic increases in flow rate may also be generated by the chaotic nature of the free surface. Flow rates that range between the film rate and the plug flow rate (the envelope between the two curves in Figure 7-9), are subject to unstable wave growth. If the wave grows sufficiently in amplitude to touch the opposing fracture wall, the system will locally saturate, resulting in a local reduction in flow that will force the fluid to back up the system or find another route. However, since the flow rate is within the specified range, fluid prefers (by energy argument) to sustain a free surface. Sudden changes in flow rate or air movement within the fracture could induce the fluid to desaturate and reform the free-surface film, which is the more efficient transport mode (Dragila and Wheatcraft, 1998). Local development of a film could drain fluid from higher in the fracture and generate sudden changes in flow rate. These peaks in transport rate would be short lived if the free surface once again became unstable. Duration of the episodic fast-flow events may be related to the time it takes for a solitary wave to form and grow. Observations of such spurious, short-lived events have been recently reported (Podgorney and Wood, 1999). A qualitative schematic of the expected behavior is shown in Figure 7-11.

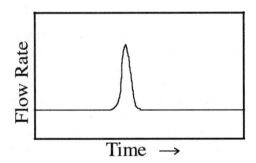

FIGURE 7-11 Conceptual schematic of changes in flow rate due to a chaotic fast-flow event within a fracture.

SEEPAGE GENERATED FILM FLOW

Free-surface films have been seen to form in fractures open to the surface such as expansion cracks in clay soils. Tokunaga and Wan (1997) showed development of a significant free-surface film along a vertically oriented fracture in a sample of Bishop tuff. Figure 7-12 shows the dramatic increase in the hydraulic conductivity of the system by the presence of the free-surface film. In a vadose environment these films could transport large fluxes down interconnected paths within the fracture system. But the observation of chlorine-36 at the ESF is hundreds of feet down from the surface, and fractures in this region may not be completely connected to the surface. A second likely method for generating free-surface films is by seepage into a fracture.

Flow through unsaturated porous material is dominated by capillary forces. Fluid moves only through the saturated-pore portions of the matrix. When matrix fluid encounters a wide-aperture air-filled fracture, the fluid stops moving until the hydrostatic pressure builds up sufficiently at the boundary. If the fracture is inclined from the horizontal, seepage from the fracture wall will flow down the wall and travel as a free-surface film to the terminus of the fracture. The algo-

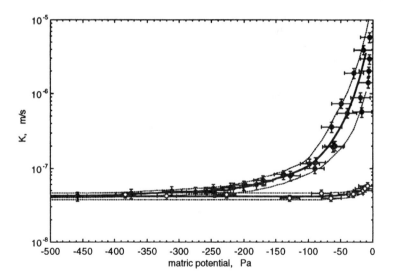

FIGURE 7-12 Hydraulic conductivity measurements on fractured tuff with fractures sealed (open circles) and with fluid allowed to flow down an unsealed fracture as a free-surface film. From Tokunaga, T. K., and J. Wan, 1997. Water film flow along fracture surfaces of porous rock. Water Resources Research 33(6): 1287-1295. Copyright by American Geophysical Union.

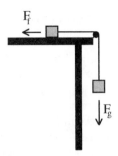

FIGURE 7-13 Sliding block conceptual model for seep generation at a fracture wall.

rithm presented below was developed to model flux rate and flow velocity for free-surface films generated by seepage into a fracture. The rate at which the matrix feeds the seep is derived using an analogous model; a block of wood sliding at constant velocity on a table is tied to a second block that is hanging freely (Figure 7-13). The energy of the system is composed of the gain in energy of the hanging block and the loss of energy due to friction by the sliding block. Similarly, the energy equation for the seep needs to include the energy gradient provided by the gravitational field and the energy loss due to the frictional characteristic of the porous medium that is accounted for by the hydraulic conductivity. The flow equation for the source of the seep becomes

$$Q(x)_{source} = K_{matrix} A(x)_{seepageface} = K_{matrix} Wx \qquad (7.25)$$

where K is the hydraulic conductivity of the porous matrix; A the cross-sectional area of the seepage face; W the lateral width of the seepage wall; and x the vertical distance over which seepage occurs. Although the source flow rate is determined by the hydraulic conductivity of the porous matrix, once the film starts to flow, the film thickness becomes a dynamic variable and different equations apply. The source of seepage is a distributed source, thus the film is being continuously fed, and as a result, continuously increasing in thickness and velocity. The film flow rate is a function of the source flow rate and the distance along the wall:

$$Q_{film}(x) = \int_{x=0}^{x=x} dQ(x)_{source}. \qquad (7.26)$$

The Nusselt film thickness, Nusselt velocity, and mean film flux rate are now a function of location down the seepage wall and a function of matrix properties,

$$h_N(x) = \left(\frac{3\eta}{g}\right)^{1/3} \left(K_{matrix} x\right)^{1/3}, \qquad (7.27)$$

$$u_N(x) = \left(\frac{g}{3\eta}\right)^{1/3} \left(K_{matrix}x\right)^{2/3}.$$

(7.28)

Since mass must be conserved, the flux carried by the film is equal to the flux exiting the seep. Thus, there is a direct relationship between film properties and seep matrix properties:

$$Q(x)_{film} = Wh_N(x)\bar{u}_N(x) = WxK_{matrix}.$$

(7.29)

After the film exits the seep area, its thickness and velocity will no longer increase. Film flow rate per unit width below the active seep area can be found by solving Equation 7.29 for $x = L$, where L is the length of the seep:

$$Q_{film} = h_N u_N = LK_{matrix}.$$

(7.30)

Figure 7-14 shows the relationship between seep matrix hydraulic conductivity and maximum film velocity for a chaotic film. Since the fastest rates are generated by solitary waves, the values were calculated by combining Equations 7.23 and 7.30 to obtain the following relationship:

$$\bar{u}_{max} = 2.1\left(\frac{g}{3\eta}\right)^{1/3} (LK_{matrix})^{2/3}$$

(7.31)

Using the range of velocity shown in the graph, the time it would take for a film to reach a depth of 100 m would be between 1 day and 35 years, depending on the saturated hydraulic conductivity of the matrix supporting the seep. The model above assumes a flat untextured surface, whereas a natural fracture is textured. Fracture surface topography could provide deeper pathways for devel-

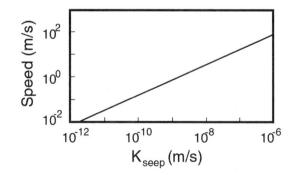

FIGURE 7-14 Graph of maximum film velocity predicted by chaotic film model as a function of seep matrix hydraulic conductivity.

opment of thicker films, providing faster conduits for the fluid. At higher seep-matric conductivity, the source flow rate would increase, resulting in thicker films. The general thickening of the film would allow lateral connection to higher-conductivity pathways. As the film progressively thickens, more and more of the flow is conducted through narrower and thicker film conduits. Noting that the predicted flow rate increases as the cube of the film thickness, deeper channels could carry a disproportionate percentage of the flow. With increasing film source flow rate, a power-law behavior may develop in the velocity profile of the film in response to possible power-law distributions in the topographic character of the natural fracture surface. It is clear that much more work needs to be done in this area.

A final comparison is made between the prediction of the algorithm above and an equivalent model using the TOUGH computer code commonly in use for fractured media transport. TOUGH is a multidimensional numerical model for simulating the transport in porous and fractured media. It assumes Darcy type flow, and fractures are modeled as porous conduits of higher permeability (Pruess and Wang, 1987). The model selected is a seepage area into a natural fracture generated by a horizontal capillary barrier. Figure 7-15 includes a schematic of the hydrologic model and shows the results of the TOUGH code model and film flow model.

Parameters used for the TOUGH model may be found in Ross et al. (1982) with some adjustment for a much longer fracture to avoid boundary problems. The cells were 1 cm square, and the fracture was a soil-filled conduit of high permeability, 10 cm in aperture and 2 m long with a seep area only 4 cm long. For the film model, flow rate used to feed the film was calculated using Equation 7.29 and the same seep permeability as used for the TOUGH model. Comparing results for the three time steps shown in Figure 7-15, it is evident that the two models address two different mechanisms in fluid flow. The mass of fluid contained in each time step in the two models is the same; however, the film redistributes the mass in a very different way than the TOUGH model. By a time of 1,000 seconds, the film has traveled 42 cm and is out of the seep area. In contrast, the TOUGH model moves the fluid laterally as well as downward due to capillary forces within the conduit.

CONCLUDING REMARKS

Free-surface films have a greater capacity for transporting fluids via unsaturated fractures than plug flow. They are minimally affected by fracture aperture characteristics and as a result do not suffer the capillary and geometric restrictions imposed on standard fracture flow models. Fractures connected to the surface can take advantage of episodic precipitation events. Fractures deeper in the vadose zone not exposed to the surface may generate seepage by creating a capillary barrier against the unsaturated porous matrix. This seepage may generate a free-surface film that will flow under the influence of gravity if the fracture

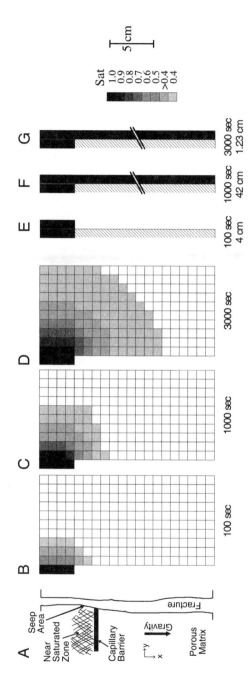

FIGURE 7-15 Comparison of results from TOUGH and free-surface film models for a seep into a fracture: (a) conceptual hydrologic model showing a seep forming from a near-saturation matrix above a capillary barrier; (b–d) results of the TOUGH model for three time steps; (e–g) results of the same problem using free-surface film model.

is inclined toward the vertical. Where the fracture system is not connected, free-surface film will transport fluid at very rapid speeds to the terminus of the fracture, where it is again reabsorbed into the matrix. By this mechanism, the fracture system becomes a short circuit to the porous matrix flow, transporting fluid rapidly across entire sections of unsaturated porous matrix. A series of fortuitously placed fractures could dramatically increase the transport rate of a contaminant.

Much work is still to be done in determining the mass exchange between film and matrix induced by the chaotic behavior of the free surface. Experimental studies have shown up to a 500 percent increase in solute transfer between matrix and fluid film when the film becomes wavy (chaotic) (Striba and Hurt, 1955). The assumption that pressure balance must be sustained across the film-matrix interface is still an open issue. Even though the free surface of the film would indicate that pressure in the film is atmospheric, tensiometer measurements in the Bishop Tuff experiment indicate that the free-surface film was subatmospheric. The hopscotch potential of a combined fracture-film transport mechanism suggests that dispersion of contaminants in a fractured vadose system may require a different approach than used in standard unsaturated porous transport.

The proposed free-surface film model represents one mechanism that may be responsible for the presence of bomb-pulse chlorine-36 at great depths at Yucca Mountain. In addition, it proposes explanations for interesting observed behavior such as episodic fast-flow events, erratic flow rate observations, and very rapid transport through air-filled fractures. Further research is required in order to comprehensively understand the extent to which this mechanism is operative in fractured vadose zones.

REFERENCES

Chu, K. J., and A. E. Dukler, 1974. Statistical characteristics of thin, wavy films: Part II. Studies of the substrate and its wave structure. AIChE Journal 20(4): 695-706.

Dragila, M. I., 1999. A New Theory for Transport in Unsaturated Fractures: Free-Surface Film Flows. Ph.D. Dissertation, University of Nevada, Reno.

Dragila, M. I., and S. W. Wheatcraft, 1997. Chaotic dynamics in unsaturated fracture flow. Abstract, Geological Society of America Meeting.

Dragila, M. I., and S. W. Wheatcraft, 1998. Non-linear characteristics of fluid flow in unsaturated fractures. Chapman Conference on Fractal Scaling, Non-linear Dynamics, and Chaos in Hydrologic Systems, Clemson, S.C., May 12-15.

Dukler, A. E., and O. P. Bergelin, 1952. Characteristics of flow in falling fluid films. Chemical Engineering Progress 48(11).

Fabryka-Martin, J. T., A. V. Wolfsberg, S. S. Levy, K. Campbell, P. Tseng, J. L. Roach, and L. E. Wolfsberg, 1998. Evaluation of flow and transport models of Yucca Mountain, based on chlorine-36 studies for FY98. Los Alamos National Laboratory, Yucca Mountain Project Milestone Report SP33DDM4.

Geller, J. T., G. Su, and K. Pruess, 1996. Preliminary studies of water seepage through rough-walled fractures. Lawrence Berkeley National Laboratory, LBNL-38810, UC-403.

Kapitza, P. L., 1965. Wave flow on thin layers of a viscous fluid (1943). In: Collected Works of P. L. Kapitza, D. Ter Haar, ed., Pergamon Press.

Nakoryakov, V. E., B. G. Pokusaev, S. V. Alekseenko, and V. V. Orlov, 1977. Instantaneous velocity profile in a wavy fluid film. Journal of Engineering Physics and Thermophysics 33(3): 1012-1016. Translated from Inzhenerno-Fizicheskii Zhurnal 33(3): 399-404.

Nicholl, M. J., R. J. Glass, and S. W. Wheatcraft, 1994. Gravity-driven infiltration instability in initially dry horizontal fractures. Water Resources Research 30(9): 2533-2546.

Nitao, J. J., and T. A. Buscheck, 1991. Infiltration of a liquid front in an unsaturated, fractured porous medium. Water Resources Research 27(8): 2099-2112.

Nitao, J. J., T. A. Buscheck, and D. A. Chesnut, 1993. Implications of episodic nonequilibrium fracture-matrix flow on repository performance. Nuclear Technology 104: 385-402.

Nusselt, W., 1916. Die Oberflocaekondensation des Easserdampfer, ZVDI, 60: 541.

Persoff, P., and K. Pruess, 1985. Two-phase flow visualization and relative permeability measurement in natural rough-walled rock fractures. Water Resources Research 31(5): 1175-1186.

Podgorney, R. K., T. R. Wood, T. M. Stoops, R. G. Taylor, and J. M. Hubbell, 1998. Basalt outcrop infiltration tests to evaluate chaotic behavior of unsaturated flow in fractured rock. Data Summary Report, 1997 Field Season, Idaho National Laboratories.

Podgorney, R. K., and T. R. Wood, 1999. Observations of water movement in variably saturated fractured basalt and its possible implications on predictive modeling. In: Proceedings of the International Symposium on Dynamics of Fluids in Fractured Rocks, Concepts and Recent Advances, B. Faybishenko, ed. Lawrence Berkeley Laboratory, pp. 300-304.

Prandtl, L., 1904. Uber Flussigkeitsbewegung bei sehr kleiner Reibung. Proc. Third Internat. Math. Cong. Heidelberg [English translation in NACA Technical memo, 452].

Pruess, K., and J. S. Y. Wang, 1987. Numerical modeling of isothermal and nonisothermal flow in unsaturated fractured rock: A review. In: Flow and Transport Through Unsaturated Fractured Rock. D. D. Evans and T. J. Nicholson, eds. Geophys. Monogr. Ser. 42, American Geophysical Union, Washington, D.C., pp. 11-21.

Pruess, K., and Y. W. Tsang, 1990. On two-phase relative permeability and capillary pressure of rough-walled rock fractures. Water Resources Research 26(9): 1915-1926.

Pumir, A., P. Manneville, and Y. Pomeau, 1983. On solitary waves running down an inclined plane. Journal of Fluid Mechanics 135: 27-50.

Ross, B., J. W. Mercer, S. D. Thomas, B. H. Lester, 1982. Benchmark Problems for Respiratory Siting Models, NRC-Report, NUREG/CR-3097.

Striba, C., and D. M. Hurt, 1955. Turbulence in falling liquid films. AIChE J. 1: 178.

Su, G., J. T. Geller, K. Pruess, and F. Wen, 1999. Experimental studies of water seepage and intermittent flow in unsaturated, rough-walled fractures. Water Resources Research, 35(4): 1019-1038.

Telles, A. S. , and A. E. Dukler, 1970. Statistical characteristics of thin, vertical, wavy liquid films. Ind. Eng. Chem. Fundam. 9(3): 412-421.

Tokunaga, T. K. and J. Wan, 1997. Water film flow along fracture surfaces of porous rock. Water Resources Research 33(6): 1287-1295.

Tsang, Y. W., 1984. The effect of tortuosity on fluid flow through a single fracture. Water Resources Research 20(9): 1209-1215.

Tsang, Y. W., and C. F. Tsang, 1987. Channel model of flow through fractured media. Water Resources Research 23(3): 467-480.

Wang, J. S. Y., and T. N. Narasimhan, 1985. Hydrologic mechanism governing fluid flow in a partially saturated, fractured, porous medium. Water Resources Research 21(12): 1861-1874.

Wasden, F. K., and A. E. Dukler, 1992. An experimental study of mass transfer from a wall into a wavy film. Chemical Engineering Science 47(17/18): 4323-4331.

White, F. M., 1991. Viscous Fluid Flow. 2nd ed., McGraw Hill, New York.

Yih, C.-S., 1963. Stability of liquid flow down an inclined plane. The Physics of Fluids 6(3).

Yu, L.-Q., 1995. Studies of wavy films in vertical gas-liquid annual flows. Ph.D. dissertation, University of Houston.

Yu, L.-Q., F. K. Wasden, A. E. Dukler, and V. Balakotaiah, 1995. Nonlinear evolution of waves on falling films at high Reynolds numbers. Physics of Fluids 7(8): 1886-1902.

8

What Do Drops Do? Surface Wetting and Network Geometry Effects on Vadose-Zone Fracture Flow

Thomas W. Doe[1]

ABSTRACT

Capillary conceptual models of vadose zone flow predict that large pores and open fractures should transmit water only when the rock as a whole approaches saturation. Observations, such as bomb-pulse radionuclides at Yucca Mountain, indicate more rapid transport in fractures at lower overall saturation values than capillary theory would suggest. These and other observations encourage a re-examination of vadose-zone conceptual models for fractures. In addition to film flow processes, this paper suggests that flow as discrete drops may also play a role in flow on fracture surfaces at lower saturation values. The study of drop flow versus film flow and capillary flow is strongly influenced by the wetting properties of the rock. Specifically, characteristics of a zero-contact angle versus a finite-contact angle may control whether flows occur as continuous films or discrete drops on fracture surfaces.

Drops in fractures may contact either one wall or both walls, in which case they may also be called capillary islands or blobs. Drops are static unless their mass exceeds a critical value to initiate sliding. The resistance to drop flow arises from wetting effects in addition to viscous effects, and in some circumstances wetting effects may dominate, especially in the initiation of flow. Drop flows appear in fracture-flow experiments described in the technical literature. The dependence of flow processes on contact angle emphasizes the importance of understanding wetting processes on natural fractures and on material used in laboratory simulations.

[1] Golder Associates, Inc., Redmond, Washington

Vadose flow in fracture networks is mainly in the direction of fracture dip. Strongly preferred nonvertical dip directions in fracture networks can divert flow from vertical directions toward the direction of fracture dip. Fracture network geometries can also lead to flow focussing into relatively small portions of the fracture network.

INTRODUCTION AND BACKGROUND

The movement of water and the transport of contaminants in the vadose zone have become matters of national importance. Vadose zone concerns are critical for radioactive waste disposal projects that involve thick unsaturated zones such as at Yucca Mountain or Ward Valley. Beyond these specific projects, the vadose zone has developed increased importance for problems of groundwater contamination, particularly in arid regions.

The dominant conceptual model for vadose zone flow can be called the capillary conceptual model. The capillary conceptual model of flow holds that water flows in pore spaces that contain continuous films held in place by capillary tension (Richards, 1931). The analytical backbone of the capillary model is Richards' equation (Richards, 1931).

According to the capillary model, capillary tension increases with decreasing pore size. Due to their higher capillary tension and lower matric potential, smaller pores will be the first to accept water on imbibition, and the last to yield water on drainage. As water preferentially resides in that portion of the rock having the lowest permeability (that is, the smaller pores), unsaturated rocks should be poor conductors of water except at saturation values of nearly one. As rocks become less saturated, the relative permeability of the rock to water also decreases. Because the capillary model conceptualizes flow in continuously connected pore water, flow ceases when the water in the pores becomes disconnected as the rock reaches its residual saturation value.

Wang and Narasimham (1985) and Peters and Klavetter (1988) extended the capillary conceptual model to fractured rock. As fractures tend to include the largest openings, particularly in fractured, consolidated rock, matrix pores should fill before fractures. The use of porous analogs and continuum models for flow in fractures has been justified by assuming that the roughness of fractures and the interactions of fracture asperities behave similarly to intergranular pores. Peters and Klavetter (1988) developed a continuum model for water movement in unsaturated fractures based on analogs of capillary tube bundles and derived equations of similar form to Richards' equation.

In Wang and Narasimhan's (1985) model, water is retained in fractures only at capillary bridges, which allow flow between the matrix blocks but not along the fracture (Figure 8-1a). According to Wang and Narasimhan (1985), "within a partially drained fracture, the presence of a relatively continuous air phase will produce practically infinite resistance to liquid flow in the direction parallel to the fracture." Within this conceptual model, fractures would contain continuous water only at saturation values that approach one. In arid regions, where infiltration rates are low and rocks may be expected to have low saturation values, these extensions of the capillary conceptual model to fractures predicted that fracture flow would be virtually nonexistent in thick vadose zones.

The capillary conceptual model has its roots in soil physics, a discipline that primarily supports agriculture and forestry concerns. For these applications, which are mainly in fine-grained soils, the capillary conceptual model has been successful and has stood the test of time. In rocks with larger voids and fractures, some questions have arisen.

The soil physics literature notes potential inconsistencies between capillary flow theory and natural behavior. Jury et al. (1991) have discussed the problems of flow in fractures and zones with large pores (macropores): "At the present time there is no complete theory describing water flow through structural voids (sometimes called macropore flow). There is uncertainty over how important subsurface voids can be in water flow, since if large they should only fill at matric potentials near saturation. Nonetheless, substantial indirect evidence of flow through structural voids has been obtained by tracer studies. These have shown that a fraction of a chemical application can migrate to substantial depths with only a small amount of water input" (Jury et al.,1991, p. 153). Hillel's (1980) soil physics textbook notes that a " very challenging problem related to the infiltration into swelling soils is how to account for the role of cracks when a field of swelling soil is allowed to dry on top. In due time, a system of regularly spaced cracks appear. Infiltration into such cracked soil, in its initial stages at least, is obviously different from the orderly, one-dimensional process described by most existing theories. Much of the applied water bypasses the surface zone as it runs directly into the cracks."

In recent years, several studies have produced evidence that vadose-zone flow in fractured rocks occurs at higher velocities than simple capillary-based models would predict. Pruess et al. (1999) summarize these observations, which include detection of bomb-pulse radionuclides in the Yucca Mountain underground test facility. Such radionuclides, which were produced by the atmospheric nuclear testing of the 1950s and early 1960s, suggest downward water movements of hundreds of meters in tens of years. These are considered "fast flows" relative to those expected from capillary theory. Pruess (1999) defines the fundamental paradox of unsaturated fracture flow: how can fast flow occur in the presence of the strong matrix imbibitions of partially saturated rock?

Laboratory testing on simulated fractures has turned up behavior that appears inconsistent with capillary flow theory. In particular, Su et al. (1999) note the formation of isolated drops on surfaces with connecting bridges that form, disconnect, and reform with time. Various spatial and temporal instabilities that are not predicted by Richards' equation have been reported in laboratory experiments. These include instabilities that lead to fingering (Glass et al., 1995), as well as cyclic pressure and flow rate variations despite constant-input sources (Persoff and Pruess, 1995). Similar cycling behavior is also reported in experiments using coarse-grained soil analogs (Prázak et al., 1992). Prázak notes that this behavior is not consistent with the expectations of Richards' equation. A later section of this paper discusses laboratory observations in greater detail.

Several conceptual models have been proposed that may resolve the inconsistency issues. Pruess (1999) notes the apparent paradox of fracture flow in the presence of large matrix imbibition potentials. A solution to this paradox, he points out, is to limit the interactions of the matrix and the fractures. This can be accomplished by using a dual-permeability model (Figure 8-1b) that views the fractures and matrix as separate, but interacting flow systems (see chapter 11). This approach may be justified on several conceptual bases, such as (1) fracture coatings that restrict flow between the fractures and matrix, or (2) focusing mechanisms that restrict flow to a small portion of the fracture network. The fracture-

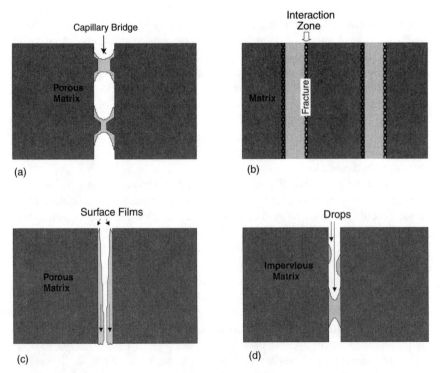

FIGURE 8-1 Conceptual models of vadose fracture flow processes: (a) model of Wang and Narasimhan (1985); (b) Dual-Permeability Model with fracture-matrix interaction zone; (c) Film-Flow Model of Tokunaga and Wan (1997); (d) Drop-Flow Model with one-walled and two-walled drops.

matrix interaction term is difficult to measure but it can be obtained by inverse modeling using an appropriate data set for calibration. The focusing mechanisms may derive from some combination of the fracture geometries and the flow processes.

One alternative model to capillary flow is the film flow model (Figure 8-1c; Tokunaga and Wan, 1997), which proposes the existence of films on fracture surfaces that form when the matrix approaches saturation and water is expelled to the fracture surfaces. Film flow requires a water source that is sufficient to maintain the film, which may be appreciable given the flow capacities of films. For situations where the water flow is not as great, we explore drop flows (Figure 8-1d) as yet another alternative conceptual model. Another difference between drop flows and film flows is the influence of wetting properties. Film flows presume a zero contact angle, while drop flows may apply when the contact angle is nonzero.

This paper places a major focus on drops. The paper was originally conceived as a review of fracture network models and the geometric considerations that might influence vadose-zone groundwater flow. In the course of preparing this paper, other questions of fundamental flow processes in fractures arose. These questions were inspired by many hours observing water flow on the glass windshield of an automobile caught in Seattle traffic. Observations of water on glass surfaces show discontinuous drops that stick to the surface and are immobile until they reach a critical size and begin to slide. At first glance this drop process does not conform to assumptions of the capillary conceptual model, which holds that flow happens in continuous films, and stops when the water becomes discontinuous, as in drops. Can one reconcile the flow of water in discontinuous drops with capillary theory? Or does drop flow constitute a fundamentally different flow process? How does flow in drops compare with other alternatives to capillary flow such as film flow (Tokunaga and Wan, 1997)?

Drop flow studies are extensive in the fluid mechanics and surface chemistry literature, but have received little attention in the soil physics or hydrogeology literature. This paper intends to develop awareness of the drop flow model as a potential alternative to capillary models for vadose zone fracture flow. The paper proposes drop flow as a means of moving water at saturation levels where water becomes discontinuous in the pore spaces and along fracture surfaces.

As such, this paper does not propose a specific practical approach to using drop flow mechanisms in analyses or numerical simulations. To become a practical tool, drop flow studies would need to deal with many of the same issues that conventional capillary flow models continue to wrestle with, such as how to deal with upscaling and heterogeneity of properties. Even if studies do not produce practical new methods, the periodic re-examination of conceptual models for their compatibility with natural observations and fundamental physical processes is desirable under any circumstances. Such a re-examination becomes necessary when the inferences of accepted conceptual models become inconsistent with observations of the natural world.

This paper defines "drop" as an isolated body of liquid that is not continuous with the rest of the liquid phase. For fractures, this definition includes drops that contact neither wall (this requires a large fracture aperture), a single wall, or both walls (Figure 8-1d). Some authors use terms other than "drop" for features that contact both walls. Kneafsey and Pruess (1998) use the term "capillary island," while Su (personal communication, 2000) uses the term "blobs." The reason for considering single-wall features and double-wall features together as "drops" lies in the similar physics of flow, as discussed below.

In the following sections, this paper reviews the fundamentals of surface-liquid interactions. This is an important starting point, as wetting properties and contact angles control, in part, whether liquids spread in films in zero contact angle systems, or isolate themselves into drops in systems with nonzero contact angles. Measurements of contact angles on quartz and on materials commonly used for laboratory simulation of fracture flow, such as glass or epoxy, are not zero. A nonzero contact angle suggests that water forms discontinuous drops and rivulets on surfaces rather than the continuous adsorbed films suggested by Richards (1931) or film flow theory (Tokunaga and Wan, 1997).

The next section reviews literature on the behavior of drops on surfaces. The movement of drops on surfaces or between surfaces for liquid-solid systems with nonzero contact angle differs significantly from conventional hydrogeologic thinking. Specifically, on single surfaces or fractures with large aperture, water forms a discontinuous network of drops or rivulets depending on the flow rate. The resistance to drop flow arises from the energy requirements for wetting the surface in advance of the drop, and viscosity enters only when the drop achieves some appreciable velocity. Drops that lack sufficient mass to allow gravity to overcome this wetting resistance are immobile. Following the discussion of the physics of drop flow, we review experiments on fracture flow in the laboratory.

Finally, we discuss how these concepts can be extended to fracture networks. This section looks at geometric factors that may lead to localization of flow in focused pathways.

SPREADING, WETTING, AND FLUID ADHESION ON SURFACES

An understanding of flow on surfaces requires an appreciation of the physics of wetting and wettability. This section briefly reviews the basics of this topic. For further discussion please refer to, for example, Berg (1993) or Hiemenz and Rajagopalan (1997); this discussion draws heavily from those references. According to Berg (1993), "wettability" refers to "the response evinced when a liquid is brought into contact with a solid surface initially in contact with a gas or another liquid." As Berg points out, several things can happen when a liquid contacts a solid surface, including:

1. the liquid may spread spontaneously, forming a film whose extent is limited only by the mass of liquid available; or

2. the liquid may spread on the surface until it achieves an equilibrium with the gas and the solid to form a three-phase interface with a contact angle; or

3. the liquid may have no interaction with the surface at all.

Which of these processes occurs depends on the wetting properties of the liquid and the surface. Depending on those properties, one may expect a range of processes that are important for vadose zone flow.

Wetting is classically viewed in terms of surface energies that are associated with interfaces between liquids, solids, and gases. These surface energies are related to the surface tensions σ_{LG} for the liquid-gas interface, σ_{LS} for the liquid-solid interface, and σ_{SG} for the solid-gas interface. If we assume as Berg (1993) does that practical differences between the free energies of the liquid and solid with the gas phase and with a vacuum are negligible, then $\sigma_{LG} \approx \sigma_L$ and $\sigma_{SG} \approx \sigma_S$.

First, let us consider the liquid itself. Liquids, particularly water, may have a strong *cohesion*, which reflect the water's attraction to itself. The work of cohesion, W_C, is defined as the work required to create a unit surface area of liquid. Separating a single mass of water into two masses creates two surfaces where there had been one. If a unit area of liquid has a surface energy reflected by σ_L, then the divided bodies will have twice that energy; hence, $W_C = 2\sigma_L$.

In contrast to cohesion, which describes interfaces in a single material, *adhesion* describes the interface between two different materials, such as a liquid and a solid. If we view the work of adhesion, W_A, as the work required to disjoin or de-wet a unit area of solid-liquid interface, then the de-wetting process creates two new surfaces with the energies associated with the liquid and the solid, and it eliminates the interface between the liquid and solid; hence, $W_A = \sigma_S + \sigma_L - \sigma_{LS}$.

The work of wetting, W_W, refers to the work required to de-wet a unit area of solid surface. As this de-wetting creates a solid-gas surface at the expense of a liquid-solid surface, the work may be expressed as the difference of the surface energies of solid-gas surface and the liquid-solid interface, or, $W_W = \sigma_S - \sigma_{SL}$. This wetting process occurs when liquid is imbibed or drained from a material such as pores or capillary tubes. A final useful wetting relationship is the work of spreading, W_S. If a surface has been coated or covered with a liquid, this is the work required to create a unit area of solid-gas interface while eliminating unit areas of liquid-gas interface and liquid-solid interface; hence $W_S = \sigma_S - \sigma_L - \sigma_{LS}$. The work of spreading can also be arrived at by taking the difference of the work of cohesion and the work of adhesion.

Thomas Young (1805) studied the contact angle at the junction of a solid, a liquid, and a gas phase. Young's law relates the contact angle to the surface tensions by:

$$\cos\theta = \frac{\sigma_{SG} - \sigma_{SL}}{\sigma_{LG}}. \qquad (8.1)$$

Substituting Young's law (Equation 8.1) into the wetting relationships ties them to contact angle as:

$$W_A = \sigma_{LG}(1 + \cos\theta),$$ (8.2)

$$W_W = \sigma_{LG}\cos\theta,$$ (8.3)

$$W_S = \sigma_{LG}(\cos\theta - 1).$$ (8.4)

Berg (1993) points out that these values of work are the negative free energies associated with these processes, hence positive values for Equations 8.2 through 8.4 indicate that adhesion, wetting, or spreading occur spontaneously.

For each process, there is a critical contact angle that separates positive values from negative values for these expressions. The ranges of these critical angles and the associated processes are shown in Figure 8-2. Adhesion (Equation 8.2) occurs provided the contact angle is less than 180°. A practical implication of this relationship is that the liquid can produce drops that are capable of "sticking" to a surface when the contact angle is less than 180°. If the contact angle is 180°, there is no adhesion between the liquid and the solid.

With respect to wetting[2] (Equation 8.3), the critical contact angle is 90°. This is the contact angle that differentiates capillary rise from capillary depression.

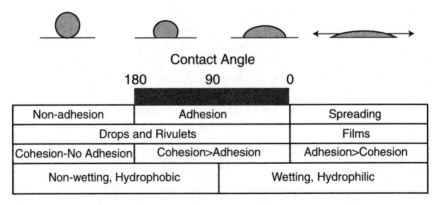

FIGURE 8-2 Relationship between adhesion, cohesion, wetting, spreading, and contact angle.

---

[2] The definition of the term "wetting" can be confusing as it has many uses depending on the circumstances. Some authors use "wetting" to refer to any system where there is adhesion, that is, contact angles less than 180° (Padday, 1993), while others, including Richards (1931), use "wetting" nearly synonymously with spreading, or a contact angle of zero. Perhaps the most common usage is for contact angles less than 90°.

This is an extremely important relationship for vadose zone flow as noted in the equations for capillary rise, H, in tubes and between parallel plates with separations of e, a fluid with density ρ_w, subject to gravitational acceleration, g, (Wang and Narasimhan, 1985):

$$H = \frac{2\sigma\cos\theta}{\rho_w g e} \tag{8.5}$$

If the contact angle is less than 90°, then liquids spontaneously imbibe into smaller pore spaces, such as pores and fracture asperities that are in contact. Such a condition is considered hydrophilic. At higher contact angles, liquids require work to be brought in contact, and thus are nonimbibing, or hydrophobic. The 90° contact angle and the surface-energy processes it represents are a fundamental separator of hydrophilic behavior and hydrophobic behavior.

As to spreading (Equation 8.4), the critical contact angle is zero, as any positive contact angle produces negative values of the spreading coefficient. When a liquid spreads, it spontaneously forms films on surfaces rather than drops. Thus the zero contact angle is critical for defining film flows. The zero contact angle thus provides a separator between regimes where film flows will dominate, from those where drops will be more prevalent at lower saturations.

Contact Angles for Geologic Materials and Surface Roughness Effects

The previous section reviewed the fundamentals of wetting, spreading, and adhesion. The properties of wettability are reflected in the contact angle. Specific values of contact angle separate regimes that exhibit differences of flow process, such as film flow versus drop flow, and hydrophilic versus hydrophobic behavior.

Literature reviews of contact angle data for geologic materials turned up relatively few references. There may be several reasons for this. Contact angles are commonly assumed to be zero as a matter of convenience due to the complexities that nonzero contact angles introduce to theoretical development (Letey et al., 1962). Furthermore, contact angles may change with time and the duration of contact with water (Carillo et al., 2000). Natural materials are heterogeneous, and may display different properties even at small scales. Finally, contact angles are notoriously irreproducible unless care is taken to keep the surface free of contamination (Berg, 1993). Such cleanliness is not a concern of nature. Despite the technical concerns with contact angle, its importance to questions of hydrophobicity and preferential flow cannot be denied, even though contact angles are hard to measure.

In reviewing contact angles, one may start with laboratory measurements on mineral surface and laboratory modeling materials. Studies by Sklodowska et al. (1999) of liquid interaction in ore processing report contact angles of water on quartz and glass to be 32.79 ± 1.12° and 51.05 ± 0.84°, respectively. Sobolev et al. (2000) determined dynamic contact angles on quartz capillaries and found that contact angles varied between approximately 30° and 70° depending on capillary radius and flow velocity. Su et al. (1999) report contact angles of 33° and 63° respectively for the glass and epoxy used in fracture visualization studies.

Contact angles have been measured in soils by a variety of methods, including times required for infiltration, time for penetration of water drop into a soil surface, and capillary rise methods (Yuan and Hammond, 1968). Pal and Varade (1971) report contact angle measurements of about 60° on untreated, sieved sands. Bachmann et al. (2000) applied sieved soils to surfaces using adhesives and measured the resulting contact angles from drop profiles. These measurements may approximate behaviors of rough, heterogeneous fractures. A portion of the sand grains were treated with hydrophobic coatings. Their measured contact angles varied from 0° to 120° depending on the percentage of hydrophobic grains in the soil.

It should be noted that the laboratory contact angles discussed above (except for the soil measurements) are given for smooth surfaces, and we will refer to these as intrinsic contact angles. The effective contact angle may be reduced by surface micro-roughness. This effect is well demonstrated in the application of paints or adhesives, where it is common to roughen a surface prior to application to assure better adhesion. Hazlett (1993) provides an overview of the literature on this effect. He concluded that the phenomenon was real, although there was not yet a consensus on its causes. Viewed thermodynamically, a rough topography has a greater effective surface area, and thus a greater surface energy, than a smooth surface with the same plan view area. Although various theories on the effects of roughness on contact angle remain open to discussion, the effect of roughness on contact angle has been verified experimentally. Bikerman (1950) demonstrated this phenomenon in experiments on the inclination angle for the sliding of drops on treated steel surfaces. The inclination angle at which drops slide is a function of the contact angle, and for the same steel material Bikerman showed a 21° decrease in the sliding angle through progressive stages of surface treatment to decrease the roughness.

The topic of roughness is very important for the design of laboratory experiments. If laboratory experiments are to reproduce realistic wetting behaviors, the simulated fracture surfaces will need to incorporate the full range of roughness effects that are present on natural fractures. Experimental approaches that use castings of natural fractures may recreate the larger scale asperities while missing the finer roughness scales that affect wetting properties.

The issue of contact angle and roughness is critical for using laboratory experiments as analogs for natural flow. Many analog experiments use smooth surfaces of glass or epoxy for which the intrinsic contact angles range between 20° and 65°. These contact angles indicate that spreading conditions do not exist and the flow will assume the form of rivulets or drops. However, roughness may contribute sufficient additional surface energy to reduce the effective contact angles to zero. This possibility is suggested by qualitative observations of wetting on roughened glass used in film flow experiments where spreading on horizontal surfaces implied a zero contact angle (Wan et al., 2000; Tokunaga et al., 2000).

FLUID MOVEMENT ON SINGLE SURFACES WITH NONZERO CONTACT ANGLES

As discussed above, the energy relationships between the liquid and the surface influence the fundamental flow processes on that surface. Current flow theories assume continuous adsorbed water films and a contact angle of zero. Film flow theories (Tokunaga and Wan, 1997) also assume a zero contact angle. If the contact angle is not zero, water may tend to form drops or rivulets. This section describes the flow processes for such systems.

Drop and Rivulet Flows

If the liquid has a contact angle with a flat surface, the cohesion of the liquid exceeds its adhesion to the surface, and the liquid will form discontinuous drops unless the flux of liquid is sufficient to keep the surface flooded. Although there is little theoretical literature on the flow of drops between two surfaces, extensive work has been done on flow on single surfaces (Sadhal et al., 1997).

This section concentrates on drop flows; however, flows on surfaces where the water has a nonzero contact angle can take several forms depending on the flow rate. At high flow rates, the water can form a curtain that completely floods the surface. As the flow proceeds down an inclined surface, the curtains that begin as uniform fronts may break up into rivulets. The rivulets have a regular spacing (Huppert, 1982; Johnson et al., 1999), with a spacing, or wavelength, λ_{max}, that is related to the thickness of the film and the ratio of viscous to capillary forces as expressed by the capillary number, Ca:

$$\lambda_{\max} = \frac{Kd_{0m}}{(3Ca)^{1/n}},$$ (8.6)

$$Ca = \frac{Q\rho v}{d_0 \sigma}$$ (8.7)

where K is an experimental constant, d_0 is the thickness, d_{0m} is a theoretical thickness, Q is the flow rate, ρ is fluid density, σ is the surface tension, n is an experimentally determined power (theoretically 1/3), and v is the kinematic viscosity.

The formation of rivulets on single surfaces is similar in form to the finger flows described by Glass et al. (1995), but it results from different processes. Glass and colleagues explain fingers as the result of a gravitational density insta-bility that arises when a lower density fluid is trapped beneath a higher density fluid. The only upward path for the lower density fluid is through the overlying fluids. Fingering also arises when such trapping does not exist, as noted by Lenormand and Zarcone (1989), in which case the origin of the fingering, like the rivulets on single surfaces, is a capillary instability rather than density instability.

Physics of Drop Sliding

Drops are familiar companions to our everyday lives. Drops are common on glass windows, mirrors, car windshields, ceramic tiles, and clear plastic dispos-able cups like those the airlines use for beverages. Aside from being a curiosity in day-to-day living, the understanding of drops has many practical applications. The encouragement of spreading and the avoidance of drops are essential for paints, adhesives, and coatings. Drop promotion is essential for waterproofing. Drops affect condenser design, as the energy transfer on the surfaces of condens-ers is more efficient for drops rather than for films, especially if the condenser surfaces are inclined and the drops can roll off to expose fresh surfaces for condensation (Dussan and Chow, 1983). Drops are important for spray retention on foliage, as chemical applications will work better if drops adhere rather than roll off the surfaces of foliage (Furmidge, 1962).

Sadhal et al. (1997) provide an extensive review of drop statics and dynamics. The basic form of drops on surfaces is that of a section of sphere distorted by gravitational effects. On a horizontal surface, the profile of the drop meets the surface at the contact angles for the vapor-solid-liquid system. On inclined sur-faces, gravity distorts the drop so that it has a larger contact angle on the downdip side of the drop than on the updip side. The updip and downdip contact angles are called the advancing and retreating contact angles, or θ_a and θ_r, respectively (Figure 8-3), and the difference between these values is the contact angle hysteresis.

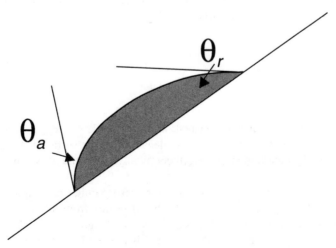

FIGURE 8-3 Advancing and receding contact angles for drops.

Furmidge (1962) experimented with drop retention of sprays on solid surfaces in the course of research to optimize the retention of agricultural applications on the surfaces of foliage. He noted that spraying spatters fluid on the surfaces with different sizes. As the spraying continues, drops coalesce and grow to a critical size at which they are no longer static and slide off the surface. Furmidge placed drops of various sizes and masses on smooth brass plates, which he tilted to induce sliding.

When a drop slides down an inclined surface, it wets an area equal to the width of the drop times the distance the drop has moved (Figure 8-4). A similar area is dewatered behind the path of the drop. Based on the differences in the work performed by the advancing, wetting surface and the retreating, de-wetting surface, Furmidge gave the sliding criterion for drops in terms of a critical angle of inclination for the surface, φ, or

$$\rho g V \sin\varphi = w\sigma(\cos\theta_r - \cos\theta_a), \tag{8.8}$$

where ρ is fluid density, σ is the surface tension, V is the volume of the drop, and w is the drop width. Note that width refers here to the length dimension of the drop perpendicular to the dip direction of the slope.

Furmidge (1962) proposed this relationship on empirical grounds; however, Dussan (1985) and Dussan and Chow (1983) showed that it could be proven by physical fundamentals. The insight afforded by this equation is that drops are held in place by various surface energies that include the fluid surface tension and the adhesive forces, which are lumped into the definition of the contact angle. Equation 8.8 shows that the resistance to sliding is based on the width of the drop and not on its total contact area with the surface. If two drops have the same

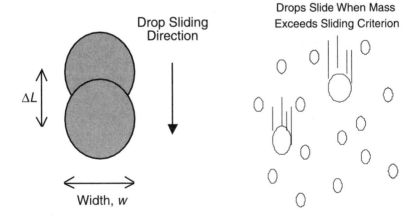

FIGURE 8-4 Sliding criteria for drops.

volume but different widths, the longer, narrower drop will resist sliding more than the wider, shorter one. Later experiments on drop sliding by Extrand and Kumagai (1995) showed that the contact angle hysteresis (the difference between the cosines of the advancing and receding contact angles) varies with drop shape, and the retentive forces could vary with the drop length. Dussan and Chow (1983) analyzed the dynamics of drop sliding theoretically and developed expressions for drop velocities. These analyses are quite complex and will not be repeated here. The approach does assume that contact angles are functions of drop velocity, and the resistance to movement changes with velocity. If the resistance increases, drops will achieve a stable velocity on surfaces rather than accelerate indefinitely.

Drop Sliding Between Parallel Plates Where the Drop Contacts Both Plates

The major portion of the work on drop flows comes from applications to single surfaces. These analyses are directly relevant to flow on fracture surfaces, provided the height of the drop is less than the fracture aperture, and the drop is in contact with only one fracture wall. Drops in contact with both walls of a fracture have been observed in experiments (Kneafsey and Pruess, 1998; Su et al., 1999).

The problem of a drop contacting both fracture walls has not been studied as rigorously as the case of a single surface. We might expect, however, that having a drop in contact with two surfaces changes the problem mainly by requiring two surfaces to be wetted rather than one. Hence a first approximation for drops in contact with both fracture walls is that the wetting resistance doubles by having to wet two surfaces instead of one.

If we assume the drop is circular with a radius, r, its approximate volume will be the area of the drop times the parallel plate aperture, e. The width of the drop will be $2r$. We also introduce an additional factor of two for the two surfaces. For a drop in contact with both parallel plates, Furmidge's sliding criterion defines a critical radius for sliding, r_c, which becomes:

$$r_c = \frac{4\sigma(\cos\theta_r - \cos\theta_a)}{e\rho g \pi \sin\varphi} \qquad (8.9)$$

Figure 8-5 shows the critical drop size for vertical parallel plates as a function of aperture and contact angle. This figure assumes the advancing and retreating contact angles are ±50 percent of the contact angle on a horizontal surface.

Drops may contact either one wall or both walls of the fracture depending on the height of the drop and aperture of the fracture. We can use Furmidge's relationships to assess whether the drop flow will be drops in contact with one wall or with both walls. If the heights of the critically-sized drops are smaller than the fracture aperture, then we should expect that flow is on one fracture wall only.

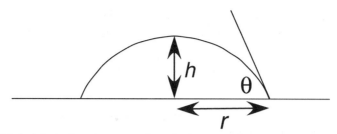

FIGURE 8-5 Spherical cap representation of a drop on a planar surface.

To make a calculation of a critically-sized drop on a single wall, we must make assumptions about the shape of the drop to calculate its height. The heights and shapes of drops are determined by a complicated interaction of surface properties and gravity; however, to get a rough approximation, we may assume that the drop has the form of a spherical cap (Figure 8-6). In this analysis we assume that the liquid-surface contact angle defines the portion of the sphere that forms that cap. For example, a contact angle of 30° would use a cap lying 30° about the top of a sphere. The volume of such a spherical cap is

$$V = \frac{1}{3}\pi \left(\frac{r}{\sin\theta}\right)^3 (2 - 3\cos\theta + \sin^3\theta), \qquad (8.10)$$

where r is the radius of the drop on the surface.

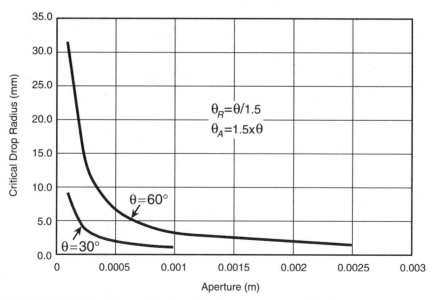

FIGURE 8-6 Critical drop radius for sliding between two parallel plates.

Inserting this volume expression into Furmidge's sliding criterion allows solution for the critical drop radius, r_c, that will induce drop sliding, which is

$$r_c = \left(\frac{6\sigma(\cos\theta_r - \cos\theta_a)\sin^3\theta}{\pi\rho g \sin\varphi(2 - 3\cos\theta + \cos^3\theta)} \right)^{1/2}. \tag{8.11}$$

The height, h_c, of a drop that is critically sized for sliding can be derived by trigonometry from the geometry of the spherical cap as:

$$h_c = \frac{r_c(\tan\theta - \sin\theta)}{\tan\theta\sin\theta}. \tag{8.12}$$

Assume that advancing and receding contact angles are a factor of 1.5 greater or less than the basic contact angle, θ, and the surfaces are vertical. Then r_c will vary from about 2 mm to 1 mm for contact angles of 30° and 90°. The corresponding drop heights will be 0.5 mm and 0.9 mm, respectively. Due to the square root relationship in calculating r, the drop size has a reduced sensitivity to the contact angles. Again, this is a rough approximation subject to experimental testing due to the use of a spherical form for the volume calculation.

These analyses suggest that the heights of critical drops on single surfaces will be on the order of a millimeter or less. Hence, we may expect that single-walled drop flow will occur for fractures larger than 1 mm in aperture, and two-walled flow will occur in fractures with less than 0.5 mm aperture for contact angles consistent with air, water, and quartz or glass.

Larger aperture fractures will be more conducive to flow by drop sliding than smaller fractures for two reasons. If we consider drops in contact with both walls, the critical volume for drop sliding will be larger for smaller aperture fractures, and the radius of a critical fracture will be larger. Larger aperture fractures are also more conducive to flow because they are more likely to contain drops in contact with single walls, and these drops have less resistance to flow because they wet only one surface.

Asperities can be expected to play a role in sliding and drop retention. With asperities, the radius of the drop required for sliding will need to be larger for at least two reasons. Asperities reduce the volume of the drop for a given drop radius; and thus the drop requires a larger radius to slide. Second, the drop will leave behind a volume of water as capillary retention at the asperity's point of contact or the point of least wall separation, if contact is not made.

Once drops begin to move they should continue to slide unless they lose volume, the fracture aperture changes, surface properties of the fracture change, or the inclination angle of the fracture changes. Drops may gain volume by coalescence with small static drops in their path. They may also lose volume by evaporation or imbibition to the matrix along the fracture walls. Drops may also lose volume to asperities, as they would leave behind some mass in capillary retention.

The movement of water in drops will differ in significant ways to the capillary conceptual model of water movement in the vadose zone and to film flow concepts. Specifically:

• The initiation of flow is related to the wetting properties of the surfaces, and once water is moving, wetting resistance may play an equal or more significant role than viscous drag along the fracture walls.

• Drops can be immobile; drops will remain in place until they accumulate sufficient volume to overcome wetting resistance, and they become immobile again if they lose mass or fracture conditions change. Once immobile, they will remain immobile until additional mass enters the drop.

• Darcy's law may not apply if the flow resistance is not primarily viscous and the concept of a potential gradient does not apply to discontinuous drops.

• Drop movement is not a diffusion process.

In terms of boundary conditions, drop flows are best described by flow rate or flux boundary conditions, as head boundary conditions do not apply. In this case, there are clear relationships based on conservation of mass for drop velocity, flux, and saturation. Subject to experimental verification, the application of a constant flux at the top of a fracture results in an accumulation, or storage, of drops or mass at the top of the fracture until a drop reaches a mass sufficient to initiate sliding. At higher fluxes, this flow might occur as a rivulet. At flow fluxes, the mass should emerge from the lower boundary as drops. The long-term average fluxes would balance the influx; however, the instantaneous outflow rate could be high or nothing depending on whether or not a drop was emerging at any particular moment.

EXPERIMENTAL EVIDENCE

Laboratory experiments of flow in simulated fractures are important for understanding flow processes. This section presents a brief review of experiments involving flow in simulated fractures.

Rasmussen (1991) performed experiments on a vertical fracture simulated by two glass panels. The test involved injecting water into a hole in the interior of the fracture under constant head conditions. The sustained injection of the constant-head condition created a region of local saturation primarily downward from the injection point, but also a slight distance upward. The main use of the experiment was to provide data to compare with numerical models of the air-water interface. The experimental results largely agreed with the analytical predictions of the position of the air-water interface, with some variability caused by air bubbles entrapped in the water-saturated zone. Because of the constant-head injection conditions, the flow occurred in locally saturated rivulets rather than as drops or films.

Fourar et al. (1993) performed two-phase flow experiments on smooth and rough fractures and correlated the results to porous media and pipe models. The experiments used smooth glass plates and plates with 1 mm beads in the aperture of the fracture to simulate roughness. The fracture materials were glass, the liquid was water, and the gas was compressed air. The experimental conditions involved the pressure injection of air into initially water-filled fractures, which would appear to be a better analog for oil reservoir processes than for vadose zone flow. Of particular interest in these experiments is the structure of the phases. The structure of the two phases varied with the gas injection rate. At low gas injection rates, the gas bubbles disperse into the water. With increasing rate, the bubbles start to become unstable and begin to finger. At yet higher gas rates, the gas occupies the major portion of the fracture. At these higher gas flow rates, the flow geometry of the water varied with the water injection rate. At lower water injection rates, the water moved as liquid drops in the gas stream. At higher liquid rates, the water flowed as unstable films on the fracture walls. The expectation for a porous medium was that each phase would occupy its own continuous network of pores, the wetting phase in the smaller pores and the nonwetting phase in the larger. The experiments, on the other hand, showed that only one phase was continuous and the other phase traveled as either bubbles or drops. The occupancy locations of the phases were constantly changing.

Nicholl et al. (1994) prepared experiments using commercial glass plates with textured surfaces to represent the fractures and deionized water with small amounts of dye to represent the liquid. The test conditions involved slug injections of water at the top of the air-filled fracture. The basic hypothesis of the experiment was that the flow process was one of density inversion, that is, the entry of higher-density water at the top of the flow system would displace the lower-density air, and the geometry of the water distribution would reflect the gravitational instability. The primary observation of the experiment was the breakup of the water invasion front into fingers (Figure 8-7). Being a slug injection, the fingers were not replenished as they moved down the steeply dipping fracture, and they left drained regions behind them. Smaller fingers were observed to have lower velocities or to stop altogether. Ultimately the liquid would form disconnected clusters, or "drops," though Nicholl et al. do not use this term. This paper also presents the results of dyed water injections in the top of a natural fracture, which showed clearly the development of fingers. One significant point of this work was to demonstrate that the fingering is primarily the result of instability in the wetting front independent of aperture heterogeneity.

Persoff and Pruess (1995) report fracture flow experiments using epoxy replicas of natural fractures. The epoxy is reported to have a contact angle of 20° on flat surfaces. Distilled water and nitrogen gas were used for the wetting and nonwetting phases, respectively. As with the experiments of Fourar et al. (1993),

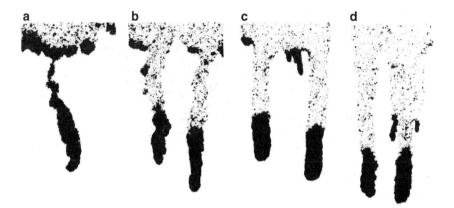

FIGURE 8-7 Fingers and drops from laboratory experiments. From Nicholl, M. J., R. J. Glass, and S. W. Wheatcraft, 1994. Gravity-driven infiltration instability in initially dry nonhorizontal fractures. Water Resources Research 30: 2533-2546. Copyright by American Geophysical Union.

these experiments involved gas injection into initially water-filled fractures. Despite the constant rate injections of gas and water, the pressure did not achieve steady values as pressures and flow oscillated with intermittent water blockage and clearing of critical pathway throats in the fracture.

Kneafsey and Pruess (1998) simulated thermally driven two-phase flow effects. Their experiments used natural fractures, epoxy casts of natural fractures, flat glass, and textured glass. The glass fractures were prepared both with and without sandblasting to enhance surface roughness. Pentane was used as the liquid phase. Although the major experimental aim was to demonstrate thermal effects, such as heat pipes, the work did record a variety of flow structures and regimes. The three types of flow were continuous rivulets, intermittent rivulets and drops, and films. Continuous rivulets occurred on the rapid introduction of the pentane or when there was sufficient liquid supply to support continuous flow. Intermittent flow occurred from quasi-stable, saturated islands. When mass was added to an island by flow or condensation, it produced an intermittent rivulet. Drops are described as extreme forms of rivulet, where liquid would accumulate at the top of a wider-aperture section of the fractures and release a drop when the mass became sufficient to overcome the capillary resistance. In some cases, drops fell freely without touching the fracture walls. Film flows occurred when the aperture of the fracture was larger than the film thickness and may have also transported fluid to the saturated islands.

Su et al. (1999) further document intermittent flow effects in rough-walled fractures using experimental conditions more closely related to vadose percolation than those of Kneafsey and Pruess (1998). These experiments used epoxy casts of natural fractures and water with a dye to aid visualization. The contact angle of water on the epoxy was 63°, which was similar to values observed for drops on smooth granite surfaces. The experiments varied the angle of inclination of the fracture. The fractures were dry prior to the introduction of the water under constant rate conditions. The fracture replica tests produced fingers of liquid rather than a uniform wetting front. The fingers broke off to form capillary islands, which were connected to the water source and to one another by thin rivulets or threads (Figure 8-8). The seepage velocities of these systems were of the order of 0.02 cm/s. Su et al. (1999) note that current analytical and numerical models of flow in unsaturated porous media do not predict this type of intermittent flow.

Su et al. (1999) followed up their fracture-cast experiments with further tests on glass parallel plates. The plates were arranged to form multiple constant-

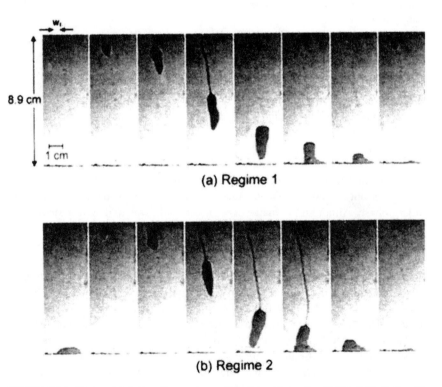

FIGURE 8-8 Drop or blob flow between parallel plates. From Su, G. W., J. T. Geller, K. Pruess, and F. Wen, 1999. Experimental studies of water seepage and intermittent flow in unsaturated, rough-walled fractures. Water Resources Research 35: 1019-1037. Copyright by American Geological Union.

aperture regions. These experiments defined three regimes under conditions of constant-rate flow:

• lower flows where islands break off from the water source and move down the fracture;
• intermediate flows where the islands remain connected to the source by threads until they reach the downstream end of the fracture, where they break off; and
• higher flow rates where the source and the sink maintained a continuous thread.

In summary, experimental work on both fractures and coarse porous media indicate conditions of discontinuous flow in the form of drops or rivulets. This behavior is not surprising given the finite contact angles of the liquid-surface systems that are used for the experiments. Another manifestation of drop flow modes is oscillations in outlet flows that occur when drops are exiting the flow system.

The occurrence of drop and rivulet flows may be an artifact of using smooth surfaces for experiments. Where experiments used casts of natural fractures, it is not clear whether or not the process reproduces the micro-roughness in addition to the macro-roughness of the visible asperities. One observation in reviewing experiments is the importance of measuring and reporting the contact angles of the fluids and the surfaces that are used.

FRACTURE NETWORK GEOMETRY AND VADOSE FLOW

As mentioned in the introduction, the occurrence of fast flow in fractures in the presence of strong matrix imbibition is a major paradox in unsaturated fracture flow systems. Pruess (1999) attempted to reconcile these observations by limiting the surface area for fracture and matrix interactions. Pruess (1999) suggested ways to limit surface area: "(1) reduction of fracture wall area available for imbibition due to spatial localization of flow, (2) reduction of time available for imbibition due to the episodic nature of seepage, and (3) reduction of fracture matrix flow due to the presence of mineral coatings of low permeability on the fractures walls." In our discussions below, we will look mainly at the localization of flow in fracture networks for reducing matrix-fracture contact.

Driving Forces and Fracture Dip

An understanding of fracture network effects in vadose flow begins with the effect of fracture dip on the driving forces. The driving force for vadose fracture flow is gravity, and gravity acts in the direction of the dip of conducting fractures. The influence of gravitational acceleration varies with the sine of the dip angle.

Within the interior of a fracture, the direction of flow should be coincident with the fracture dip angle. A key issue for flow direction involves how effectively asperities and variability of surface properties can affect flow direction. If asperities occur as random islands of contact, they may serve as locations of storage, but would not affect the flow direction in the fracture. However, an asperity pattern that creates linear ridges could act as a "wick" and divert flow off the fracture dip direction. Such linear asperity zones could arise from either mechanical processes, such as arrest lines formed during tensile propagation, or geochemical processes that preferentially deposit or dissolve material along channels or ridges. Variability of surface wettability can also affect flow direction. Variations in surface materials or roughness could create regions with lower wetting resistance, and thus create pathways that are not strictly in the direction of fracture dip.

The edges of fractures may have special properties and processes compared with the rest of the fracture. The edge of the fracture forms a boundary, which will have a lower dip inclination than the rest of the fracture. However, an edge of a fracture likely has a smaller aperture and higher capillary suction, hence a continuous water film may occupy the fracture edge and be a path for preferential flow.

If we assume that the driving forces for flow are dominantly gravitational, and flow is in the dip direction of fractures, we can develop several mechanisms for flow localization arising from the geometries of fracture networks. These include flow diversion, flow focusing, and flow connectivity.

Flow Diversion

A common assumption of vadose zone flow is the assumption of vertical hydraulic gradient. When flow is confined to fracture planes, this assumption is correct only if the fractures are vertical. If the fractures are not vertical the hydraulic gradient will be reduced by the sine of the fracture dip. Furthermore, if the fractures have a dominant dip direction, the dominant flow direction may be skewed away from vertical toward the direction of fracture dip.

Dominant dip directions may arise in bedded rocks such as sedimentary or volcanic rocks where there are fracture sets that follow the weakness along the bedding planes. Exfoliation joints or other topographically controlled fracture systems could serve a similar purpose. The overall effect of a preferred dip direction would be a diversion of the flow from vertical toward the dip direction. Figure 8-9 illustrates this effect.

Flow Focusing

Flow focusing is another potential consequence of fracture network geometry. Pruess (1999) presented an extensive discussion of focusing effects, where

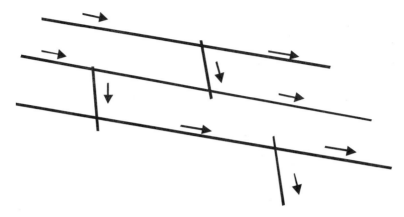

FIGURE 8-9 Flow diversion by fractures with strongly preferred orientations.

he divided focusing mechanisms into "external" and "internal." External focusing is the localization of flow on the external boundaries of the flow system. This may involve localization at surface drainages or at the surface exposures of fractures.

Internal flow focusing is controlled by the fracture network geometry. Pruess (1999) showed how barriers to flow can focus the flow to small portions of the fracture system. When a flow encounters a barrier, the flow is diverted to the down-dip edge of the barrier. The flow stream from this point is considerably more focused than it was upstream of the barrier. Numerical simulations using a Richards' equation-based flow simulator show that these focused flows do not spread as gravitational forces overwhelm dispersive effects. The sorts of features that may serve as barriers include zones of fracture closure due to mineralization or stress. Fracture intersections in sparsely fractured rock also may serve as focusing agents (Figure 8-10).

Two other conceptual models involving focused flow in fracture networks deserve mention. The first is the "weeps" model (Gauthier et al., 1992), which assumes fracture flow in local regions of saturation. A form of weeps model, conditioned to the fracture-matrix interaction effects in numerical models, was developed by Ho and Wilson (1998). The second involves numerical methods to validate the active fracture area in dual-permeability models (Liu et al., 1998).

Gauthier et al. (1992) developed a simplified, "weeps" approach to vadose fracture flow for the Yucca Mountain project. They viewed fracture flow as occurring in locations of local fracture saturation, or weeps, distributing the flow based on the parallel plate approximations for fracture transmissivity. The basis of the method is not fundamental flow processes, but rather a mass balance approach. The model uses the assumed infiltration rate to the fractured rock, and determines the number of saturated fractures that will accommodate the assumed

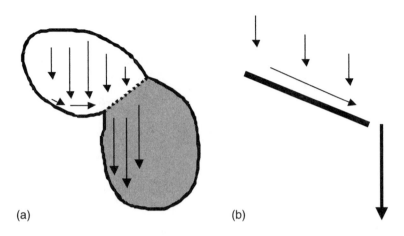

(a)   (b)

FIGURE 8-10 Flow focusing: (a) focussing at intersection; (b) diversion at barrier. After
Pruess (1999).

rate. The model also assumes little or no matrix interaction based on the rationale
that fracture coatings impede matrix-fracture interaction. Gauthier (1994) extended the model to allow weeps of different sizes. The weeps approach was
specifically designed to assess the number and distribution of inflow points to a
repository. As such, it attempts to be a bounding calculation. For systems with
low infiltration rates, as in arid environments, the weeps model provides insights
into how a few fractures may be sufficient to accommodate the entire flux. For
example, an infiltration rate of 15 mm/year over a 1-km^2 area could be accommodated theoretically by a single, 1-m-long fracture with an aperture of slightly less
than a millimeter.

Ho and Wilson (1998) combined the "weeps" approach with the results of
dual-permeability modeling to arrive at the wetted and active portions of a fracture network at Yucca Mountain. They started with the fracture-matrix reduction
factor that comes out of a dual-permeability model that had been calibrated to
matrix saturation and geochemical data. The fracture-matrix reduction factor is
the portion or the percentage of the total fracture surface area that is available for
fracture-matrix mass transfers. They then assumed that the product of the total
fracture area and the fracture-matrix reduction factor was the active flowing area
of the fractures. Using values of fracture spacing obtained by fracture mapping to
obtain the total fracture area, they applied the fracture-matrix interaction term to
get weep spacings. The flow rate per weep is the total percolation flux (L/T)
multiplied by the area associated with each weep. The approach used several
assumptions including homogeneous, vertical fractures with constant spacing.

Liu et al. (1998) developed another method for estimating the portion of the fracture system that is actively flowing water. Their conceptual model assumes that the active portion of the connected fracture space, f_a, is a number between 0 and 1. This active portion is assumed to be related to the effective fracture saturation by a power of γ, which is a positive constant related to the network properties. The saturation is "effective" in that it is scaled to make the residual saturation zero:

$$f_a = \left(\frac{S_f - S_r}{1 - S_r} \right)^\gamma . \qquad (8.13)$$

Liu goes on to derive van Genuchten relationships for capillary pressures and relative permeability in terms of γ. The γ term cannot be measured directly. It becomes a parameter that comes out of an inverse modeling procedure. This procedure adjusts parameter values using forward models that successively iterate until it obtains optimized estimates of parameter values. In the case of Liu et al. (1998), the inversion of fracture properties primarily is conditioned to reproduce measured matrix saturation and water potential data. Using data from Yucca Mountain, Liu et al. (1998) constrained their inversion using fracture trace maps and fracture permeability data from pneumatic tests.

Connectivity Considerations

Perhaps the most significant distinction of fracture networks from porous continua is the question of connectivity. Fractures force fluids to flow on discrete pathways, which may be highly discontinuous in sparsely fractured rock masses. In order to have a fracture pathway from the surface to the water table, the fracture pathway must have continuity. Within sparse fracture networks, fracture pathways from the surface may dead end, and could potentially host local perched water bodies (Stephens, 1996). Conversely, paths that connect with the water table may not have connection to water sources at the surface, in which case, the fracture pathway will be preferentially drained. Variable connectivity of fracture networks, therefore, is a potential agent for focusing flow.

SUMMARY AND CONCLUSIONS

The observation of fast flow in vadose fracture networks encourages a re-examination of the capillary model of vadose-zone flow for fractures. One important aspect of capillary flow models is the assumption that flow ceases when the saturation drops to the point where the liquid no longer forms a continuous phase. This paper develops an alternative conceptual model involving the flow of drops and to a lesser extent rivulets. The existence of drops implies a nonzero contact angle, which is counter to a common, though not universal, assumption in vadose flow studies.

Measurements of contact angles on rock materials suggest that contact angles with water are greater than zero. Such a wetting relationship encourages the formation of drops and rivulets, whose flow physics is fundamentally different from conventional hydrogeologic models. While gravity drives flow in both conventional and drop flows, for drops and rivulets on open fractures, the resistance arises from wetting phenomena in addition to viscosity. Drops are static unless their mass provides sufficient energy to overcome wetting resistances.

Drops may occur as either single-wall or two-wall drops, if both fracture walls are contacted. Single-wall drops are more likely in larger fractures, and they will be more likely to slide because contact with the second wall increases resistance to flow. Single-wall drops are more likely in fractures with apertures greater than 1 mm.

Laboratory experiments on simulated fractures show drop and rivulet effects. These effects may arise from the use of modeling materials that have nonzero contact angles with the water. A critical question in the use of laboratory results is whether or not these artificial surfaces simulate real fractures. Real fractures may have significant micro-roughness that can reduce the apparent contact angle to zero; thus changing the flow process from drops to films. The current understanding of wetting properties of fractures is incomplete, but essential if appropriate processes are to be identified and used in simulations and analyses.

Fracture network geometry affects vadose-zone fracture flow in several ways. First, gravity encourages flow in the dip direction of the fracture. Gravitational effects are dip-angle dependent varying by the sine of the dip angle. A directional consequence of this phenomenon is that preferred orientations of fractures can create diversion effects that direct flow in vertical directions.

Fracture network geometries serve to focus flow to a portion of the fracture network that is small. Barriers within fractures, fracture edges, and fracture intersections all confine flow to portions of networks. Fracture networks may be discontinuous. Discontinuous pathways that are preferentially connected to the surface will tend to pond or act as perched zones. Discontinuous paths that preferentially connect to the water table will drain. Fast-path flow will be enhanced by the presence of continuous, large-aperture pathways.

REFERENCES

Bachman, J., R. Horton, R. van der Pleog, and S. Woche, 2000. Modified sessile drop method for assessing initial soil-water contact angle of sandy loam. Soil Science Society of America Journal 64: 564-567.

Berg, J. C., 1993. Role of acid-base interactions in wetting and related phenomena. In J. Berg, ed. Wettability. New York: Decker, pp. 75-148.

Bikerman, J. J., 1950. Sliding of drops from surfaces of different roughnesses. Journal of Colloidal Science 5: 349-359.

Carillo, M. L. K., J. Letey, and S. Yates, 2000. Measurement of initial soil-water contact angle of water repellent soils. Soil Science Society of America Journal 63: 433-436.

Dussan V. E. B., 1985. On the ability of drops of bubbles to stick to non-horizontal surfaces of solids. Part 2. Small drops or bubbles having contact angles of arbitrary size. Journal of Fluid Mechanics 151: 1-20.

Dussan V. E. B., and R. T. Chow, 1983. On the ability of drops or bubbles to stick to non-horizontal surfaces of solids. Journal of Fluid Mechanics 137: 1-29.

Extrand, C. W., and Y. Kumagai, 1995. Liquid drops on an inclined plane: The relationship of contact angles, drop shape, and retentive force. Journal of Colloid and Interface Science 170: 515-521.

Fourar, M., S. Bories, R. Lenormand, and P. Persoff, 1993. Two-phase flow in smooth and rough fractures: Measurement and correlation by porous-medium and pipe flow models. Water Resources Research 29: 3699-3708.

Furmidge, C. G. L., 1962. Studies at phase interfaces. 1. The sliding of liquid drops on solid surfaces and a theory for spray retention. Journal of Colloidal Science 17: 309-324.

Gauthier, J. H., M. L. Wilson, and F. C. Lauffer, 1992. Estimating the consequences of significant fracture flow at Yucca Mountain. High Level Radioactive Waste Management, Proceedings of the Third Annual International Conference, American Nuclear Society, pp. 891-898.

Gauthier, J. H., 1994. An updated fracture flow model for total-system performance assessment of Yucca Mountain. High Level Radioactive Waste Management, Proceedings of the Fifth Annual International Conference, American Nuclear Society, pp. 1663-1670.

Glass, R., M. J. Nicholl, and V. C. Tidwell, 1995. Challenging models for flow in unsaturated rock through exploration of small-scale processes. Geophysical Research Letters 22: 1457-1460.

Hazlett, R. D., 1993. On surface roughness effects in wetting phenomena. In: K. L. Mittal, ed. Contact Angle, Wettability, and Cohesion. VSP Utrecht, Netherlands, pp. 173-181.

Hiemenz, P. C., and R. Rajagopalan, 1997. Principles of Colloid and Surface Chemistry. 3rd ed. Marcel Dekker, Inc., New York, 650 p.

Hillel, D., 1980. Fundamentals of Soil Physics. Academic Press, New York, 413 p.

Ho, C. K., and M. Wilson, 1998. Calculation of discrete fracture flow paths in dual continuum models. High Level Radioactive Waste Management, Proceedings of the Seventh International Conference, American Nuclear Society, pp. 375-377.

Huppert, H. E., 1982. Flow and instability of a viscous current down a slope. Nature 300: 427-429.

Johnson, M. F. G., R. A. Schluter, M. J. Miksis, and S. G. Bankoff, 1999. Experimental study of rivulet formation on an inclined plate by fluorescent imaging. Journal of Fluid Mechanics 394: 339-354.

Jury, W. A., W. R. Gardner, and W. H. Gardner, 1991. Soil Physics. 5th ed. Wiley, New York.

Kneafsey, T. J., and K. Pruess, 1998. Laboratory experiments on heat-driven two-phase flows in natural and artificial rock fractures. Water Resources Research 34: 3349-3367.

Lenormand, R., and C. Zarcone, 1989. Capillary fingering: Percolation and fractal dimension. Transport in Porous Media 4: 52-61.

Letey, J., J. Osvorn, and R. Pelishek, 1962. Measurement of liquid-solid contact angles in soil and sand. Soil Science 93: 149-153.

Liu, H. H., C. Doughty, and G. Bodvarsson, 1998. An active fracture model for unsaturated flow and transport in fractured rocks. Water Resources Research 34: 2633-2646.

Nicholl, M. J., R. J. Glass, and S. W. Wheatcraft, 1994. Gravity-driven infiltration instability in initially dry nonhorizontal fractures. Water Resources Research 30: 2533-2546.

Padday, J. F., 1993. Spreading, wetting, and contact angles. In K. L. Mittal, ed. Contact Angle, Wettability, and Cohesion. VSP Utrecht, Netherlands, pp. 97-108.

Pal, D., and S. B. Varade, 1971. Measurement of contact angle of water in soils and sand. Journal of Indian Society of Soil Science 19: 339-446.

Persoff, P., and K. Pruess, 1995. Two-phase flow visualization and relative permeability measurement in natural rough-walled rock fractures. Water Resources Research 31: 1175-1186.

Peters, R. R., and E. A. Klavetter, 1988. A continuum model for water movement in an unsaturated fractured rock mass. Water Resources Research 24: 416-430.

Prázak, J., M. Sír, F. Kubik, J. Tywoniak, and C. Zarcone, 1992. Oscillation phenomena in gravity-driven drainage in coarse porous media. Water Resources Research 28: 1849-1855.

Pruess, K, 1999. A mechanistic model for water seepage through thick unsaturated zones in fractured rocks of low matrix permeability. Water Resources Research 35: 1039-1051.

Pruess, K., B. Faybishenko, and G. Bodvarsson, 1999. Alternative concepts and approaches for modeling flow and transport in thick unsaturated zones of fractured rocks. Journal of Contaminant Hydrology 38: 281-322.

Rasmussen, T.C., 1991. Steady fluid flow and travel times in partially saturated fractures using a discrete air-water interface. Water Resources Research 27: 66-77.

Richards, L. A., 1931. Capillary conduction of liquids through porous mediums. Physics 1: 318-333.

Sadhal, S. S., P. S. Ayyaswamy, and J. N. Chung, 1997. Transport phenomena with drops and bubbles. Springer Verlag, Berlin.

Sklodowska, A., M. Wozniak, and R. Matlakowska, 1999. The method of contact angle measurements and estimation of work of adhesion in bioleaching of metals. Biological Procedures Online 1(3): (www.science.uwaterloo.ca/ bpo/).

Sobolev, V., N. Churaev, M. Velarge, Z. Zorin. 2000. Surface tension and dynamic contact angle of water in quartz capillaries. Journal of Colloid and Interface Science 222: 51-54.

Stephens, D., 1996. Vadose Zone Hydrology. CRC Press, Boca Raton.

Su, G. W., J. T. Geller, K. Pruess, and F. Wen, 1999. Experimental studies of water seepage and intermittent flow in unsaturated, rough-walled fractures. Water Resources Research 35: 1019-1037.

Tokunaga, T. K., and J. Wan, 1997. Water film flow along fracture surfaces in porous rock. Water Resources Research 33: 1287-1295.

Tokunaga, T. K., J. Wan, and S. Sutton, 2000. Transient film flow on rough fracture surfaces. Water Resources Research 36: 1737-1746.

Wan, J., T. K. Tokunaga, T. Orr, J. O'Neill, and R. W. Conners, 2000. Glass casts of rock fracture surfaces: A new tool for studying flow in a partially saturated, fractured, porous medium. Water Resources Research 36: 355-360.

Wang, J. S., and T. N. Narasimhan, 1985. Hydrologic mechanisms governing fluid flow in saturated, fractured porous media. Water Resources Research 21: 1861-1874.

Young, T., 1805. On the cohesion of fluids. Philos. Transactions Royal Society of London A84.

Yuan, T. L., and L. Hammond, 1968. Evaluation of available methods for soil wettability measurement with particular reference to soil-water contact angle determination. Soil and Crop Science Society of Florida Proc. 28: 56-63.

9

Investigating Flow and Transport in the Fractured Vadose Zone Using Environmental Tracers

Fred M. Phillips[1]

ABSTRACT

Environmental tracers have been applied to a variety of settings that are closely related to fractured vadose zone hydrology, such as structured agricultural soils, unstable wetting fronts, macropores in watersheds, and caves. Results from these tracer applications can give insight into hydrological processes in more traditionally defined fractured vadose zones. The most generally useful tracers appear to be the stable isotopes of oxygen and hydrogen in the water molecule, tritium, halides, and chlorine-36, but there are a wide variety of potentially applicable environmental tracers. Environmental tracers can provide information on the integrated response of actual hydrological systems to real-world boundary conditions over long time scales. Environmental tracers should be considered a primary means of investigating fractured vadose zones in all studies where understanding the integrated system response is an important objective.

INTRODUCTION

In the late 1960s the issue of groundwater contamination, along with other environmental problems, began to receive worldwide attention. One example of this concern was the Chalk aquifer in southern England, a resource of particular national importance inasmuch as it provides approximately 15 percent of the

---

[1] Department of Earth and Environmental Science, New Mexico Tech, Socorro

country's water supply (Foster, 1975). Several lines of evidence indicated that this aquifer was particularly vulnerable to pollution. Extensive nitrate contamination had been detected under farmed areas. Bacterial contamination had also been detected, even under areas with thick vadose zones. Also, it was well known that areas underlain by chalk usually generated little or no runoff, even after intense rain. This evidence was interpreted to indicate that pollutants could be rapidly transported to the water table through interconnected channels in the heavily fractured chalk.

The United Kingdom Atomic Energy Authority decided to take advantage of the pulse of elevated tritium then raining out after the recent series of United States and Soviet atmospheric nuclear weapons tests in order to better understand solute transport through the Chalk vadose zone. The results of Smith et al. (1970) were surprisingly at variance with the previous understanding. The measured Chalk vadose zone profiles showed that the tritium peak had penetrated only about 5 m in the past decade (Figure 9-1), indicating slow, relatively uniform infiltration rates between 0.5 and 1 m yr^{-1}. This apparent rate of transport was supported by the results of follow-up studies. Instead of elucidating the contaminant transport problem, the new data only created confusion. How could bacteria be advected to the water table in a matter of hours or days, but tritium take 25 years?

The insight of Foster (1975) solved the dilemma within a few years. Downward transport was almost entirely within the pervasive open fractures. The matrix of the Chalk had very high microporosity, but very low permeability, and hence transport (largely lateral) was mainly by diffusion. The transient pulse of tritium was rapidly diffused into the matrix blocks and retained, while bacteria were too large to enter the matrix pores. The nitrate input was relatively steady and predated that of the tritium by many years, and hence was close to equilibrium with the matrix, permitting unimpeded transport to the water table. This conceptual model was borne out by subsequent studies (Barker and Foster, 1981; Foster and Smith-Carrington, 1980).

The Chalk tritium anomaly was one of the earliest studies in which the importance of matrix diffusion was recognized. It was entirely independent of the better-known discovery of the same phenomenon while assessing fractured, saturated rock as a potential setting for high-level nuclear waste repositories (Neretnieks, 1980). It illustrates well the power of environmental tracers for assessing solute transport in fractured rock, as well as the dangers of inadequate conceptual models causing misinterpretation of the data.

SCOPE

This chapter will focus on the application of environmental tracers to understanding the hydrology of fractured vadose zones. Environmental tracers are defined as solutes that are introduced into the hydrological cycle by either natural

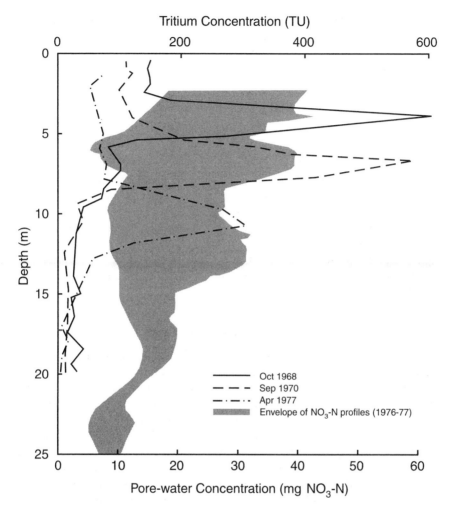

FIGURE 9-1 Profiles of tritium and nitrate concentration with depth, measured at sites in the Chalk of southern England. Tritium is a transient pulse moving downward, whereas nitrate is close to steady state. Data from Foster and Smith-Carrington (1980) and Barker and Foster (1981).

processes or as an inadvertent result of human activities. Deliberately introduced tracers are of great utility for hydrological investigations in both the laboratory and the field, but these are most commonly employed in conjunction with hydraulic testing and are best considered as a component of such tests. There exists a large body of literature on the use of environmental tracers in nonfractured vadose zones, but a comprehensive review of this topic is beyond the scope of this volume. Two fractured vadose zone sites, Apache Leap and Yucca Mountain, have undergone extensive environmental tracer investigations, but these are treated individually elsewhere in this volume and hence will not be described in detail in this chapter.

COMPARATIVE STUDIES

Environmental tracers have received comparatively little attention in the hydrological literature as a means of understanding flow and transport in fractured rock. For example, the recent National Research Council report *Rock Fractures and Fluid Flow: Contemporary Understanding and Applications* (National Research Council, 1996) contained within 500 pages of text only 3 pages dealing with environmental tracers, and these described the results of a single case study. Environmental tracers were not mentioned as a possible avenue for better understanding of fractured-rock hydrology. In contrast, such tracers have been extensively used for related studies in other disciplines. It is worthwhile to examine what techniques or findings can be transferred to the fractured vadose-zone problem.

Structured Agricultural Soils

Farmers have been conducting environmental tracer tests in the vadose zone on a grand scale for many years. A variety of agricultural chemicals of environmental concern are routinely applied to fields and then allowed to infiltrate, under either irrigation or natural precipitation. These chemicals have commonly been observed to migrate to the groundwater at rates much faster than would be predicted by uniform one-dimensional flow (Jury and Flühler, 1992). Numerous field tracer investigations (Bowman and Rice, 1986; Flury et al., 1994; Jaques et al., 1998; Johnston et al., 1998; Jørgensen et al., 1998) have revealed that this fast transport can often be attributed to preferential flow along high-permeability pathways, commonly referred to as "structured soils" in agriculture.

Agricultural soils are typically at the low end of the scale in terms of permeability contrast between fast pathways and matrix. In most studies there has been little attempt to sample the fast pathways directly; instead, high-spatial-resolution sampling of the subsoil is employed and inferences about pathways are drawn based on the measured breakthrough curves. In some cases identification of the fast pathways may be attempted by adding dyes and excavating after the experiment is complete.

One significant advantage of the agricultural soil experiments is that the shallow depths and loose nature of the soil permit much more detailed sampling than is usually possible for fractured consolidated rock. The relatively low permeability contrasts and high matrix permeability permit detection of preferential flow by sampling the matrix, rather than needing to directly sample fracture flow, which is much more difficult. One important result of the agricultural soil studies is that, when sampled at this kind of resolution, nearly all soils show significant indications of preferential flow. Easily visible macropores or other fast paths are not required to produce preferential flow. The breakthrough curves observed in these tracer studies are typically highly skewed with very long tails (Figure 9-2). Continuum models usually adequately reproduce these curves (e.g., Bowman and Rice, 1986; Jaques et al., 1998).

Unstable Wetting Fronts

When large amounts of water infiltrate into dry soils, unstable wetting fronts may develop if the soil has water-repellent surfaces or if the surficial layer is coarse (Glass et al., 1989; Hendrickx et al., 1993). The wetting front changes from a relatively uniform, planar shape to one with elongated fingers that drain the high water-content layer at the soil surface. The flow through such fingers is some-what analogous to fracture flow. The fingers may, under some circumstances, be persistent, in the sense that during repeated infiltration and drying cycles the

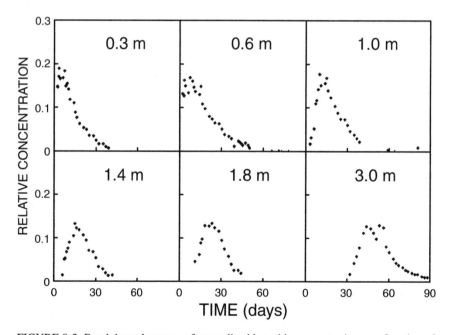

FIGURE 9-2 Breakthrough curves of normalized bromide concentration as a function of time, for different depths, under a semiweekly irrigation regime. Data are from Bowman and Rice (1986). The strongly skewed and tailing breakthrough curves are characteristic of solute transport in strongly structured soils.

same finger paths will be reoccupied (Glass et al., 1989; Ritsema et al., 1998). In real soils there is undoubtedly a continuum between true unstable wetting front propagation (which can be observed even in a completely homogeneous soil) and preferential flow, which requires a high-permeability pathway.

Most studies of unstable wetting fronts in soils have measured water content rather than employing tracers because the contrast between the low initial water content of the soils (which is required for the fingers to form) and the high water content after the fingers have penetrated makes this measurement simple, as does the ease of excavation of the soil. However, anionic tracers have been applied to a limited extent to verify solute transport predictions in the presence of unstable wetting (Van Dam et al., 1990). The extent to which wetting fronts will become unstable in consolidated rocks is at present uncertain, but it does seem likely that under conditions of high water application rates to rock surfaces, preferential paths will develop and may tend to be reoccupied during subsequent wettings. Given the difficulty of excavating consolidated rock and of measuring water content within such rock, tracers may provide an attractive alternative for evaluating related flow effects in more typical fractured rocks.

Watershed Hydrology

Environmental tracers have provided critical evidence for the importance of fracture-analogous flow in the generation of runoff from watersheds. At one time the sudden response in streamflow that closely follows precipitation events was thought to result largely from overland flow of the rainfall. Studies employing $\delta^{18}O$ and δ^2H in the water molecule, and chloride, demonstrated that in many cases a large fraction of the storm runoff was not from the current precipitation event, but was in fact water that had been stored in the subsurface system for some period of time (McDonnell et al., 1991; Rodhe, 1981; Sklash, 1990; Sklash et al., 1976). Rapid infiltration of precipitation through macropores causes a pressure response of the water table that results in an increase in stream discharge through groundwater discharge into the streambed. This type of lumped system response is surely a close analog to the behavior of many fractured vadose zone/ aquifer systems.

Numerous watershed studies have used environmental tracers to directly study preferential flow phenomena (Hammermeister, 1982; Mulholland et al., 1990; Leany et al., 1993). One of the most striking examples of flow partitioning between matrix and fast paths has been provided by Newman et al. (1998). In this study of a hillslope in a ponderosa pine forest in New Mexico, flow interceptors were emplaced on top of the B horizon (dominated by clay) and the C horizon (bedrock), and the interflow generated from these horizons was compared with matrix water sampled by coring. The matrix water in the B horizon had chloride concentrations in the range of 200 to >300 mg L^{-1}, while water collected in the interceptors during the same period had concentrations ranging from 2 to 30 mg

L^{-1}. The very high matrix chloride concentrations resulted from transpiration of matrix water during prolonged periods of low precipitation. The dramatic contrast in concentrations indicates that the macropores were acting almost independently of the matrix. However, after high moisture contents were maintained for more than a week, the chloride concentration of the macropore flow began to rise as solute exchange between the pores and matrix became appreciable (Figure 9-3). This response nicely illustrates the highly transient nature of the matrix/macropore interaction that also characterizes more typical fractured rock in the vadose zone.

Cave Hydrology

It is the desire of every vadose fractured rock hydrologist to have a tunnel drilled beneath the research site in order to actually see and sample in detail the

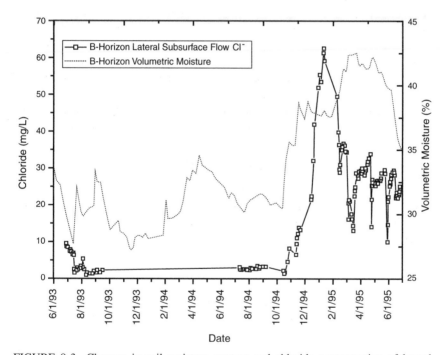

FIGURE 9-3 Changes in soil moisture content and chloride concentration of lateral subsurface flow from the B horizon with time in a semiarid hillslope in New Mexico, from Newman, B. D., A. R. Campbell, and B. P. Wilcox, 1998. Lateral subsurface flow pathways in a semiarid ponderosa pine hillslope. Water Resources Research 34: 3485-3496. Copyright by American Geophysical Union. The increase in chloride concentration beginning in November 1994 illustrates increased exchange between macropores and matrix as moisture content rises.

fractured vadose zone. This has been accomplished (generally at great expense) at a few sites, notably Yucca Mountain. However, an extensive natural network of such tunnels already exists, and their hydrology has been investigated for many years. They are called caves.

Sampling drips into caves provides a direct means of accessing flow passing through fractured and solutioned vadose zones (Even et al., 1986; Harmon, 1979; Ingraham et al., 1990; Yonge et al., 1985). Much effort has been directed toward tracing the transit time from the land surface to the cave depth. These times vary greatly: two weeks for a cave in Kentucky (Harmon, 1979), several months for caves in England (Atkinson et al., 1985), decades at Carlsbad Caverns, New Mexico (Chapman et al., 1992). Both stable isotopes and tritium have proved useful tracers in these systems. Some cave drips show the isotopic signature of nearly every precipitation event, others show no temporal variation at all, indicating either very long flow paths along which infiltrating waters are homogenized by mixing, or large water volumes stored in the matrix (Chapman et al., 1992; Yonge et al., 1985). Transit times and compositions can vary markedly between fractures separated by only a short distance (Even et al., 1986).

Cave seeps represent an underutilized resource for understanding the hydrology of fractured vadose zones. To date, results of these studies have been treated largely as aids to the paleoenvironmental interpretation of cave precipitates, and in relation to the hydrology of particular localities. A more comprehensive and systematic investigation of environmental tracers in cave seeps could add greatly to our understanding of flow and transport in fractured vadose zones.

SUITABLE TRACERS

A very wide variety of environmental tracers have been applied to hydrological problems. Some of these are particularly suitable for work in the fractured vadose zone and others are not. The following tracers appear to be among the most suitable.

Stable Isotopes of Oxygen and Hydrogen in the Water Molecule

Deuterium (^2H) and oxygen-18 (^{18}O) are stable isotopes that are incorporated into the water molecule itself. They can thus be considered in some senses to be the "perfect tracer" for water movement. In natural, near-surface hydrological systems these heavy isotopes are fractionated relative to the more abundant light forms of their elements mainly by two influences: evaporation and variations in the isotopic composition of precipitation. Evaporation enriches the heavy isotopes. Fractionation in precipitation depends mainly on temperature (Dansgaard, 1964), and thus carries a seasonal signature. Precipitation falling in the summer tends to be isotopically heavy, that falling in the winter light. These fractionations can be applied to various types of tracing, such as determining transit times from the surface to a point at depth by comparing time trends of composition with the seasonal precipitation cycle (Saxena, 1984), or distinguishing long-residence matrix water from recent fracture water by the heavy evaporation signature in the former. For example, Thoma et al. (1979) were able to discern the annual cycle in δ^2H of recharge through a sand dune in Bordeaux down to 25 m depth. In addition, because temperature decreases with increasing elevation, the lighter isotopic composition of higher-altitude precipitation can be used to help determine the source of recharge (Scholl et al., 1996). The application of stable isotopes to hydrological problems has been reviewed by IAEA (1981) and Coplen et al. (1999).

Tritium

Tritium (^3H) is an unstable isotope of hydrogen with a half-life of 12.45 years. It is naturally produced at fairly low levels in the atmosphere by cosmic radiation. A very large amount of tritium, illustrated in Figure 9-4, was also

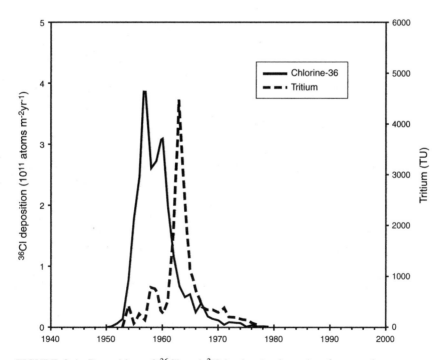

FIGURE 9-4   Deposition of ^{36}Cl and ^3H in the Arctic region from nuclear weapons testing. Tritium fallout (not decay-corrected) was reconstructed by Östlund (1982), and ^{36}Cl was measured in Greenland ice by Synal et al. (1990).

released into the atmosphere during the atmospheric thermonuclear weapons testing period (1954-1964). Radioactive decay, rainout, and dilution have by now reduced this pulse down nearly to prebomb levels. During the atmospheric testing period, and for about two decades subsequent to it, this pulse served as a very useful transient tracer (IAEA, 1983c). At present, however, radioactive decay and mixing in the subsurface have reduced the tritium levels of even peak-fallout-period infiltration to the point where it is often difficult to distinguish such water from more recent infiltration. This ambiguity will only increase as the bomb pulse decays.

In saturated zone settings this problem can be remedied by measuring the concentrations of both ^3H and its decay product, ^3He (Cook and Solomon, 1997). The tritium content at the time of recharge can be inferred from the added concentrations of parent and daughter, and the time since recharge from the ratio of the two. This technique is not generally applicable to fractured vadose zones, however, since the ^3He can readily escape by volatilization. As a result, the utility of tritium as a vadose zone tracer is likely to decline in the future.

Tritium is presently most useful in vadose zones where residence times are long. In such settings, relatively high tritium concentrations may be indicative of fast flow paths (e.g., fractures), whereas low tritium indicates matrix water. In this case it may not be necessary to distinguish the bomb-pulse tritium; the knowledge that measurable tritium corresponds to residence times shorter than 50 years and at- or below-detection tritium to longer residence times may be sufficient. In short-residence-time vadose zones (shallow ones or those with uniformly high permeability and large recharge rates), the tritium content is likely to be spatially uniform and yield little information on flow paths.

Halides

Chloride, bromide, and iodide share similar geochemical characteristics in hydrologic systems. In the reduced, anionic form (i.e., Cl^-, Br^-, and I^-) they do not tend to form insoluble minerals and are repelled from negatively charged mineral surfaces, and hence act conservatively in groundwater and soil water. They tend to be left behind when water is removed by evaporation or transpiration, and thus increases in concentration can help to indicate the proportion of water that has infiltrated relative to that lost to the atmosphere (Allison et al., 1994). This evapotranspirative concentration can thus also "tag" long-residence-time matrix water, compared to freshly infiltrated fracture or macropore water (Newman et al., 1998; Scanlon, 1992).

Cl^-/Br^- ratios can serve to distinguish atmospheric salts carried by surface infiltration from subsurface salts (Davis et al., 1998). This information can be of importance when the fractured rock matrix consists of high-chloride lithologies such as rhyolite or basalt. Shifts in the Cl/Br ratio of porewater may then provide another qualitative indicator of flow path and residence time. In addition, Cl/Br ratios may be useful to infer the nature of the infiltration process. Because bromide is somewhat less conservative than chloride due to a greater propensity for reaction with plants and other organic matter, the Cl/Br ratio of infiltrating water tends to increase with residence time in the soil zone (Davis et al., 1998). Iodine is even less conservative than bromine because of its stronger affinity for organic matter.

Chlorine-36

Chlorine-36 is a radioactive isotope of Cl with a half-life of 301 ka (Bentley et al., 1986). It is produced at relatively high levels in the atmosphere by cosmic ray reactions on Ar, and in the subsurface at much lower levels from reaction of stable Cl with neutrons derived from U and Th decay. The atmospheric deposition of ^{36}Cl has fluctuated at several time scales. Variations at the $\sim 10^4$ year scale have been observed by Plummer et al. (1997) and Baumgartner et al. (1998). A major pulse of ^{36}Cl was released during the late 1950s by atmospheric nuclear weapons testing (Elmore et al., 1982; Zerle et al., 1997), shown in Figure 9-4. The systematics of ^{36}Cl applications to hydrology have recently been reviewed by Phillips (1999).

The most straightforward application of ^{36}Cl to fractured-rock hydrology is mapping the depth of penetration of the bomb-pulse ^{36}Cl. Due to decay of the bomb-pulse ^3H, this radionuclide is now in many cases the most easily distinguishable signal in the vadose zone (Bentley et al., 1982). The most notable use of this tracer thus far in fractured vadose zones has been at Yucca Mountain (Fabryka-Martin et al., 1997). As with tritium, ^{36}Cl is most applicable to long-residence-time vadose zones. Cook and Walker (1996) have noted that in arid soils where the bomb-pulse ^{36}Cl has not penetrated below the root zone, its depth cannot be used to infer the groundwater recharge rate. Nevertheless, this information on depth of penetration may be very important for assessing potential contaminant transport.

Other Tracers

Many tracers that have proved very useful in saturated-zone studies over short time scales are not generally applicable to studying the liquid phase of vadose-zone problems because they are too volatile. These include ^{85}Kr and ^3He in the $^3H/^3He$ pair (Cook and Solomon, 1997) and chlorofluorocarbons, refrigerant compounds that have built up in the atmosphere over the past four decades (Busenberg and Plummer, 1991). These volatile tracers may either be lost from vadose-zone water, or diffuse into the water, through the gas phase in the vadose zone. Carbon-14 is another common hydrological tracer that is too volatile for vadose-zone applications. [Note that, although these volatile tracers are not generally applicable to tracing liquid water movement in the vadose zone, they may be useful for studying gas-phase behavior (Thorstenson et al., 1998)].

A wide range of chemical compounds may form suitable tracers, depending on circumstances. Anionic chemicals are particularly useful because they usually do not interact strongly with the solid phase. Nitrate has been frequently applied in this way because it is a common agricultural contaminant (Barker and Foster, 1981; Johnston et al., 1998). The stable isotopes of many elements can also be used as tracers. Among the most commonly employed are isotopes of carbon (Bar-Matthews et al., 1996), sulfur (IAEA, 1983a), nitrogen (Böhlke and Denver, 1995), and boron (Leenhouts et al., 1998). Strontium isotopes have been applied to reconstructing ancient flow regimes through the fractured vadose zone at Yucca Mountain (Johnson and DePaolo, 1994). Although the principles are well established, the application of these isotopic techniques to problems of fractured vadose zone hydrology has been very limited.

LIMITATIONS AND ADVANTAGES

Environmental tracers are particularly suitable for answering some kinds of questions, but not others. These are briefly summarized below.

Limitations

• *Investigating perturbed flow systems.* Environmental tracers typically yield information on flow systems under fluctuating natural boundary conditions. They are thus not optimal for investigating situations where the circumstances vary significantly from the natural system. For example, environmental tracers would probably give little information about the infiltration of gasoline beneath a leaking storage tank. Similarly, studies of environmental tracers beneath an undisturbed fractured desert vadose zone would be of minimal value in predicting the migration of herbicides beneath a golf course established in the same area.

• *Detailed mechanistic studies.* Many studies are directed toward establishing a fundamental mechanistic understanding of flow and transport (e.g., relation between fracture roughness and flow rate, or fracture coatings and matrix imbibition). One motivation for this type of study is that numerical models are frequently constructed using such mechanistic equations as the building blocks. Environmental tracers are generally of limited value for such investigations because the tracer signals are the result of the combined influence of a wide range of environmental variables that include the mechanistic flow and transport effects, but that are integrated over a range of spatial scales and also include effects of fluctuations in the environmental boundary conditions. For this type of goal, either laboratory experiments or field experiments in which a portion of the system can be isolated and stressed in a controlled fashion are generally preferable alternatives.

• *Natural limitations.* In some circumstances, suitable environmental tracer signals may not be present at detectable levels in the system, or may be too subtle to be of much use. In other cases, the difficulty of obtaining high quality samples for environmental tracers in fractured vadose zones may make the cost prohibitive. In these situations, artificial tracer experiments will often be the best alternative.

Advantages

• *Integration of natural processes.* In many cases the desired goal is to predict flow and transport through fractured vadose zones under conditions approximating the natural situation. This prediction is usually attempted using numerical models that have been developed and calibrated using a limited number of short-term observations and artificial hydraulic and tracer tests. In this circumstance, the extent to which a model assembled from such components can successfully predict the behavior of an integrated system under fluctuating natural boundary conditions is frequently uncertain. Testing against environmental isotope data that represent an integrated response of the system can serve to greatly increase confidence in the modeling results.

• *Long temporal or spatial scales.* Predictions over long time scales are frequently desired, particularly for waste disposal problems. The validity of extrapolating model simulations far into the future is often questioned. Environmental tracers can often provide information on solute transport over time scales that are much longer than would ever be feasible for artificial experiments. Analogously, tracers can provide direct information on transport times through large hydrological systems where obtaining an adequate database of model parameter values is difficult.

• *Minimal disturbance of natural systems.* In some cases, extensive drilling and hydraulic and tracer testing cannot be performed, either to prevent compromising sites that are supposed to be "leak tight," or to avoid damaging environmentally sensitive areas. In such cases it may still be possible to obtain samples from natural sources that can be analyzed for environmental tracers.

The bottom line is that environmental tracers can provide a form of "ground truth" with regard to the natural behavior of complex fractured flow systems that is not obtainable in any other way. Unless the system is under conditions that are far from natural, or no useful environmental tracer can be identified, environmental tracers should always be considered a primary investigation technique.

EXAMPLES

Flow Beneath Mont-Blanc

Determining the pattern of flow within a large mountain is a formidable task, and it is difficult to imagine how it could be accomplished using only conventional physical hydrology methods. However, Fontes et al. (1978) were able to obtain a good understanding of the flow regime beneath Mount-Blanc in the French/Italian Alps merely by sampling water outflows into the 12-km-long tunnel beneath the mountain and analyzing them for environmental tracers. Part of their results is shown in Figure 9-5. The close correspondence between the profile of $\delta^{18}O$ and the topographic profile of the mountain above the tunnel is good evidence for nearly vertical downward flow (the lighter isotope values under the peaks reflect colder condensation temperatures with higher altitude). The data preclude arguments that permanent freezing conditions at the highest altitudes do not permit recharge. The anomalous interval of heavy $\delta^{18}O$ between

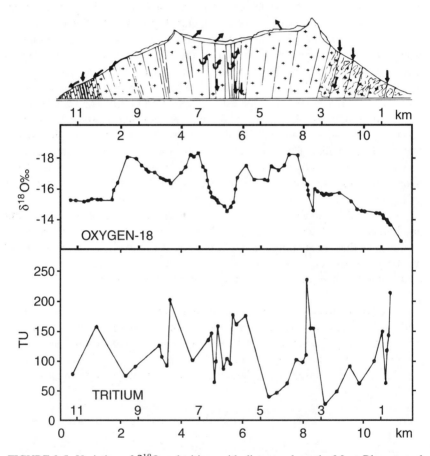

FIGURE 9-5 Variation of $\delta^{18}O$ and tritium with distance along the Mont-Blanc tunnel, from Fontes et al. (1978). Note the correspondence between overlying topography and $\delta^{18}O$ values, but lack of correlation with tritium. Reproduced with permission from the International Atomic Energy Agency.

5 and 6 km was shown to be due to evaporation of the water during transit through the fracture system, demonstrating that the fractures are not in a fully saturated state. The variable, but frequently high tritium values indicate that there is little interconnection between fractures (and that they do not draw from any common reservoir), and that while some fractures are transmitting infiltration to the tunnel level in a matter of months, others take more than a decade. These conclusions are similar to those obtained by Rauert and Stichler (1974) from a tracer investigation of the 7-km-long Tuxer Hauptkamm tunnel in Austria, except that the tritium concentrations were very low throughout the central section of that tunnel, indicating uniformly slower rates of downward flow.

Hydraulics of Fractured Tuff in New Mexico

Los Alamos National Laboratory recently drilled several boreholes through the Bandelier Tuff at the laboratory site. The fractured rhyolite tuff caps long, fingerlike mesas that project eastward from the flanks of the Jemez Mountains. Deep canyons bound the mesas, and the water table is over 100 m below the land surface. Vertical profiles of stable-isotope composition and chloride concentration were measured with depth and are illustrated in Figure 9-6. The strongly enriched $\delta^{18}O$ and very high chloride concentrations in the top 30 m are striking. They are accompanied by very low water contents in the rhyolite matrix. Newman and Birdsell (1997) attributed this profile to geothermally induced convection of air through the mesas. Air moves inward through outcrops of highly permeable basal-surge pumice deposits that are exposed on the canyon walls. Geothermal heating in the centers of the mesas causes the air to rise through fractures that pervade the low-permeability tuff, carrying water vapor with it. Oxygen-18 and chloride are concentrated in the residual pore water. Geothermally stimulated convection of air is undoubtedly a common phenomenon in fractured vadose zones (Weeks, 1987), and this example illustrates how environmental tracers can identify circumstances in which it exerts a profound influence on vadose-zone hydraulics and porewater chemistry.

Tension Fractures in Desert Vadose Zones

In general, the hydraulics of vadose zones of the American Southwest do not appear to be strongly affected by the presence of vertical fractures or other vertical macropores. One exception to this is narrow vertical fractures, sometimes extending as much as several hundred meters deep and many kilometers in length, that spontaneously appear in piedmont deposits or basin fills. In some cases these fissures seem to be a response to groundwater pumping, but in others the cause is not apparent (Larson and Péwé, 1986). Although these fissures are initially only a few millimeters wide, diversion of surface flows can erode the upper portions into chasms over 10 m in both width and depth.

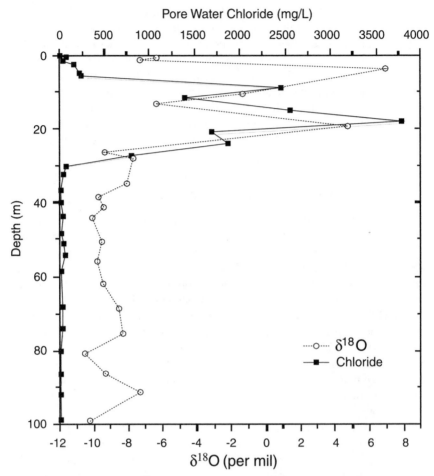

FIGURE 9-6 Variation of $\delta^{18}O$ and chloride concentration with depth in fractured rhyolite tuff at Los Alamos, New Mexico (Newman and Birdsell, 1997; B. Newman, Los Alamos National Laboratory, personal communication, 1999). Enriched portion of profile above 30 m overlies a pyroclastic unit that allows air circulation through the fractured vadose zone.

Dramatic erosion of the fissure tops is evidence that large amounts of water have been diverted from ephemeral surface drainages or overland flow to depth. How does the drainage of water into the fissure affect the subsurface hydrology? Scanlon (1992) and Scanlon et al. (1997) have addressed this issue by measurement of chloride, ^{36}Cl, and stable isotopes of oxygen and hydrogen (as well as physical hydrologic parameters) in the vicinity of such fissures. Results are illustrated in Figures 9-7 and 9-8. Variations in water content were small and difficult

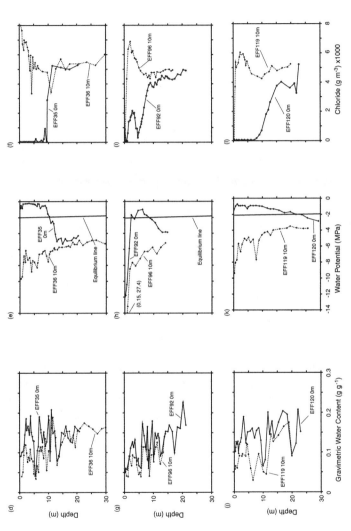

FIGURE 9-7 Profiles of water content, soil-water potential, and chloride concentration in boreholes adjacent to an earth fissure at Eagle Flat, Texas. From Journal of Hydrology, 203, Scanlon, B. R., R. S. Goldsmith, and J. G. Paine, Analysis of focused unsaturated flow beneath fissures in the Chihuahuan Desert, Texas, USA. Pp. 58-78, 1997, with permission from Elsevier Science. Lengths indicated on the figures ("0 m" and "10 m") refer to distances of the boreholes from the fissure. All boreholes immediately adjacent to the fissure (0 m) show low chloride concentrations and less negative soil-water potentials than at 10 m, due to interception of runoff by the fissure.

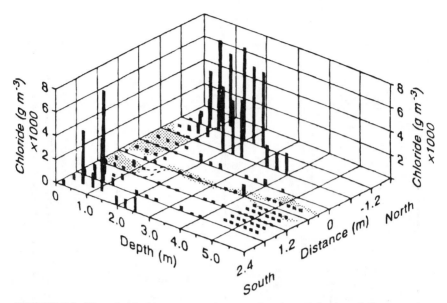

FIGURE 9-8 Plot of chloride concentration as a function of depth and distance away from an earth fissure at Eagle Flat, Texas. From Journal of Hydrology, 203, Scanlon, B. R., R. S. Goldsmith, and J. G. Paine, Analysis of focused unsaturated flow beneath fissures in the Chihuahuan Desert, Texas, USA. Pp. 58-78, 1997, with permission from Elsevier Science. Low chloride concentrations adjacent to the fissure delineate the zone of leaching by runoff captured by the fissure. The configuration of the fissure is indicated by the stippled pattern.

to interpret. Chloride concentration measurements, however, showed a pronounced leached zone surrounding the fissures. The zone was typically 5 to 10 m in width and extended downward 10 to 20 m. This pattern shows that there is a strong tendency for imbibition of the fracture water into the soil matrix. The rate of imbibition is apparently rapid enough that most fracture flow is depleted within 20 m of the surface, a significant finding with respect to contaminant transport and aquifer recharge.

Flow Through Unsaturated Fractured Rhyolite
at Yucca Mountain, Nevada

Yucca Mountain is at present the nation's only site under investigation for its suitability as a potential high-level nuclear waste repository. An 8-km-long U-shaped tunnel called the Exploratory Studies Facility (ESF) has been drilled beneath the mountain to provide access to the interior of the mountain for hydrogeologic and engineering research. The thickness of rhyolite tuff over the tunnel is approximately 300 m, although it is thinner close to the portals. The rhyolite consists of interbedded dense welded tuff and less welded pyroclastic deposits. Fractures in the welded units were recognized as potential fast paths for flow from the surface. The more porous and less cohesive pyroclastic strata were anticipated to provide a capillary barrier to fast-path flow.

Such hypotheses are very difficult to test at the mountain scale using conventional methods. Measurement of dissolved ^{36}Cl, however, can provide a tracer for infiltration on time scales ranging from decades to millennia. Rock samples were collected during drilling of the tunnel (no liquid water seeps were found). Sample points were located both by systematic collection at 200-m intervals and by "feature-based" sampling of observed faults and fractures (Fabryka-Martin et al., 1997, 1998). Many samples showed $^{36}Cl/Cl$ ratios that were elevated above the contemporary value of $\sim500 \times 10^{-15}$ (Figure 9-9). Most of these, however, were below the limit of $\sim1,200 \times 10^{-15}$ that has been demonstrated by measurements on

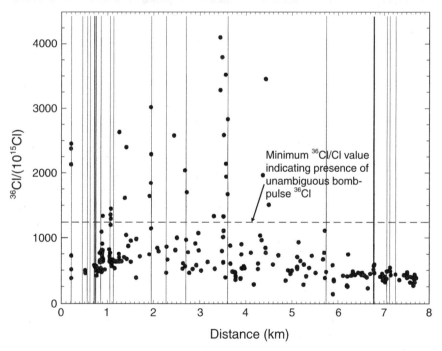

FIGURE 9-9 Distribution of $^{36}Cl/Cl$ ratios as a function of distance along the Exploratory Studies Facility tunnel beneath Yucca Mountain (Fabryka-Martin et al., 1998. Copyright 1998 by the American Nuclear Society, La Grange Park, Illinois.). Vertical lines illustrate the position of faults within the tunnel. The current meteoric $^{36}Cl/Cl$ ratio is $\sim500 \times 10^{-15}$. Values higher than $1,200 \times 10^{-15}$ indicate presence of bomb-pulse ^{36}Cl that infiltrated from the surface in the past 50 years.

fossil rat urine to be the atmospheric deposition value during the late Pleistocene (Plummer et al., 1997). All of the systematic samples were below $1,200 \times 10^{-15}$. Nine locations, however, contained ratios above $1,200 \times 10^{-15}$, and as high as $4,000 \times 10^{-15}$, that are most probably ascribed to infiltration of bomb-fallout ^{36}Cl. All except two of these were associated with known faults.

These results indicate a remarkable variability in vertical transport rates. Infiltration of the ^{36}Cl in the samples with ratios between 500×10^{-15} and $1,200 \times 10^{-15}$ was probably prior to ~15,000 years ago. Infiltration of the ^{36}Cl in the samples with ratios greater than $1,200 \times 10^{-15}$ was probably less than 50 years ago. The fast paths are almost uniformly associated with faults, indicating that the displacements have ruptured the less-welded units and greatly increased their fracture connectivity and vertical permeability. The presence of presumably Pleistocene-age ^{36}Cl in fractured, but not faulted, sections of the ESF indicates that the less-welded units greatly retard downward fracture transport where the pyroclastics are not faulted.

This environmental tracer study has yielded critically important information on the hydrology of this fractured vadose zone. It illustrates well the ability of environmental tracers to provide information at a wide range of time scales, under fluctuating natural boundary conditions. The results should be of great utility in either evaluating or designing a high-level nuclear waste repository at Yucca Mountain.

SUMMARY

Environmental tracers have been applied for many years to a variety of hydrologically related flow-and-transport problems. Many of these are variants on the fractured-vadose setting, including structured agricultural soils, unstable wetting fronts, macropores in watersheds, and caves. Actual applications to classical fractured vadose zones have not been numerous in the published literature, but this situation is more likely to reflect a general under-investigation of this environment rather than a lack of suitable tracers. The most generally applicable tracers would appear to be the stable isotopes of O and H in the water molecule, tritium, halides, and ^{36}Cl.

The major advantage of environmental tracers is that they provide information on the integrated response of actual hydrological systems to real-world boundary conditions, over long time scales. They are thus particularly useful for problems where long-time or large-scale responses are important. They are also useful for formulating conceptual models of fractured vadose zones, calibrating models, and building confidence in them. They are less useful for detailed mechanistic investigations aimed at understanding the response of small portions of the fractured vadose system to carefully controlled stresses.

Environmental tracers should be considered a primary means of investigating fractured vadose zones in all studies where understanding the integrated system response is an important objective. Fractured vadose zones are complex hydrological environments where understanding and predicting flow and transport is inherently a very difficult task. Environmental tracers must be incorporated along with mechanistic studies and numerical modeling in order to produce results that can be applied and interpreted with confidence.

REFERENCES

Allison, G. B., G. W. Gee, and S. W. Tyler, 1994. Vadose-zone techniques for estimating groundwater recharge in arid and semiarid regions. Soil Sci. Soc. Am. J. 58: 6-14.

Atkinson, T. C., J. W. Hess, and R. S. Harmon, 1985. Stable isotope variations in recharge to a karst aquifer, Yorkshire Dales, England: Preliminary work. Ann. Soc. Géol. Belg 108: 225.

Bar-Matthews, M., A. Ayalon, A. Matthews, E. Sass, and L. Halicz, 1996. Carbon and oxygen isotope study of the active carbonate-water system in a karstic Mediterranean cave: Implications for paleoclimate research in semiarid regions. Geochim. Cosmochim. Acta 60: 337-347.

Barker, J. A., and S. S. D. Foster, 1981. A diffusion-exchange model for solute movement in fissured porous rock. Quart. J. Eng. Geol. (London) 1: 17-24.

Baumgartner, S., J. Beer, J. Masarik, G. Wagner, L. Meynadier, and H.-A. Synal, 1998. Geomagnetic modulation of the ^{36}Cl flux in the GRIP ice core, Greenland. Science 279: 1330-1332.

Bentley, H. W., F. M. Phillips, S. N. Davis, S. Gifford, D. Elmore, L. E. Tubbs, H. E. Gove, 1982. Thermonuclear ^{36}Cl pulse in natural water. Nature 300: 737-740.

Bentley, H. W., F. M. Phillips, and S. N. Davis, 1986. Chlorine-36 in the terrestrial environment. In: Handbook of Environmental Isotope Geochemistry, Vol. 2B (P. Fritz and J.-C. Fontes, eds.). Elsevier, Amsterdam, pp. 427-480.

Böhlke, J. K., and J. M. Denver, 1995. Combined use of groundwater dating, chemical, and isotopic analyses to resolve the history and fate of nitrate contamination in two agricultural watersheds, Atlantic coastal plain, Maryland. Water Resour. Res. 31: 2319-2340.

Bowman, R. S. and R. C. Rice, 1986. Transport of conservative tracers in the field under intermittent flood irrigation: Water Resour. Res. 22: 1531-1536.

Busenberg, E., and L. N. Plummer, 1991. Chlorofluorocarbons (CCl_3F and CCl_2F_2): Use as an age dating tool and hydrologic tracer in shallow ground-water systems. U.S. Geol. Surv., Water Resour. Invest. Rept. 91-4034.

Chapman, J. B., N. L. Ingraham, and J. W. Hess, 1992. Isotopic investigation of infiltration and unsaturated flow processes at Carlsbad Cavern, New Mexico. J. Hydrol. 133: 343-363.

Cook, P. G., and D. K. Solomon, 1997. Recent advances in dating young groundwater: Chlorofluorocarbons, $^3H/^3He$ and ^{85}Kr. J. Hydrol. 191: 245-265.

Cook, P. G., and G. Walker, 1996. Evaluation of the use of ^3H and ^{36}Cl to estimate groundwater recharge in arid and semi-arid environments. Symposium on Isotopes in Water Resources Management, Vienna, International Atomic Energy Agency, pp. 397-403.

Coplen, T. B., A. L. Herczeg, and C. Barnes, 1999. Isotope engineering: Using stable isotopes of the water molecule to solve practical problems. In: Environmental Tracers in Subsurface Hydrology (P. Cook and A. L. Herczeg, eds.). Kluwer Academic, Boston, pp. 79-110.

Dansgaard, W., 1964. Stable isotopes in precipitation. Tellus 16: 436-468.

Davis, S. N., D. O. Whittemore, and J. Fabryka-Martin, 1998. Uses of chloride/bromide ratios in studies of potable water. Ground Water 36: 338-350.

Elmore, D., L. E. Tubbs, D. Newman, X. Z. Ma, R. Finkel, K. Nishiizumi, J. Beer, H. Oeschger, and M. Andree, 1982. The ^{36}Cl bomb pulse measured in a shallow ice core from Dye 3, Greenland. Nature 300: 735-737.

Even, H., I. Carmi, M. Magaritz, and R. Gerson, 1986. Timing the transport of water through the upper vadose zone in a karstic system above a cave in Israel. Earth Surf. Processes Landforms 11: 181-191.

Fabryka-Martin, J. T., A. V. Wolfsberg, P. R. Dixon, S. S. Levy, J. A. Musgrave, and H. J. Turin, 1997. Summary report of chlorine-36 studies: Sampling, analysis, and simulation of chlorine-36 in the Exploratory Studies Facility. Los Alamos National Laboratory, LA-13352-MS, Los Alamos.

Fabryka-Martin J. T., A. V. Wolfsberg, S. S. Levy, J. L. Roach, S. T. Winters, L. E. Wolfsberg, D. Elmore and P. Sharma, 1998. Distribution of fast hydrologic paths in the unsaturated zone at Yucca Mountain. 8th Annual International High-Level Radioactive Waste Management Conference, Las Vegas, Nevada. American Nuclear Society, La Grange Park, Illinois, pp. 93-96.

Flury, M., H. Flühler, W. A. Jury, and J. Leuenberger, 1994. Susceptibility of soils to preferential flow of water: A field study. Water Resour. Res. 30: 1945-1954.

Fontes, J.-C., G. C. Bortolami, and G. M. Zuppi, 1978. Hydrologie isotopique du massif du Mont-Blanc: In: Isotope Hydrology (Proc. Symp., Neuherberg, 19-23 June). International Atomic Energy Agency, Vienna, Vol. 1, pp. 411-440.

Foster, S. S. D., 1975. The chalk groundwater tritium anomaly: A possible explanation: J. Hydrol. 25: 159-165.

Foster, S. S. D., and A. Smith-Carrington, 1980. The interpretation of tritium in the chalk unsaturated zone: J. Hydrol. 46: 343-364.

Glass, R. J., T. S. Steenhuis, and J.-Y. Parlange, 1989. Mechanism for finger persistence in homogeneous, unsaturated porous media: Theory and verification. Soil Sci. 148: 60-70.

Hammermeister, D. P., 1982. Perched water tables on hillsides in western Oregon. II. Preferential downslope movement of water and anions. Water Resour. Res. 46: 819-826.

Harmon, R. S., 1979. An isotopic study of groundwater seepage in the central Kentucky karst. Water Resour. Res. 15: 476-480.

Hendrickx, J. M. H., L. W. Dekker, and O. H. Boersma, 1993. Unstable wetting fronts in water repellent field soils. J. Env. Qual. 22: 109-118.

IAEA (International Atomic Energy Agency), 1981. Stable isotope hydrology: Deuterium and oxygen-18 in the water cycle. Technical Report Series No. 210, Vienna, 337 pp.

IAEA (International Atomic Energy Agency), 1983a. Aqueous sulfur. Ch. 8.4 in: Isotope techniques in the hydrogeological assessment of potential sites for the disposal of high level radioactive wastes. Technical Report Series No. 228, Vienna, pp. 76-86.

IAEA (International Atomic Energy Agency), 1983b. Isotope techniques in the hydrogeological assessment of potential sites for the disposal of high level radioactive wastes: IAEA Technical Report Series No. 228, 151 pp.

IAEA (International Atomic Energy Agency), 1983c. Tritium. Ch. 7 in: Isotope techniques in the hydrogeological assessment of potential sites for the disposal of high level radioactive wastes. Technical Report Series No. 228, Vienna, pp. 57-61.

Ingraham, N. L., J. B. Chapman, and J. W. Hess, 1990. Stable isotopes in cave pool systems: Carlsbad Caverns, New Mexico. Chem. Geol. (Isot. Geosci. Sect.) 86: 65-74.

Jaques, D., D. J. Kim, J. Diels, J. Vanderborght, H. Vereecken, and J. Feyen, 1998. Analysis of steady-state chloride transport through two heterogenous field soils. Water Resour. Res. 34: 2539-2550.

Johnson, T. M., and D. J. DePaolo, 1994. Interpretation of isotopic data in groundwater-rock systems: Model development and application to Sr isotope data from Yucca Mountain. Water Resour. Res. 30: 1571-1588.

Johnston, C. T., P. G. Cook, S. K. Frape, L. N. Plummer, E. Busenberg, and R. J. Blackport, 1998. Ground water age and nitrate distribution within a glacial aquifer beneath a thick unsaturated zone. Ground Water 36: 171-180.

Jørgensen, P. R., L. D. McKay, and N. H. Spliid, 1998. Evaluation of chloride and pesticide transport in a fractured clayey till using large undisturbed columns and numerical modeling. Water Resour. Res. 34: 539-554.

Jury, W. A., and H. Flühler, 1992. Transport of chemicals through soil: Mechanisms, models, and field applications. Adv. Agron. 47: 141-201.

Larson, M. K., and T. L. Péwé, 1986. Origin of land subsidence and earth fissuring, northeast Phoenix, Arizona. Assoc. Eng. Geol. Bull. 23: 139-165.

Leany, F. W. J., K. R. J. Smettem, and D. J. Chittleborough, 1993. Estimating the contribution of preferential flow to subsurface runoff from a hillslope using deuterium and chloride. J. Hydrol. 147: 83-103.

Leenhouts, J. M., R. L. Bassett, and T. I. Maddock, 1998. Utilization of intrinsic boron isotopes as co-migrating tracers for identifying potential nitrate contamination sources. Ground Water 36: 240.

McDonnell, J. M., M. K. Stewart, and I. F. Owens, 1991. Effect of catchment-scale subsurface mixing on stream isotopic response. Water Resour. Res. 27: 3065-3073.

Mulholland, P. J., G. V. Wilson, and P. M. Jardine, 1990. Hydrogeological response of a forested watershed to storms: Effects of preferential flow along shallow and deep pathways. Water Resour. Res. 26: 3021-3036.

National Research Council, 1996. Rock Fractures and Fluid Flow: Contemporary Understanding and Applications. National Academy Press, Washington, D.C., 551 pp.

Neretnieks, I., 1980. Diffusion in the rock matrix: An important factor in radionuclide retardation. J. Geophys. Res. 85: 4379-4397.

Newman, B. D., and K. H. Birdsell, 1997. Characterization of deep evaporation in mesas at Los Alamos, New Mexico, using environmental tracers and numerical modeling. Eos 78(46):F318.

Newman, B. D., A. R. Campbell, and B. P. Wilcox, 1998. Lateral subsurface flow pathways in a semiarid ponderosa pine hillslope. Water Resour. Res. 34: 3485-3496.

Östlund, H. G., 1982. The residence time of the freshwater component in the Arctic Ocean. J. Geophys. Res. 87: 2035-2043.

Phillips, F. M., 1999. Chlorine-36. In: Environmental Tracers in Subsurface Hydrology (P. G. Cook and A. L. Herczeg, eds.). Kluwer Academic, Boston, pp. 299-348.

Plummer, M. A., F. M. Phillips, J. Fabryka-Martin, H. J. Turin, P. E. Wigand, and P. Sharma, 1997. Chlorine-36 in fossil rat urine: An archive of cosmogenic nuclide deposition over the past 40,000 years. Science 277: 538-541.

Rauert, W., and W. Stichler, 1974. Groundwater investigations with environmental isotopes. In: Isotope Techniques in Groundwater Hydrology. International Atomic Energy Agency, Vienna, Vol. 2, pp. 431-443.

Ritsema, C. J., L. W. Dekker, J. L. Nieber, and T. S. Steenhuis, 1998. Modeling and field evidence of finger formation and finger recurrence in a water repellent sandy soil. Water Resour. Res. 34: 555-567.

Rodhe, A., 1981. Spring flood: Meltwater or groundwater? Nordic Hydrol. 12: 21-30.

Saxena, R. K., 1984. Seasonal variations of oxygen-18 in soil moisture and estimation of recharge in esker and moraine formations. Nordic Hydrol. 15: 235-242.

Scanlon, B. R., 1992. Moisture and solute flux along preferred pathways characterized by fissured sediments in desert soils. J. Cont. Hydrol. 10: 19-46.

Scanlon, B. R., R. S. Goldsmith, and J. G. Paine, 1997. Analysis of focused unsaturated flow beneath fissures in the Chihuahuan Desert, Texas, USA. J. Hydrol. 203: 58-78.

Scholl, M. A., S. E. Ingebritsen, C. J. Janik, and J. P. Kauahikaua, 1996. Use of precipitation and groundwater isotopes to interpret regional hydrology on a tropical volcanic island: Kilauea volcano area, Hawaii. Water Resour. Res. 32: 3525-3538.

Sklash, M. G., 1990. Environmental isotope studies of storm and snowmelt runoff generation. In: Process Studies in Hillslope Hydrology (M. G. Anderson and T. P. Burt, eds.). John Wiley, New York, pp. 401-435.

Sklash, M. G., R. N. Farvolden, and P. Fritz, 1976. A conceptual model of watershed response to rainfall, developed through the use of oxygen-18 as a natural tracer. Can. J. Earth Sci. 13: 271-283.

Smith, D. B., P. L. Wearn, H. J. Richards, and P. C. Rowe, 1970. Water movement in the unsaturated zone of high and low permeability strata by measuring natural tritium. In: Isotope Hydrology 1970, International Atomic Energy Agency, Vienna, pp. 73-87.

Synal, H.-A., J. Beer, G. Bonani, M. Suter, and W. Wölfli, 1990. Atmospheric transport of bomb-produced ^{36}Cl. Nucl. Instrum. Meth. Phys. Res. B52: 483-488.

Thoma, G., N. Esser, C. Sonntag, W. Weiss, J. Rudolf, and P. Leveque, 1979. New technique of in-situ soil moisture sampling for environmental isotope analysis applied at Pilat sand dune near Bordeaux: In: Isotopes in Hydrology. International Atomic Energy Agency, Vienna, pp. 753-768.

Thorstenson, D. C., E. P. Weeks, H. Haas, E. Busenberg, L. N. Plummer, and C. A. Peters, 1998. Chemistry of unsaturated zone gases sampled in open boreholes at the crest of Yucca Mountain, Nevada. Data and basic concepts of chemical and physical processes in the mountain. Water Resour. Res. 34: 1507-1530.

Van Dam, J. C., J. M. H. Hendrickx, H. C. van Ommen, M. H. Bannik, M. T. van Genuchten, and L. W. Dekker, 1990. Water and solute movement in a coarse-textured water-repellent field soil. J. Hydrol. 120: 359-379.

Weeks, E. P., 1987. Effect of topography on gas flow in unsaturated fractured rock: Concepts and observations. In: Flow and Transport Through Unsaturated Fractured Rock. Geophysical Monograph Series, Volume 42 (D. D. Evans and T. J. Nicholson, eds.) American Geophysical Union, Washington, D.C., pp. 165-170.

Yonge, C. J., D. C. Ford, J. Gray, and H. P. Schwarcz, 1985. Stable isotope studies of cave seepage water. Chem. Geol. (Isot. Geosci. Sect.) 58: 97-105.

Zerle, L., T. Faestermann, K. Knie, G. Korschinek, and E. Nolte, 1997. The ^{41}Ca bomb pulse and atmospheric transport of radionuclides. J. Geophys. Res. 102: 19517-19527.

10

Lessons from Field Studies at the Apache Leap Research Site in Arizona

Shlomo P. Neuman,[1] Walter A. Illman,[2] Velimir V. Vesselinov,[3] Dick L. Thompson,[4] Guoliang Chen,[5] and Amado Guzman[6]

ABSTRACT

This paper summarizes lessons learned from single-hole and cross-hole pneumatic injection tests recently completed by The University of Arizona in unsaturated fractured tuffs at the Apache Leap Research Site (ALRS) near Superior, Arizona. The research was designed to investigate, test, and confirm methods and conceptual-mathematical models that can be used to determine the role of fractures and fracture zones in flow and transport through partially saturated porous rocks, with emphasis on the characterization of fracture connectivity, permeability, and porosity, and their dependence on location, direction, and scale. Over 270 single-hole tests have been conducted in six shallow vertical and inclined boreholes at the site by Guzman et al. (1996). These authors used a steady-state analysis to obtain permeability values for borehole test intervals of various lengths, based solely on late pressure data from each test. We summarize briefly the results of this earlier work and discuss more recent pressure and pressure-derivative type-curve interpretations, as well as numerical inverse analyses, of transient data from some of the single-hole tests. Our transient analyses of single-hole tests yield information about

[1] University of Arizona in Tucson
[2] CNWRA, Southwest Research Institute, San Antonio, Texas
[3] Geoanalysis Group, Los Alamos National Laboratory, New Mexico
[4] City of Tucson Water Department, Arizona
[5] Pinnacle West, Phoenix, Arizona
[6] Water Management Consultants, Tucson, Arizona

air permeability, air-filled porosity, skin factor, borehole storage, phenomenology, and dimensionality of the flow regime on a nominal scale of 1 m in the immediate vicinity of each test interval. Transient air permeabilities agree well with previously determined steady-state values, which, however, correlate poorly with fracture density data. We used the single-hole test results, together with borehole televiewer data, to help design and conduct 44 cross-hole pneumatic tests in 16 boreholes at the site (one test included 22 boreholes), including those used previously for single-hole testing. In each cross-hole test, air was injected at a constant mass flow rate into a relatively short borehole interval of length 1-2 m while monitoring (a) air pressure and temperature in the injection interval; (b) barometric pressure, air temperature, and relative humidity at the surface; and (c) air pressure and temperature in 13 short (0.5-2 m) and 24 longer (4-20 m) intervals within the injection and surrounding boreholes. We focus here on one of these cross-hole tests, labeled PP4. During this test, pressure responses were detected in 12 of the 13 short monitoring intervals and 20 of the 24 longer intervals. We used two methods to analyze the test results: a graphical matching procedure of data against newly developed pressure and pressure-derivative type-curves, and an automatic parameter estimation method based on a three-dimensional finite volume code (FEHM) coupled with an inverse code (PEST). The type-curve approach treats short and longer intervals as points or lines, depending on distance between injection and monitoring intervals. The type-curve approach accounts indirectly for storage effects in monitoring intervals due to the compressibility of air. The finite volume code allows representing borehole geometry and storage more realistically, and directly, by treating each borehole as a high-permeability cylinder of finite length and radius. Analyses of pressure data from individual monitoring intervals by the two methods, under the assumption that the rock acts as a uniform and isotropic fractured porous continuum, yield comparable results. These results include information about pneumatic connections between the injection and monitoring intervals, corresponding directional air permeabilities, and air-filled porosities. All of these quantities are found to vary considerably from one monitoring interval to another on scales ranging from a few meters to over 20 m. Together with the results of earlier site investigations, our single- and cross-hole test analyses reveal that, at the ALRS, (a) the pneumatic pressure behavior of fractured tuff is amenable to analysis by methods that treat the rock as a continuum on scales ranging from meters to tens of meters; (b) this continuum is representative primarily of interconnected fractures; (c) its pneumatic properties vary strongly with location, direction and scale—in particular, the mean of pneumatic permeabilities increases, and their variance decreases, with scale; (d) this scale effect is most probably due to the presence in the rock of various size fractures that are interconnected on a variety of scales; and (e) given a sufficiently large sample of spatially varying pneumatic rock properties on a given scale of measurement, these properties are amenable to analysis by geostatistical methods, which treat them as correlated random fields defined over a continuum.

INTRODUCTION

Issues associated with the site characterization of fractured rock terrains, the analysis of fluid flow and contaminant transport in such terrains, and the efficient handling of contaminated sites are typically very difficult to resolve. A major source of this difficulty is the complex nature of the subsurface "plumbing systems" of pores and fractures through which flow and transport in rocks take place. There is at present no well-established field methodology to characterize the fluid flow and contaminant transport properties of unsaturated fractured rocks. In order to characterize the ability of such rocks to conduct water, and to transport dissolved or suspended contaminants, one would ideally want to observe these phenomena directly by conducting controlled field hydraulic injection and tracer experiments within the rock. In order to characterize the ability of unsaturated fractured rocks to conduct nonaqueous-phase liquids such as chlorinated solvents, one would ideally want to observe the movement of such liquids under controlled conditions in the field. In practice, there are severe logistical obstacles to the injection of water into unsaturated geologic media, and logistical as well as regulatory obstacles to the injection of nonaqueous liquids. There also are important technical reasons why the injection of liquids, and dissolved or suspended tracers, into fractured rocks may not be the most feasible approach to site characterization when the rock is partially saturated with water. Injecting liquids and dissolved or suspended tracers into an unsaturated rock would cause them to move predominantly downward under the influence of gravity, and would therefore yield at best limited information about the ability of the rock to conduct liquids and chemical constituents in directions other than the vertical. It would further make it difficult to conduct more than a single test at any location because the injection of liquid modifies the ambient saturation of the rock, and the time required to recover ambient conditions may be exceedingly long.

Many of these limitations can be overcome by conducting field tests with gases rather than with liquids, and with gaseous tracers instead of chemicals dissolved in water. Experience with pneumatic injection and gaseous tracer experiments in fractured rocks is limited. Much of this experience has been accumulated in recent years by The University of Arizona at the Apache Leap Research Site (ALRS) near Superior, Arizona, and by the U.S. Geological Survey (USGS) near the ALRS (LeCain, 1995) and at Yucca Mountain in Nevada (LeCain, 1996; LeCain and Walker, 1996). To our knowledge, the earliest pneumatic injection tests were conducted by Boardman and Skrove (1966) to determine fracture permeability following a contained nuclear explosion. Their analysis was based on the steady-state, isothermal, radial flow equation for ideal gases. Other earlier work includes air injection tests conducted by Montazer (1982) in unsaturated fractured metamorphic rocks, and injection methods developed for fractured formations containing natural gas of the kind considered by Mishra et al. (1987).

This paper focuses on single- and cross-hole pneumatic injection tests conducted by our group at the ALRS. The site is situated near Superior in central Arizona. It consists of a cluster of 22 vertical and inclined (at 45°) boreholes that have been completed to a maximum vertical depth of 30 m within a layer of slightly welded unsaturated tuff. The boreholes span a surface area of 55 m by 35 m and a volume of rock on the order of 60,000 m^3. The upper 1.8 m of each borehole was cased, and a surface area of 1,500 m^2 was covered with a plastic sheet to minimize infiltration and evaporation. Core data and borehole television images are available for many of the boreholes.

Early work related to our area of study at the ALRS is described by Evans (1983), Schrauf and Evans (1984), Huang and Evans (1985), Green and Evans (1987), Rasmussen and Evans (1987, 1989, 1992), Tidwell et al. (1988), Yeh et al. (1988), Weber and Evans (1988), Chuang et al. (1990), Rasmussen et al. (1990, 1996), Evans and Rasmussen (1991), and Bassett et al. (1994). It included drilling 16 boreholes and conducting numerous field and laboratory investigations. Information about the location and geometry of fractures in the study area has been obtained from surface observations, the examination of oriented cores, and borehole televiewer records. Fracture density, defined by Rasmussen et al. (1990) as number of fractures per meter in a 3-m borehole interval, ranges from 0 to a maximum of 4.3 per meter. Though the fractures exhibit a wide range of inclinations and trends, most of them are near vertical, strike north-south, and dip steeply to the east. Surface fracture traces reveal a steeply dipping east-west set. An experimental study of aperture distributions in a large natural fracture at the ALRS was published by Vickers et al. (1992).

Single-hole pneumatic injection tests were conducted in 87 intervals of length 3 m in nine boreholes by Rasmussen et al. (1990, 1993). The tests were conducted by injecting air at a constant mass rate between two inflated packers while monitoring pressure within the injection interval. Pressure was said to have reached stable values within minutes in most test intervals, allowing the calculation of air permeability by means of steady-state formulae. Figure 5b of Rasmussen et al. (1993) suggests a good correlation ($r = 0.876$) between pneumatic and hydraulic permeabilities at the ALRS. Figure 10-1 shows a scatter plot of pneumatic permeability versus fracture density for 3-m borehole intervals based on the data of Rasmussen et al. (1990). It suggests a lack of correlation between fracture density and air permeability.

The single-hole tests of Rasmussen et al. (1990, 1993) were of relatively short duration and involved relatively long test intervals. Guzman et al. (1994, 1996) and Guzman and Neuman (1996) conducted a much larger number of single-hole pneumatic injection tests of considerably longer duration over shorter intervals in six boreholes. A total of 184 borehole segments were tested by setting the packers 1 m apart as shown in Figure 10-2. Additional tests were conducted in segments of lengths of 0.5, 2.0, and 3.0 m, bringing the total number of tests to over 270. The tests were conducted by maintaining a constant injection rate until

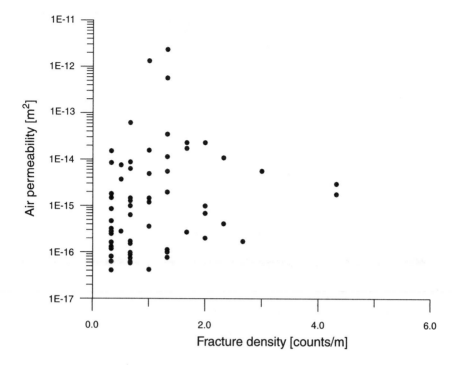

FIGURE 10-1 Air permeability versus fracture density. Data from Rasmussen et al., 1990.

air pressure became relatively stable and remained so for some time. The injection rate was then incremented by a constant value and the procedure repeated. In most tests, three or more such incremental steps were conducted in each borehole segment while recording the air injection rate, pressure, temperature, and relative humidity. For each relatively stable period of injection rate and pressure, air permeability was estimated by treating the rock around each test interval as a uniform, isotropic continuum within which air flows as a single phase under steady state, in a pressure field exhibiting prolate spheroidal symmetry.

The results of these steady-state interpretations of single-hole air injection tests are listed in Guzman et al. (1996). They reveal that:

1. air permeabilities determined in situ from steady-state single-hole test data are much higher than those determined on core samples of rock matrix in the laboratory, suggesting that the in situ permeabilities represent the properties of fractures at the site;

Locations of 1.0 m Air Permeability Measurements
(Circles do not indicate sphere of influence)

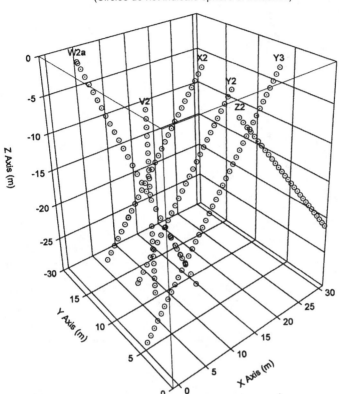

FIGURE 10-2 Perspective toward the northeast showing center locations of 1-m single-hole pneumatic test intervals; overlapping circles indicate retested locations. After Guzman et al. (1996).

2. it is generally not possible to distinguish between the permeabilities of individual fractures and the bulk permeability of the fractured rock in the immediate vicinity of a test interval by means of steady-state single-hole test data;

3. the time required for pressure in the injection interval to stabilize typically ranges from 30 to 60 min, increases with flow rate, and may at times exceed 24 h, suggesting that steady-state permeability values published in the literature for this and other sites, based on much shorter air injection tests, may not be entirely valid;

4. steady-state interpretation of single-hole injection tests, based on the assumption of radial flow, corresponds closely to prolate spheroidal in intervals of

length equal to or greater than 0.5 m in boreholes having a radius of 5 cm, as is the case at the ALRS;

5. pressure in the injection interval typically rises to a peak prior to stabilizing at a constant value, possibly due to a two-phase flow effect whereby water in the rock is displaced by air during injection;

6. in most test intervals, pneumatic permeabilities show a systematic increase with applied pressure as air apparently displaces water under two-phase flow;

7. in a few test intervals, intersected by widely open fractures, air permeabilities decrease with applied pressure due to apparent inertial effects;

8. air permeabilities exhibit a hysteretic variation with applied pressure;

9. the pressure-dependence of air permeability suggests that it is advisable to conduct single-hole air injection tests at several applied flow rates and/or pressures;

10. enhanced permeability due to slip flow [the Klinkenberg (1941) effect] appears to be of little relevance to the interpretation of single-hole air injection tests at the ALRS;

11. local-scale air permeabilities vary by orders of magnitude between test intervals across the site;

12. spatial variability is much greater than that due to applied pressure and lends itself to meaningful statistical and geostatistical analysis;

13. air permeabilities are poorly correlated with fracture densities, as is known to be the case for hydraulic conductivities at many water-saturated fractured rock sites worldwide (Neuman, 1987), providing further support for Neuman's conclusion that the permeability of fractured rocks cannot be reliably predicted from information about fracture geometry (density, trace lengths, orientations, apertures and their roughness) but must be determined directly by means of hydraulic and/or pneumatic tests; and

14. air permeabilities vary systematically with the scale of measurement as represented nominally by the distance between packers in an injection interval.

The work of Guzman et al. (1994, 1996) and Guzman and Neuman (1996) strongly suggests that air injection tests yield properties of the fracture system that are relevant to both unsaturated and saturated conditions. In particular, numerical simulations by these authors show that, whereas the intrinsic permeability one determines from such tests is generally lower than the intrinsic permeability to water of fractures that surround the test interval, it nevertheless approaches the latter as the applied pressure goes up. This is so because, under ambient conditions, capillary forces tend to draw water from fractures into the porous (matrix) blocks of rock between the fractures, thereby leaving the fractures saturated primarily with air. Water saturation in the matrix blocks is therefore typically much higher than that within the fractures, making it relatively difficult for air to flow through such blocks. It follows that, during a pneumatic injection test, the air moves primarily through fractures (most of which contain relatively little water), and the test therefore yields flow and transport parameters that reflect the intrinsic properties of these largely air-filled fractures.

Core and single-hole measurements, conducted over short segments of a borehole, provide information only about a small volume of rock in the immediate vicinity of each measurement interval. Available data from the ALRS indicate that rock properties, measured on such small scales, vary erratically in space in a manner that renders the rock randomly heterogeneous and pneumatically anisotropic. Our analyses to date (Bassett et al., 1994, 1997; Guzman et al., 1996) suggest that it is possible to interpolate some of the core and single-hole measurements at the ALRS between boreholes by means of geostatistical methods, which view the corresponding variables as correlated random fields. This is especially true about air permeability, porosity, fracture density, water content, and the van Genuchten water retention parameter α, for each of which we possess enough measurements to constitute a workable geostatistical sample. A geostatistical analysis of these site variables has been conducted by Chen et al. (1997) and Illman et al. (1998).

Steady-state analyses of single-hole pneumatic test data yield only air permeability values (Guzman et al., 1996). In this paper, we discuss transient interpretations of the same data, which provide additional information about air-filled porosity, skin factor, and dimensionality of the flow regime.

Single-hole air injection tests provide information only about a small volume of rock in the close vicinity of the injection interval. Fractured rock properties measured on such small scales tend to vary rapidly and erratically in space so as to render the rock strongly and randomly heterogeneous. To determine the properties of the rock on a larger scale, cross-hole interference tests were conducted by Illman et al. (1998; see also Illman, 1999) by injecting air into an isolated interval within one borehole, while monitoring pressure responses in isolated intervals within this and other boreholes. Of the 16 boreholes used for cross-hole testing, 6 were previously subjected to single-hole testing. The results of the single-hole tests (primarily spatial distribution of air permeabilities and local flow geometry) together with other site information (primarily borehole televiewer images) served as a guide in designing the cross-hole tests.

A total of 44 cross-hole pneumatic interference tests of various types (constant injection rate, multiple step injection rates, instantaneous injection) have been conducted during 1995-1997 using various configurations of injection and monitoring intervals. The tests were conducted using modular straddle packer systems that were easily adapted to various test configurations and allowed rapid replacement of failed components, modification of the number of packers, and adjustment of distances between them in both the injection and monitoring boreholes. A typical cross-hole test consisted of packer inflation, a period of pressure recovery, air injection, and another period of pressure recovery. Once packer inflation pressure had dissipated in all (monitoring and injection) intervals, air injection at a constant mass flow rate began. It generally continued for several days, until pressure in most monitoring intervals appeared to have stabilized. In some tests, injection pressure was allowed to dissipate until ambient conditions had been recovered. In other tests, air injection continued at incremental flow rates, each lasting until the corresponding pressure had stabilized, before the system was allowed to recover. In this paper we discuss results from one of these tests, labeled PP4.

TYPE-CURVE INTERPRETATION OF
PNEUMATIC INJECTION TESTS

Most type-curve models currently available for the interpretation of single-hole and cross-hole fluid injection (or withdrawal) tests in fractured rocks fall into three broad categories: (1) those that treat the rock as a single porous continuum representing the fracture network; (2) those that treat the rock as two overlapping continua of the dual-porosity type; and (3) hybrid models that embed a major discrete fracture in a porous continuum so as to intersect the injection (or withdrawal) test interval at various angles. The prevailing interpretation of dual continua is that one represents the fracture network and the other embedded blocks of rock matrix. We take the broader view that multiple (including dual) continua may represent fractures on a multiplicity of scales, not necessarily fractures and matrix. When a dominant fracture is present in a type-curve model, it is usually pictured as a high-permeability slab of finite or infinitesimal thickness. To allow developing analytical solutions in support of type-curve models, the continua are taken to be uniform and either isotropic or anisotropic. The test interval is taken to intersect a dominant fracture at its center. Either flow across the walls of such a fracture, or incremental pressure within the fracture, is taken to be uniform in most models. Flow is usually taken to be transient with radial or spherical symmetry, which may transition into near-uniform flow as one approaches a major fracture that intersects the test interval. Some models account for borehole storage and skin effects in the injection (or withdrawal) interval.

In this paper, we interpret transient pressure data from the single-hole air injection tests previously conducted at the ALRS by Guzman et al. (1994, 1996) and Guzman and Neuman (1996) by means of modified single-continuum type-curve models developed for spherical flow by Joseph and Koederitz (1985), for radial flow by Agarwal et al. (1970), for a single horizontal fracture by Gringarten and Ramey (1974), and for a single vertical fracture by Gringarten et al. (1974). Our modifications, detailed by Illman et al. (1998) and Illman (1999), consist of recasting these models in terms of pseudopressure and developing corresponding expressions and type-curves in terms of (pseudo)pressure-derivatives. Pseudopressure is defined as (Al-Hussainy and Ramey, 1966; Raghavan, 1993)

$$w(p) = 2 \int_{P_0}^{p} \frac{p}{\mu(p)Z(p)} dp \qquad (10.1)$$

where p is pressure, μ is dynamic viscosity, $Z = pV/nRT$ is compressibility factor, V is volume, n is mass in mols, R is universal gas constant, and T is absolute temperature. We take p_0 to represent barometric pressure. Under conditions of testing at the ALRS, μZ is constant and so

$$w(p) = \frac{2}{\mu Z} \int_{P_0}^{P} p \, dp = \frac{p^2 - p_0^2}{\mu Z}. \qquad (10.2)$$

Type-curves of pressure derivative versus the logarithm of time have become popular in recent years because they accentuate phenomena that might otherwise be missed, help diagnose the prevailing flow regime, and aid in constraining the calculation of corresponding flow parameters. The type-curves are derived analytically for single-phase gas flow by linearizing the otherwise nonlinear partial differential equations, which govern such flow in uniform, isotropic porous continua. Included in the type-curves are effects of gas storage in the injection interval (known as borehole storage effect) and reduced or enhanced permeability in the immediate vicinity of this interval (known as positive or negative skin effects). Our type-curve analyses of single-hole pneumatic test data yield information about air permeability, air-filled porosity, skin factor, and dimensionality of the flow regime on a nominal scale of 1 m in the immediate vicinity of each test interval.

For purposes of cross-hole test analysis by means of type-curves, we represent the fractured rock by an infinite three-dimensional uniform, anisotropic continuum as was done by Hsieh and Neuman (1985), and linearize the airflow equations in terms of pressure (Illman et al., 1998) or pseudopressure (Illman, 1999). Hsieh and Neuman treat injection and observation intervals as points or lines; we consider the special case where injection takes place at a point and observation along a line. However, we modify their solution to account for the effects of storage and skin on pressure, and its derivative (not considered by these authors), in the observation interval.

Type-Curve Interpretation of Single-Hole Tests

We have interpreted over 40 sets of 1-m scale single-hole pneumatic injection test data by means of the spherical, radial, vertical, and horizontal fracture flow models described in the previous section (Illman et al., 1998; Illman, 1999). The majority of these data conform to the spherical flow model regardless of number or orientation of fractures in a test interval. We interpret this to mean that flow around most test intervals is controlled by a single continuum, representative of a three-dimensional network of interconnected fractures, rather than by discrete planar features. Only in a small number of test intervals, known to be intersected by widely open fractures, have planar features dominated flow as evidenced by the development of an early half-slope on logarithmic plots of pressure versus time; unfortunately, the corresponding data do not fully conform to available type-curve models of fracture flow. Some pressure records conform to the radial flow model during early and intermediate times but none do so fully at late time.

Figure 10-3 shows visual fits between incremental squared pressure (circles) and corresponding derivative (triangles) data from test CAC0813 (in a 1-m interval) and type-curves corresponding to a spherical flow model expressed in terms of dimensionless pseudopressure w_D and its derivative $\partial w_D/\partial \ln t_D$; we recall that pseudopressure, w, is proportional to incremental squared pressure when μZ is constant, as is the case at the ALRS. The data exhibit a good match with type curves that correspond to zero skin ($s = 0$); indeed, most test data from the ALRS show little evidence of a skin effect. The match yields an air permeability value of 1.56×10^{-15} m^2. The early-time data fall on a straight line with unit slope, indicative of compressible air storage within the test interval.

Figure 10-4 shows a similar type-curve match for test CHB0617 (in a 1-m interval). The data exhibit a fair match with type curves that correspond to zero skin for early to intermediate data, yielding an air permeability of 6.01×10^{-17} m^2. Late data do not match the type curves due to apparent displacement of water by air, which manifests itself as a gradually decreasing skin effect. Early-time data are strongly affected by compressible air storage within the test interval. A pressure peak, possibly due to two-phase flow, is discernible on the logarithmic scale of Figure 10-4.

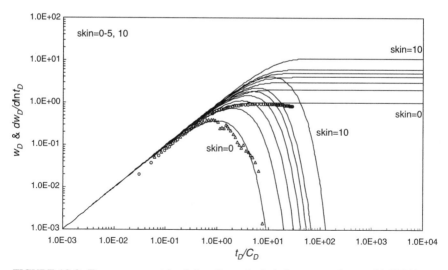

FIGURE 10-3 Type-curve match of data from single-hole pneumatic test CAC0813 to the gas, spherical flow model.

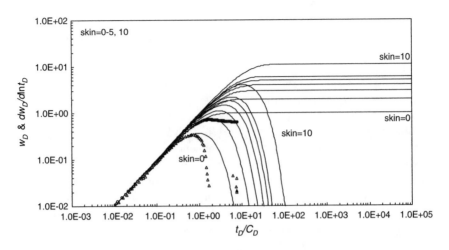

FIGURE 10-4 Type-curve match of data from single-hole pneumatic test CHB0617 to the gas, spherical flow model.

Figure 10-5 compares permeabilities obtained by steady-state and spherical transient analyses, showing that they agree reasonably well.

Figure 10-6 shows type-curve matches for single-hole test JGA0605 in a 1-m interval. In this case, the early and intermediate data appear to fit the radial flow model but the late pressure data stabilize, and the late pressure derivative data

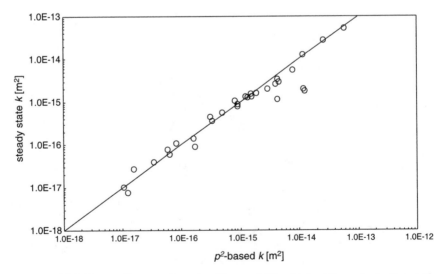

FIGURE 10-5 Scatter diagram of results of permeability obtained from steady-state and p^2-based spherical models.

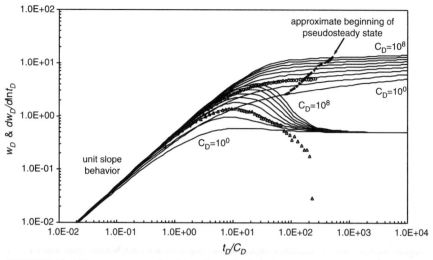

FIGURE 10-6 Type-curve match of data from single-hole pneumatic test JGA0605 to the gas, radial flow model.

drop, in manners characteristic of three-dimensional flow. The same is seen to happen when we consider Figure 10-7 data from single-hole test JJA0616 in a 1-m interval. We take this to indicate that the flow regime evolves from radial to spherical with time.

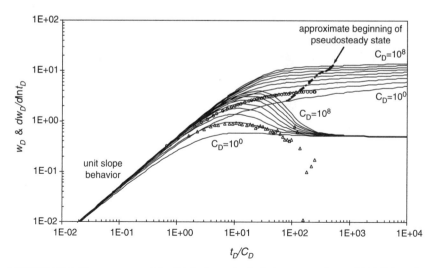

FIGURE 10-7 Type-curve match of data from single-hole pneumatic test JJA0616 to the gas, radial flow model.

The 1-m test interval JHB0612 intersects a fracture, which on televiewer (Figure 10-8) appears to be widely open. Figure 10-9 depicts an attempt on our part to match the corresponding incremental squared pressure data to type curves of pseudopressure based on the horizontal fracture flow model described in the previous section. Only the early time data appear to match one of these curves. We suspect that deviation of the late data from the type curves is due to the fact that whereas in reality the flow evolves with time to become three-dimensional, in the model it evolves to become radial. Upon ignoring the late data and considering only the early match, we obtain an air permeability of 1.3×10^{-13} m^2. This is about four times the value of 4.8×10^{-14} m^2 obtained by Guzman et al. (1996) on the basis of a steady-state analysis of the late data.

An unsuccessful attempt to match the same data with a type-curve corresponding to a vertical fracture flow model, described in the previous section, is depicted in Figure 10-10.

Incremental pressure and, to a much greater extent, derivative data from several single-hole pneumatic injection tests, one of which is illustrated in Figure 10-11, exhibit inflections that are suggestive of dual or multiple continuum behaviors. If so, we ascribe such behavior not to fractures and rock matrix as is common in the literature (Warren and Root, 1963; Odeh, 1965; Gringarten, 1979, 1982), but to fractures associated with two or more distinct length scales. However, these inflections show some correlation with barometric pressure fluctua-

FIGURE 10-8 BHTV image taken in borehole Y2 (high permeability zone).

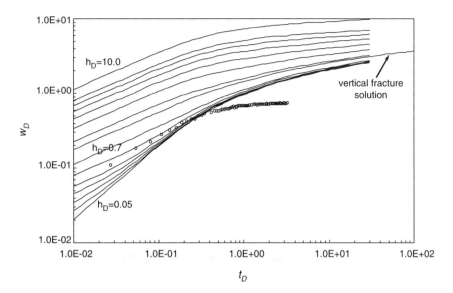

FIGURE 10-9 Type-curve match of data from single-hole pneumatic test JHB0612 to the horizontal fracture model.

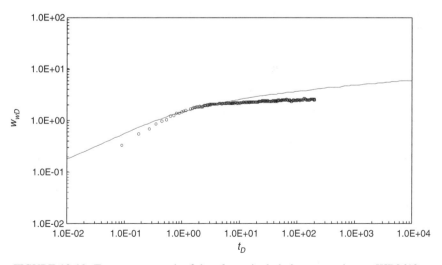

FIGURE 10-10 Type-curve match of data from single-hole pneumatic test JHB0612 to the vertical fracture model.

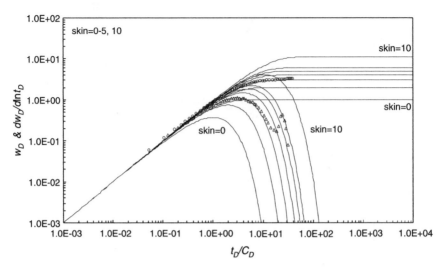

FIGURE 10-11 Type-curve match of data from single-hole pneumatic test CHB0617 to the gas, spherical flow model.

tions, implying that they might have been caused by the latter rather than by dual or multiple continuum phenomena.

Our type-curve interpretations of single-hole tests appear to be insensitive to how one linearizes the airflow equations; writing these equations in terms of pressure or pseudopressure (incremental pressure squared) leads to very similar results (Illman et al., 1998; Illman, 1999).

Type-Curve Interpretation of Cross-Hole Test PP4

A total of 44 cross-hole pneumatic interference tests of various types (constant injection rate, multiple step injection rates, instantaneous injection) were conducted during the years 1995-1997 using various configurations of injection and monitoring intervals (Illman et al., 1998; Illman, 1999). The tests were conducted using modular straddle packer systems that were easily adapted to various test configurations and allowed rapid replacement of failed components, modification of the number of packers, and adjustment of distances between them in both the injection and monitoring boreholes. A typical cross-hole test consisted of packer inflation, a period of pressure recovery, air injection and another period of pressure recovery. Once packer inflation pressure had dissipated in all (monitoring and injection) intervals, air injection at a constant mass flow rate began. It generally continued for several days, until pressure in most monitoring intervals appeared to have stabilized. In some tests, injection pressure was allowed to dissipate until ambient conditions had been recovered. In other tests, air injection continued at incremental flow rates, each lasting until the corresponding pressure had stabilized, before the system was allowed to recover.

In this paper we focus on the analysis of test PP4 conducted during the latter phase of our program. This test was selected because it involved:

1. injection into a high-permeability zone, which helped pressure to propagate rapidly across much of the site;

2. injection at a relatively high flow rate, which led to unambiguous pressure responses in a relatively large number of monitoring intervals;

3. the largest number of pressure and temperature monitoring intervals among all tests;

4. a complete record of relative humidity, battery voltage, atmospheric pressure, packer pressure, and injection pressure;

5. the lowest number of equipment failures among all tests;

6. flow conditions (such as injection rate, fluctuations in barometric pressure, battery voltage, and relative humidity) that were better controlled, and more stable, than in all other tests;

7. minimum boundary effects due to injection into the central part of the tested rock mass;

8. a relatively long injection period;

9. rapid recovery; and

10. a test configuration that allowed direct comparison of test results with those obtained from two line-injection/line-monitoring tests, and a point-injection/line-monitoring test, at the same location. Stable flow rate and barometric pressure made type-curve analysis of test PP4 results relatively straightforward.

Test PP4 was conducted by injecting air at a rate of 50 slpm into a 2-m interval located 15-17 m below the lower lip of casing in borehole Y2, as indicated by a large solid circle in Figure 10-12. The figure also shows a system of Cartesian coordinates x, y, z with origin at the center of the injection interval, which we use to identify the placement of monitoring intervals relative to this center. Responses were monitored in 13 relatively short intervals (0.5-2 m) whose centers are indicated in the figure by small white circles, and 24 relatively long intervals (4-42.6 m) whose centers are indicated by small solid circles, located in 16 boreholes. Several of the short monitoring intervals were designed to intersect a high permeability region that extends across much of the site at a depth comparable to that of the injection interval.

Type-curve interpretation of pressure data from cross-hole test PP4 included 31 intervals; pressure data from 5 intervals were not amenable to type-curve interpretation and have therefore been excluded. A special set of type-curves was developed for each pressure monitoring interval by treating the medium as if it was pneumatically isotropic. Indeed, our type-curve analysis additionally treats

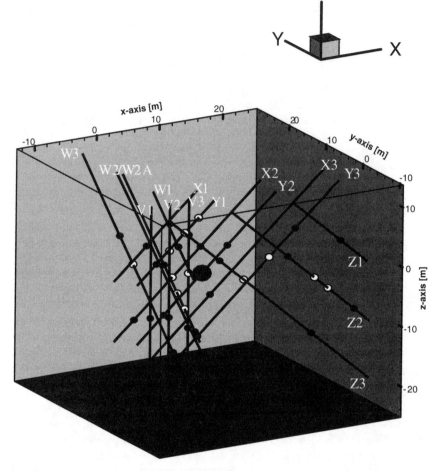

FIGURE 10-12 Locations of centers of injection and monitoring intervals. Large solid circle represents injection interval, small solid circles represent short monitoring intervals, and open circles represent long monitoring intervals.

the rock as if it was pneumatically uniform. However, since the analyses of pressure data from different monitoring intervals yield different values of pneumatic parameters, our analysis ultimately yields information about the spatial and directional dependence of these parameters.

Figures 10-13 through 10-16 show how we matched typical records of pressure buildup, and pressure derivatives, from cross-hole injection test PP4 to corresponding type-curves on logarithmic paper. Most of the matches are excellent to good, but a few are poor. Fluctuations in barometric pressure affect some

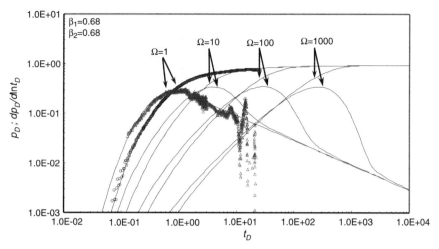

FIGURE 10-13 Type-curve match of pressure and its derivative from monitoring interval V1.

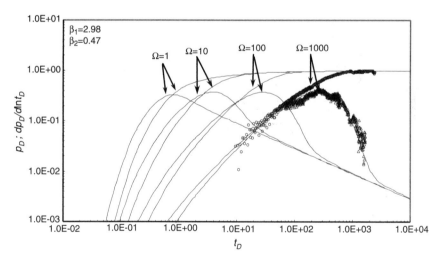

FIGURE 10-14 Type-curve match of pressure and its derivative from monitoring interval Z2U.

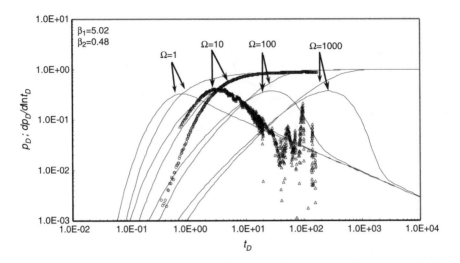

FIGURE 10-15 Type-curve match of pressure and its derivative from monitoring interval X2M.

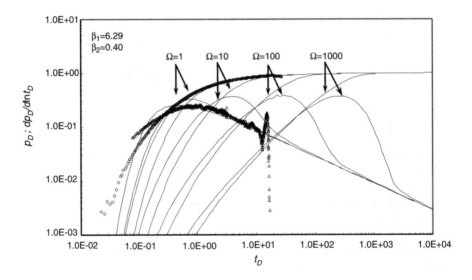

FIGURE 10-16 Type-curve match of pressure and its derivative from monitoring interval Z3M.

of the late pressure buildup data and cause their derivatives to fluctuate, at times wildly.

Our finding that most pressure buildup data match the type-curves well is a clear indication that the majority of cross-hole test PP4 results are amenable to interpretation by means of a continuum model, which treats the rock as being pneumatically uniform and isotropic while describing airflow by means of linearized, pressure-based equations. The fact that some of our data do not fit this model shows that the latter does not provide a complete description of pneumatic pressure behavior at the site. That the site is not pneumatically uniform or isotropic on the scale of cross-hole test PP4 is made evident by pneumatic parameters derived from our type-curve matches. Values of pneumatic permeability and air-filled porosity derived from these matches represent bulk properties of the rock between the corresponding monitoring interval and the injection interval. The permeabilities additionally represent directional values along lines that connect the centers of these intervals. The directional permeabilities range from $1.1 \times 10^{-16} m^2$ to $4.6 \times 10^{-13} m^2$ with a mean of -13.5 for $\log_{10}$-based k, while a corresponding (antilog) value is $2.9 \times 10^{-14} m^2$. The corresponding variance and coefficient of variation (CV) are 5.3×10^{-1} and -5.4×10^{-1}, respectively. Air-filled porosities range from 1.7×10^{-5} to 2.2×10^{-1} with a geometric mean of 3.5×10^{-3}, while their variance and coefficient of variation are 8.0×10^{-1} and -3.6×10^{-1}, respectively. Permeabilities derived from cross-hole tests have a much higher mean, and lower variance, than those from the smaller-scale single-hole tests.

NUMERICAL INVERSE INTERPRETATION OF PNEUMATIC INJECTION TESTS

We have also interpreted data from several single-hole and cross-hole tests at the ALRS by means of a three-dimensional finite-volume numerical simulator, FEHM (Zyvoloski et al., 1988, 1996, 1997), coupled to a numerical inverse code, PEST (Doherty et al., 1994), based on the assumption of single-phase airflow through a uniform, isotropic porous continuum. Our numerical model accounts directly for the ability of all boreholes, and packed-off borehole intervals, to store and conduct air through the system. The model does so by treating these as high-permeability and high-porosity cylinders of finite length and radius. It solves the airflow equations in nonlinear form, and is able to account for atmospheric pressure fluctuations at the soil surface. Our numerical model thus accounts more fully and accurately for nonlinear pressure propagation and storage through both the rock and the boreholes than do our type curves, which rely on linearized airflow equations and ignore the effect of boreholes (other than the injection interval) on pressure distribution through the system. Both allow one (in principle) to interpret multiple injection-step and recovery data simultaneously. The numerical inverse model yields information about air permeability, air-filled porosity, and effective porosity of the injection interval.

There is little information in the literature about the effect that open borehole intervals may have on pressure propagation and response during interference tests. Paillet (1993) noted that the drilling of an additional observation borehole had an effect on drawdowns created by an aquifer test. We likewise found through numerical simulations (Illman et al., 1998) that the presence of open borehole intervals has a considerable impact on pressure propagation through the site, and on pressure responses within monitoring borehole intervals during cross-hole air injection tests.

As we consider only single-phase airflow, the saturation of air and associated pneumatic properties of the rock remain constant during each simulation. The only initial condition we need to specify is air pressure, which we take to be the average barometric pressure of 0.1 MPa. The side and bottom boundaries of the flow model are impermeable to airflow. Our results suggest that these boundaries have been placed sufficiently far from injection intervals to have virtually no effect on simulated single-hole tests. The top boundary coincides with the soil surface and is maintained at a constant and uniform barometric pressure of 0.1 MPa. Though barometric pressure fluctuated during each single-hole test, these fluctuations were small and are ignored in our analysis.

The air permeability k and air-filled porosity ϕ are taken to be uniform within the computational region. In our inverse analyses, these two parameters are adjusted simultaneously with the effective porosity ϕ_w of the injection interval. The latter parameter is allowed to take on values in excess of 1 as a way to account for effective borehole volumes larger than those originally built into the computational grid.

We used the same computational grid for the analysis of single-hole and cross-hole tests. The grid measures 63 m in the x direction, 54 m in the y direction, and 45 m in the z direction, encompassing a rock volume of 153,090 m^3 (Figure 10-17). The computational grid is illustrated for the case of single-hole tests, during which injection takes place into various packed-off intervals along borehole Y2, by means of two-dimensional images in Figures 10-18 and 10-19. Boreholes are treated as porous media having much higher permeability and porosity than the surrounding rock. Figure 10-18 shows three views of the grid perpendicular to the x-y, x-z, and y-z planes. As the grid in the vicinity of boreholes is relatively fine, the corresponding areas appear dark in the figures. Figure 10-19 shows four cross-sectional views of the grid along vertical planes that contain selected boreholes. Since the grid is three-dimensional, its intersections with these planes do not necessarily occur along nodal points (i.e., what may appear as nodes in the figure need not be such). The grid includes 39,264 nodes, 228,035 tetrahedral elements, and is divided into three parts: a regular grid at the center of the modeled area, which has a node spacing of 1 m; a surrounding regular grid having a node spacing of 3 m; and a much finer and more complex

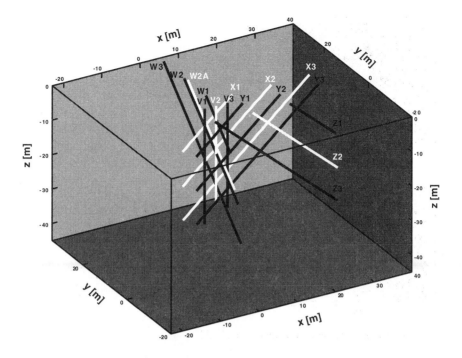

FIGURE 10-17 Boundaries of computational region.

unstructured grid surrounding each borehole. The three-dimensional grid represents quite accurately the geometry, flow properties, and storage capabilities of vertical and inclined boreholes at the site; is capable of resolving medium heterogeneity on a support scale of 1 m; is able to represent, with a high degree of resolution, steep gradients around the injection test interval, as well as pressure interference between boreholes, no matter how closely spaced; and assures smooth transition between fine borehole grids having radial structures and surrounding coarser grids having regular structures.

Inverse Analysis of Single-Hole Tests

A steady-state interpretation of pressure buildup data recorded during the first injection step (labeled A) of single-hole test JG0921 by means of an analytical formula gave a pneumatic permeability of 2.8×10^{-14} m^2 (Guzman et al., 1996), and a transient type-curve analysis based on the spherical flow model gave 2.6×10^{-14} m^2 (Illman et al., 1998); neither of these two analyses was able to provide air-filled porosity estimates. When open borehole intervals (including that used for injection) are not considered in the simulation, our numerical in-

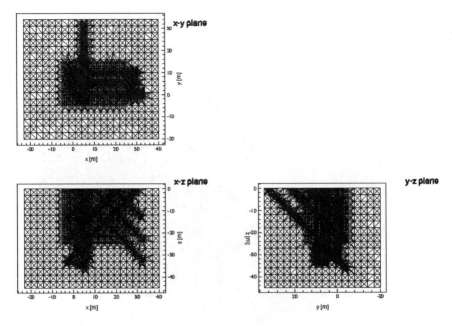

FIGURE 10-18 Side views of computational grid.

verse model yields a match that is less than satisfactory (Figure 10-20), with air permeability $k = 2.3 \times 10^{-14} \pm 2.6 \times 10^{-16}$ m^2 and air-filled porosity $\phi = 4.5 \times 10^{-1} \pm 1.9 \times 10^{-3}$, where the ± range represents 95% confidence intervals identified by PEST. The porosity estimate is much too high for fractures, suggesting that the corresponding confidence interval computed by PEST is overly optimistic. When only steady-state pressure data are included in the inverse analysis, the match at early time is poor, yielding $k = 2.8 \times 10^{-14}$ m^2 and $\phi = 4.6 \times 10^{-3}$, respectively. Here the estimate of porosity is based entirely on our specification of the time at which steady state commences (approximately 0.008 days, as indicated by open circles in Figure 10-20), which renders model sensitivity to ϕ so low as to preclude it from computing confidence intervals for either parameter. When the effect of all open borehole intervals is included and the effective porosity of the injection interval is allowed to vary simultaneously with k and ϕ, the match improves, yielding $k = 2.2 \times 10^{-14} \pm 4.4 \times 10^{-16}$ m^2, $\phi = 6.7 \times 10^{-3} \pm 4.7 \times 10^{-3}$, and $\phi_w = 7.0 \times 10^{-1} \pm 6.7 \times 10^{-2}$. The relatively large confidence interval associated with ϕ (of the same order as its estimate) reflects the fact that transient data are dominated by borehole storage, which is the reason type-curve analysis has been unable to identify this parameter (Illman et al., 1998).

Air storage in the injection interval has theoretically no effect on early recovery data, which should therefore be ideal for the estimation of air-filled porosity.

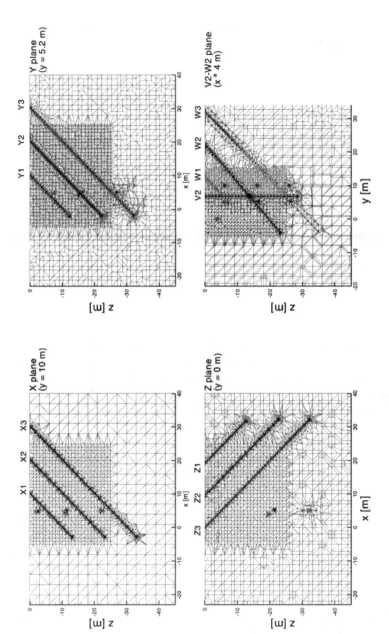

FIGURE 10-19 Vertical cross-sections through computational grid.

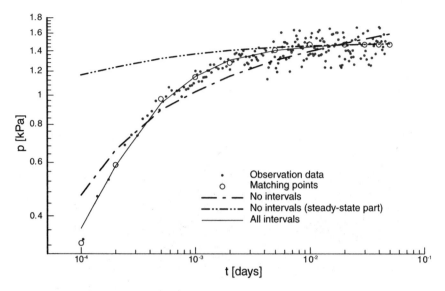

FIGURE 10-20 Pressure data from single-hole test JG0921A interpreted by various inverse models. Adapted from Illman et al. (1998).

We in fact find that such storage effects are most pronounced during the first step of a test, and less so during subsequent steps. This is seen in Figure 10-21, where pressure during each step and recovery is plotted relative to pressure established during the previous step, versus time measured relative to the end of the previous step. The figure demonstrates that pressure data from the first injection step, labeled A, exhibit the most pronounced one-to-one slope at early time, which is indicative of borehole storage; there is no discernible storage effect during the second step, labeled B, or the recovery phase, labeled R. We therefore expect a simultaneous analysis of pressure data from the entire test to yield a more reliable estimate of parameters, especially air-filled porosity, than is possible based only on data from the first step.

A reasonably good fit of our model (which now accounts for all open bore-hole intervals) to the entire two-step pressure record, including recovery data, is shown in Figure 10-22. The corresponding parameter estimates are $k = 2.4 \times 10^{-14} \pm 7.1 \times 10^{-16}$ m^2, $\phi = 1.4 \times 10^{-2} \pm 1.7 \times 10^{-3}$, and $\phi_w = 8.0 \times 10^{-1} \pm 4.6 \times 10^{-2}$. We consider these values reasonable and reliable.

It is of interest to note that a 2-m borehole interval used for injection during cross-hole test PP4 (discussed below) lay 0.1 m above that of the 1-m injection interval during single-hole test JG0921, so that the two overlapped. Injection rate during the cross-hole test (1×10^{-3} kg/s or 5×10^4 cm^3/s) was over 100 times higher than that during JG0921A. Pressure data recorded in the injection interval

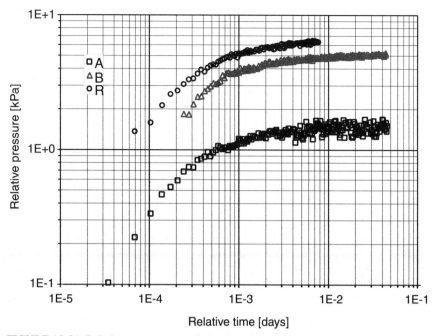

FIGURE 10-21 Relative pressure vs. relative time for all injection steps and recovery of JG0921.

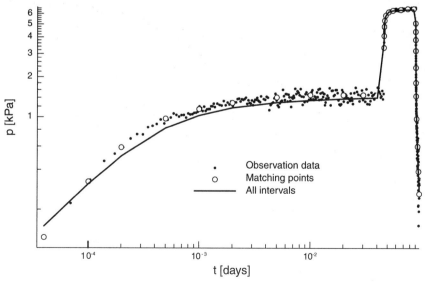

FIGURE 10-22 Pressure data from single-hole test JG0921 interpreted by inverse model incorporating all boreholes.

during the cross-hole test varied little with time and were therefore not suitable for a reliable estimation of air-filled porosity. They were, however, suitable for the estimation of air permeability, giving a value of $k = 2.2 \times 10^{-14} \pm 2.7 \times 10^{-16}$ m^2, which is very close to those we derived from the single-hole test data.

Steady-state analysis of pressure buildup data from the first injection step (A) of single-hole test JGC0609 gave a pneumatic permeability of $k = 2.0 \times 10^{-15}$ m^2 (Guzman et al., 1996), and transient type-curve analysis with a spherical flow model yielded $k = 2.9 \times 10^{-15}$ m^2 (Illman et al., 1998). In the absence of open borehole intervals, the inverse model yields a poor fit with $k = 1.8 \times 10^{-15} \pm 3.9 \times 10^{-15}$ m^2 and $\phi = 5.0 \times 10^{-1} \pm 4.2 \times 10^{-2}$. Upon including the injection interval, the fit improves dramatically to yield $k = 1.6 \times 10^{-15} \pm 2.6 \times 10^{-17}$ m^2, $\phi = 4.8 \times 10^{-3} \pm 9.4 \times 10^{-3}$, and $\phi_w = 1.3 \pm 2.6 \times 10^{-2}$. Incorporating the effects of all open boreholes results in an equally good fit with $k = 1.6 \times 10^{-15} \pm 1.3 \times 10^{-17}$ m^2, $\phi = 5.5 \times 10^{-3} \pm 4.7 \times 10^{-3}$, and $\phi_w = 1.3 \pm 1.7 \times 10^{-2}$ (Figure 10-23). Despite such good fits, the above porosity estimates are evidently unreliable due to the dominance of borehole storage effects during the transient period of the test. A borehole porosity in excess of 1 is plausible, implying that the effective volume V_S of the injection interval exceeds its nominal volume V_w.

Figure 10-24 depicts relative pressure versus relative time for injection steps 1-4 (labeled A-D, respectively) and recovery (labeled R). Here again we see that

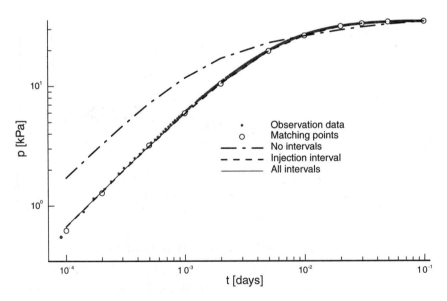

FIGURE 10-23 Pressure data from single-hole test JGC0609A interpreted by various inverse models. Adapted from Illman et al. (1998).

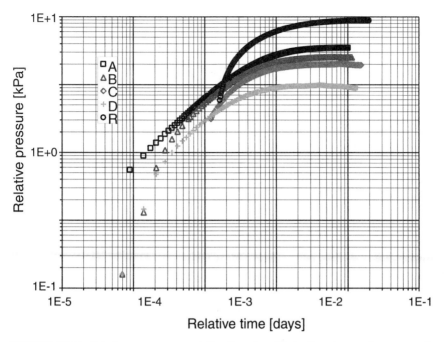

FIGURE 10-24 Relative pressure vs. relative time for all injection steps and recovery of JGC0609.

only data corresponding to step A exhibit a pronounced one-to-one slope at early time, while all other data appear to be free of borehole storage influence. We therefore expect a simultaneous analysis of pressure data from the entire test to yield a more reliable estimate of parameters, especially air-filled porosity, than is possible based only on data from the first step.

A fit of our model to the entire four-step pressure record, including recovery data, yields a good match (Figure 10-25) with $k = 1.7 \times 10^{-15} \pm 4.1 \times 10^{-17}$ m^2, ϕ $= 3.6 \times 10^{-3} \pm 2.8 \times 10^{-4}$, and $\phi_w = 1.5 \pm 6.3 \times 10^{-2}$. The model appears sensitive to all three parameters whose estimates seem reasonable to us: both k and ϕ are smaller, by about one order of magnitude, than they were in the case of single-hole test JG0921.

Inverse Analysis of Cross-Hole Test PP4

In the inverse analysis we describe here, pressure data from each monitoring interval are considered separately while pneumatic permeability and air-filled porosity are treated as if they were uniform across the site, in a manner similar to our type-curve analyses. Hence any major difference between results we obtain

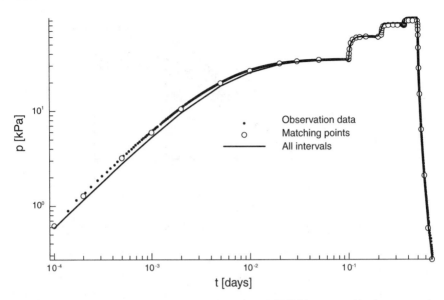

FIGURE 10-25 Pressure data from single-hole test JGC0609 interpreted by inverse model incorporating all boreholes.

by these two methods of analysis can be attributed to differences between the ways they handle nonlinearity of the governing airflow equations (without or with linearization, numerically or analytically), boundary conditions (at finite or infinite distances), and boreholes (directly or indirectly, completely or incompletely).

Matches between computed and recorded pressures are shown in Figure 10-26. Some of these matches are very poor due to barometric pressure effects, some are of intermediate quality, and some are good. Corresponding estimates of air permeability range from 3.8×10^{-15} to 3.0×10^{-12} m^2 with a mean of 2.8×10^{-13} m^2, variance of 4.9×10^{-25}, and coefficient of variation equal to 2.5. The mean, variance, and coefficient of variation of log-transformed permeability are -13.5 (corresponding to 2.9×10^{-14} m^2), 6.9×10^{-1}, and 6.0×10^{-2}, respectively. The mean, variance, and coefficient of variation of corresponding type-curve results are -13.5 (corresponding to 3.5×10^{-14} m^2), 3.2×10^{-1}, and -2.4×10^{-2}, respectively. The type-curve analysis excluded intervals X3 and Y2M but included interval Y1U, which was not considered in the inverse analysis. The two sets of air permeability values are compared in Figure 10-27.

Air-filled porosity estimated by our inverse model on the basis of pressure data from injection interval Y2M during cross-hole test PP4 is highly uncertain, due to a very rapid pressure buildup in this interval. The large air-filled porosity

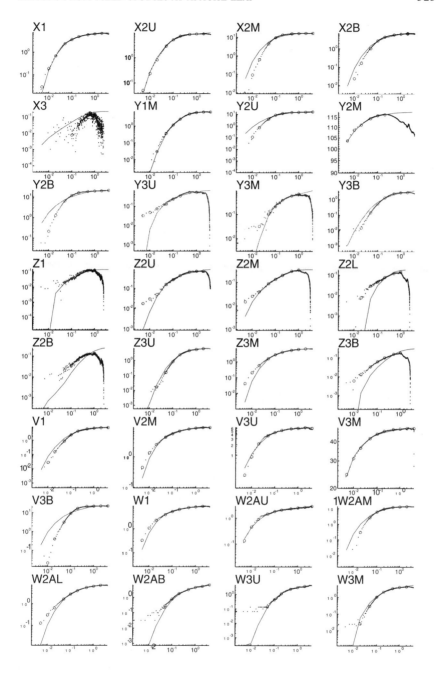

FIGURE 10-26 Matches between simulated and observed pressures (kPa, vertical axes) versus time (days, horizontal axes) in individual intervals during cross-hole test PP4.

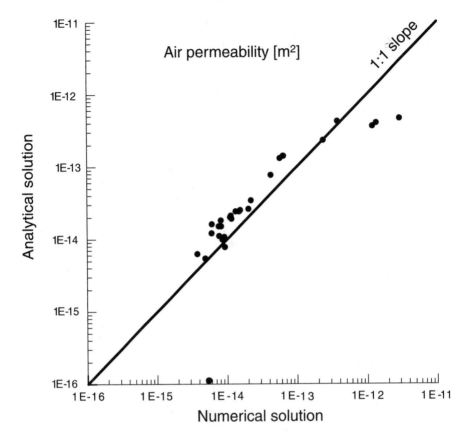

FIGURE 10-27   Analytically vs. numerically derived air permeabilities for cross-hole test PP4.

estimates (0.5) obtained on the basis of pressure data from monitoring intervals X3, Z2L, Z2B, and Z3B are equal to their specified upper bound; we consider them highly unlikely due to poor fits between calculated and observed pressure responses in these intervals (Figure 10-26), which appear to be pneumatically well connected to the atmosphere and therefore strongly influenced by barometric pressure fluctuations. Upon excluding air-filled porosity values obtained from these five borehole intervals, the range of this parameter narrows down to 5.1×10^{-3}–1.0×10^{-1} with arithmetic mean, variance, and coefficient of variation equal to 3.6×10^{-2}, 7.1×10^{-4}, and 7.3×10^{-1}, respectively. The corresponding mean, variance, and coefficient of variation of log-transformed air-filled porosities are -1.6 (corresponding to 2.7×10^{-2}), 1.4×10^{-1}, and -2.3×10^{-1}. Log-transformed air-filled porosities from type-curve analyses have mean –2.2 (corresponding to 6.6×10^{-3}), variance 3.0×10^{-1}, and coefficient of variation -4.0×10^{-1}. The two sets of air-filled porosities, compared in Figure 10-28, are seen to agree poorly; values obtained by the inverse method are consistently larger than those from type-curve analyses.

LESSONS LEARNED

The following are major lessons we have learned from our Apache Leap Research Site studies:

1. Issues associated with the site characterization of fractured rock terrains, the analysis of fluid flow and contaminant transport in such terrains, and the efficient handling of contaminated sites are typically very difficult to resolve. A

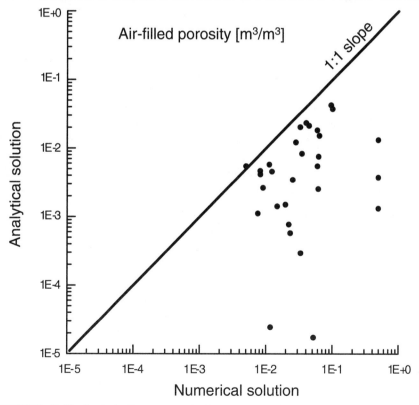

FIGURE 10-28 Analytically vs. numerically derived air-filled porosities for cross-hole test PP4.

major source of this difficulty is the complex nature of the subsurface "plumbing systems" of pores and fractures through which flow and transport in rocks take place. There is at present no well-established field methodology to characterize the fluid flow and contaminant transport properties of unsaturated fractured rocks.

2. In order to characterize the ability of unsaturated fractured rocks to conduct water, and to transport dissolved or suspended contaminants, one would ideally want to observe these phenomena directly by conducting controlled field hydraulic injection and tracer experiments within the rock. In order to characterize the ability of unsaturated fractured rocks to conduct nonaqueous phase liquids such as chlorinated solvents, one would ideally want to observe the movement of such liquids under controlled conditions in the field. In practice, there are severe logistical obstacles to the injection of water into unsaturated geologic media, and logistical as well as regulatory obstacles to the injection of nonaqueous liquids. There also are important technical reasons why the injection of liquids, and dissolved or suspended tracers, into fractured rocks may not be the most feasible approach to site characterization when the rock is partially saturated with water. Many of these limitations can be overcome by conducting field tests with gases rather than with liquids, and with gaseous tracers instead of chemicals dissolved in water.

3. The University of Arizona has successfully conducted numerous single-hole and cross-hole pneumatic injection tests in unsaturated fractured tuffs at the ALRS near Superior, Arizona, under the auspices of the U.S. Nuclear Regulatory Commission (NRC). These tests were part of confirmatory research in support of NRC's role as the licensing agency for a potential high-level nuclear waste repository in unsaturated fractured tuffs at Yucca Mountain. However, unsaturated fractured porous rocks similar to tuffs are found at many locations, including some low-level radioactive waste disposal sites, nuclear decommissioning facilities, and sites contaminated with radioactive as well as other hazardous materials. The test methodologies we have developed, and the understanding we have gained concerning the pneumatic behavior and properties of tuffs at the ALRS, are directly relevant to such facilities and sites.

4. We found it possible to interpret both single-hole and cross-hole pneumatic injection tests at the ALRS by means of analytically derived type-curves and a numerical inverse model, which account only for single-phase airflow through the rock while treating water as if it was immobile. Our type-curves are based on linearized versions of the nonlinear partial differential equations that govern single-phase airflow in uniform, isotropic porous continua under three regimes: three-dimensional flow with spherical symmetry, two-dimensional flow with radial symmetry, and flow in a continuum with an embedded high-permeability planar feature (a major fracture). The particular method of linearization appears to have only a minor impact on the results of our type-curve analyses. Included in our type-curves are effects of compressible air storage and skin in the injection interval during single-hole tests, and in monitoring intervals during

cross-hole tests. Our analytical tools include type-curves of pressure derivative versus the logarithm of time, which accentuate phenomena that might otherwise be missed, help diagnose the prevailing flow regime, and aid in constraining the calculation of corresponding flow parameters. Our numerical inverse model does not require linearizing the governing airflow equations and represents pneumatic test conditions at the site more realistically than do our type-curves.

5. Despite these differences between them, type-curve and inverse methods yield comparable values of air permeability for both single-hole and cross-hole tests. On the other hand, the inverse method yields consistently higher air-filled porosities than do type-curve analyses for cross-hole tests. A key to obtaining reliable air-filled porosities from single-hole pneumatic injection tests is to include multiple step and/or recovery data in the analysis. This is so because storage of air in the injection interval dominates pressure transients during the first step of each test, but appears to do so less in subsequent steps, and has relatively small impact on pressure recovery data.

6. At the ALRS, air permeabilities obtained from steady state and transient type-curve interpretations of single-hole pneumatic injection tests, conducted in borehole intervals of 1 m, agree closely with each other but correlate poorly with fracture density data. Airflow around the vast majority of these relatively short test intervals appears to be three-dimensional; borehole storage due to air compressibility is pronounced at early time; and skin effects are minimal.

7. During a pneumatic injection test, air moves primarily through fractures, most of which contain relatively little water, and the test therefore yields permeabilities and porosities that reflect closely the intrinsic properties of the surrounding fractures. This is so because capillary forces tend to draw water from fractures into porous (matrix) blocks of rock, leaving the fractures saturated primarily with air, and making it difficult for air to flow through matrix blocks. Since the fractures contain some residual water, the corresponding pneumatic permeabilities and air-filled porosities tend to be somewhat lower than their intrinsic counterparts. The former nevertheless approach the latter as the rate of injection goes up. This is due to displacement of water by air, which, under a constant rate of injection, appears to manifest itself in a rapid increase in pressure within the injection interval, followed by a gradual decrease. Two-phase flow of water and air additionally causes air permeabilities from single-hole pneumatic injection tests to exhibit a hysteretic variation with applied pressure.

8. In most single-hole pneumatic injection tests at the ALRS, pneumatic permeabilities increase systematically with applied pressure as air appears to displace water under two-phase flow. In a few single-hole tests, where the injection intervals are intersected by widely open fractures, air permeabilities decrease with applied pressure due to inertial effects. This pressure-dependence of air permeability suggests that it is advisable to conduct single-hole air injection tests at several applied flow rates and/or pressures. Pneumatic parameters derived from pressure data recorded in monitoring intervals during cross-hole tests ap-

pear to be much less sensitive to the rate of injection, suggesting that two-phase flow and inertial phenomena decay rapidly with distance from the injection interval. Enhanced permeability due to slip flow (the Klinkenberg effect) appears to be of little relevance to the interpretation of single-hole or cross-hole air injection tests at the ALRS.

9. Flow in the vicinity of most 1-m single-hole pneumatic test intervals at the ALRS appears to be three-dimensional regardless of the number or orientation of fractures in the surrounding rock. We interpret this to mean that such flow is controlled by a single continuum, representative of a three-dimensional network of interconnected fractures, rather than by discrete planar features. Indeed, most single-hole and cross-hole pneumatic test data at the ALRS have proven amenable to analysis by means of a single fracture-dominated continuum representation of the fractured porous tuff at the site. Only in a small number of single-hole test intervals, known to be intersected by widely open fractures, have such features dominated flow as evidenced by the development of an early half-slope on logarithmic plots of pressure versus time; unfortunately, the corresponding data do not fully conform to available type-curve models of fracture flow. Some pressure records conform to the radial flow model during early and intermediate times, but none do so fully at late time.

10. It is generally not possible to distinguish between the permeabilities of individual fractures and the bulk permeability of the fractured rock in the immediate vicinity of a test interval by means of pneumatic injection tests. Hence there is little justification for attempting to model flow through individual fractures at the site. The explicit modeling of discrete features appears to be justified only when one can distinguish clearly between layers, faults, fracture zones, or major individual fractures on scales not much smaller than the domain of interest.

11. Air permeabilities obtained from single-hole tests are poorly correlated with fracture densities, as is known to be the case for hydraulic conductivities at many water-saturated fractured rock sites worldwide (Neuman, 1987). This provides further support for Neuman's conclusion that the permeability of fractured rocks cannot be reliably predicted from information about fracture geometry (density, trace lengths, orientations, apertures and their roughness) but must be determined directly by means of hydraulic and/or pneumatic tests.

12. Core and single-hole measurements, conducted over short segments of a borehole, provide information about only a small volume of rock in the immediate vicinity of each measurement interval. Available data from the ALRS indicate that rock properties, measured on such small scales, vary erratically in space in a manner that renders the rock randomly heterogeneous and pneumatically anisotropic. Local-scale air permeabilities from single-hole tests vary by orders of magnitude between test intervals across the site; their spatial variability is much more pronounced than their dependence on applied pressure. We found it possible to interpolate some of the core and single-hole measurements at the ALRS between boreholes by means of geostatistical methods, which view the corre-

sponding variables as correlated random fields defined over a continuum. This was especially true about air permeability, porosity, fracture density, water content, and the van Genuchten water retention parameter α, for each of which we possess enough measurements to constitute a workable geostatistical sample. To differentiate between geostatistical models that appear to fit these data equally well, we used formal model discrimination criteria based on maximum likelihood and the principle of parsimony (which places a premium on simplicity; Illman et al., 1998). Standard geostatistical analysis provides best (minimum variance) linear unbiased estimates of how each such quantity varies in three-dimensional space, together with information about the quality of these estimates. Our finding supports the application of continuum flow and transport theories and models to unsaturated fractured porous tuffs at the ALRS on scales of 1 m or more.

13. Cross-hole pneumatic injection test data from individual monitoring intervals at the ALRS have proven amenable to analysis by type-curve and numerical inverse models that treat the rock as a uniform and isotropic fractured porous continuum. Analyses of pressure data from individual monitoring intervals by the two methods provided information about pneumatic connections between injection and monitoring intervals, corresponding directional air permeabilities, and air-filled porosities. All of these quantities were found to vary considerably from one monitoring interval to another in a given cross-hole test on scales ranging from a few to over 20 meters. Thus, even though the analyses treat the rock as if it was pneumatically uniform and isotropic, they ultimately yield information about the spatial and directional dependence of pneumatic connectivity, permeability, and porosity across the site.

14. Some single-hole pressure records reveal an inflection that is characteristic of dual-continuum behavior. It is possible that this inflection is caused by barometric pressure fluctuations and not by a dual-continuum phenomenon. The prevailing interpretation of dual continua is that one represents the fracture network and the other embedded blocks of rock matrix. We take the broader view that multiple (including dual) continua may represent fractures on a multiplicity of scales, not necessarily fractures and matrix.

15. The pneumatic permeabilities of unsaturated fractured tuffs at the ALRS vary strongly with location, direction, and scale. In particular, the mean of pneumatic permeabilities increases, and their variance decreases, with distance between packers in a single-hole injection test, and with distance between injection and monitoring intervals in cross-hole injection tests. This scale effect is most probably due to the presence in the rock of various size fractures that are interconnected on a variety of scales.

ACKNOWLEDGMENTS

This work was supported by the U.S. Nuclear Regulatory Commission under contracts NRC-04-95-038 and NRC-04-97-056. We wish to acknowledge with gratitude the support, advice, and encouragement of our NRC Project Manager, Thomas J. Nicholson. Walter Illman was supported in part by a National Science Foundation Graduate Traineeship during 1994-1995, a University of Arizona Graduate College Fellowship during 1997-1998, the Horton Doctoral Research Grant from the American Geophysical Union during 1997-1998, and the John and Margaret Harshbarger Doctoral Fellowship from the Department of Hydrology and Water Resources at The University of Arizona during 1998-1999. Velimir (Monty) Vesselinov conducted part of his simulation and inverse modeling work during a summer internship with the Geoanalysis Group at Los Alamos National Laboratory in 1997. We are grateful to Dr. George A. Zyvoloski for his help in the implementation of FEHM, and to Dr. Carl W. Gable for his assistance in the use of the X3D code to generate the corresponding computational grid. All pneumatic cross-hole tests at the ALRS were conducted by Walter Illman with the help of Dick Thompson, our talented and dedicated technician. Type-curve development and analyses were performed by Walter Illman, geostatistical analyses by Guoliang Chen and Velimir Vesselinov, and the development and implementation of our inverse model by Velimir Vesselinov.

REFERENCES

Agarwal, R. G., R. Al-Hussainy, H. J. Ramey Jr., 1970. An investigation of wellbore storage and skin effect in unsteady liquid flow. 1. Analytical treatment. AIME Trans. 279-290.

Al-Hussainy R., and H. J. Ramey Jr., 1966. The flow of real gases through porous media. JPT 624-636.

Bassett, R. L., S. P. Neuman, T. C. Rasmussen, A. Guzman, G. R. Davidson, and C. L. Lohrstorfer, 1994. Validation Studies for Assessing Unsaturated Flow and Transport Through Fractured Rock. NUREG/CR-6203.

Bassett, R. L., S. P. Neuman, P. J. Wierenga, G. Chen, G. R. Davidson, E. L. Hardin, W. A. Illman, M. T. Murrell, D. M Stephens, M. J. Thomasson, D. L. Thompson, and E. G. Woodhouse, 1997. Data Collection and Field Experiments at the Apache Leap Research Site, May 1995-1996. NUREG/CR-6497.

Boardman C. R., and J. Skrove, 1966. Distribution in fracture permeability of a granitic rock mass following a contained nuclear explosion. JPT 619-623.

Chen G., S. P. Neuman, and P. J. Wierenga, 1997. Infiltration tests in fractured porous tuffs at the ALRS. In: Bassett, R. L., S. P. Neuman, P. J. Wierenga, G. Chen, G. R. Davidson, E. L. Hardin, W. A. Illman, M. T. Murrell, D. M Stephens, M. J. Thomasson, D. L. Thompson, and E. G. Woodhouse. Data Collection and Field Experiments at the Apache Leap Research Site, May 1995-1996. NUREG/CR-6497.

Chuang, Y., W. R. Haldeman, T. C. Rasmussen, and D. D. Evans, 1990. Laboratory Analysis of Fluid Flow and Solute Transport Through a Variably Saturated Fracture Embedded in Porous Tuff. NUREG/CR-5482, 328 pp.

Doherty, J., L. Brebber, and P. Whyte, 1994. PEST: Model Independent Parameter Estimation. Watermark Computing; Corinda, Australia.

Evans, D. D., 1983. Unsaturated Flow and Transport Through Fractured Rock: Related to High-Level Waste Repositories. NUREG/CR-3206, 231 pp.

Evans, D. D., and T. C. Rasmussen, 1991. Unsaturated Flow and Transport Through Fractured Rock Related to High-Level Waste Repositories, Final Report: Phase III. NUREG/CR-5581, 75 pp.

Green, R. T., and D. D. Evans, 1987. Radionuclide Transport as Vapor Through Unsaturated Fractured Rock. NUREG/CR-4654, 163 pp.

Gringarten, A. C., 1979. Flow Test Evaluation of Fractured Reservoirs. Symposium On Recent Trends in Hydrology. Berkeley, Calif., February 8-11.

Gringarten, A. C., 1982. Flow Test Evaluation of Fractured Reservoirs. Geological Society of America, Special Paper 189, 237-263.

Gringarten, A. C., and H. J. Ramey Jr., 1974. Unsteady-state pressure distributions created by a well with a single horizontal fracture, partial penetration, or restricted entry. SPEJ, 413-426, Trans. AIME 257.

Gringarten, A. C., H. J. Ramey Jr., and R. Raghavan, 1974. Unsteady-state pressure distributions created by a well with a single infinite-conductivity vertical fracture. SPEJ, 347-360, Trans. AIME 257.

Guzman, A. G., and S. P. Neuman, 1996. Air injection experiments. In: Rasmussen, T. C., S. C. Rhodes, A. Guzman, and S. P. Neuman. Apache Leap Tuff INTRAVAL Experiments: Results and Lessons Learned. NUREG/CR-6096.

Guzman, A. G., S. P. Neuman, C. Lohrstorfer, and R. Bassett, 1994. In: Bassett, R. L., S. P. Neuman, T. C. Rasmussen, A. G. Guzman, G. R. Davidson, and C. L. Lohrstorfer. Validation Studies for Assessing Unsaturated Flow and Transport Through Fractured Rock. NUREG/CR-6203.

Guzman, A. G., A. M. Geddis, M. J. Henrich, C. Lohrstorfer, and S. P. Neuman, 1996. Summary of Air Permeability Data From Single-Hole Injection Tests in Unsaturated Fractured Tuffs at the Apache Leap Research Site: Results of Steady-State Test Intepretation. NUREG/CR-6360.

Hsieh, P. A., and S. P. Neuman, 1985. Field Determination of the three-dimensional hydraulic conductivity tensor of anisotropic media. 1. Theory. Water Resour. Res. 21(11): 1655-1665.

Huang, C., and D. D. Evans, 1985. A 3-Dimensional Computer Model to Simulate Fluid Flow and Contaminant Transport Through a Rock Fracture System. NUREG/CR-4042, 109 pp.

Illman, W. A., 1999. Ph.D. dissertation, The University of Arizona, Tucson.

Illman, W. A., D. L. Thompson, V. V. Vesselinov, G. Chen, and S. P. Neuman, 1998. Single- and Cross-Hole Pneumatic Tests in Unsaturated Fractured Tuffs at the Apache Leap Research Site: Phenomenology, Spatial Variability, Connectivity and Scale. NUREG/CR-5559, 188 pp..

Joseph, J. A., and L. F. Koederitz, 1985. Unsteady-state spherical flow with storage and skin. SPEJ 804-822.

Klinkenberg, L. J., 1941. The permeability of porous media to liquids and gases. American Petroleum Institute, Drilling and Production Practice, 200-213.

LeCain, G. D., 1995. Pneumatic testing in 45-degree-inclined boreholes in ash-flow tuff near Superior, Arizona. U.S. Geological Survey Water-Resources Investigations Report 95-4073.

LeCain, G. D., 1996a. Air-injection testing in vertical boreholes in welded and nonwelded tuff, Yucca Mountain, Nevada. U.S. Geological Survey Milestone Report 3GUP610M, Denver, Colorado.

LeCain, G. D., and J. N. Walker, 1996b. Results of air-permeability testing in a vertical borehole at Yucca Mountain, Nevada. Radioactive Waste Management, 2782-2788.

Mishra, S., G. S. Bodvarsson, and M. P. Attanayake, 1987. Injection and falloff test analysis to estimate properties of unsaturated fractured fractures. In: Flow and Transport Through Unsaturated Fractured Rock. D. D. Evans and T. J. Nicholson, eds. American Geophysical Union, Geophysical Monograph 42.

Montazer, P. M., 1982. Permeability of Unsaturated, Fractured Metamorphic Rocks Near An Underground Opening. Ph.D. thesis, Colorado School of Mines, Golden, Colorado.

Neuman, S. P., 1987. Stochastic continuum representation of fractured rock permeability as an alternative to the REV and fracture network concepts. In: I. W. Farmer, J. J. K. Dalmen, C. S. Desai, C. E. Glass, and S. P. Neuman, eds. Rock Mechanics, Proceedings of the 28th U.S. Symposium. Balkema, Rotterdam, 533-561.

Odeh, A. A., 1965. Unsteady-state behavior of naturally fractured reservoirs. SPEJ 5:60-65.

Paillet, F. L., 1993. Using borehole geophysics and cross-borehole flow testing to define connections between fracture zones in bedrock aquifers. Journal of Applied Geophysics 30:261-279.

Raghavan, R., 1993. Well Test Analysis. Prentice Hall Petroleum Engineering Series, Prentice-Hall, Inc., 558 pp.

Rasmussen, T. C., and D. D. Evans, 1987. Unsaturated Flow and Transport Through Fractured Rock-Related to High-Level Waste Repositories, NUREG/CR-4655, 474 pp.

Rasmussen, T. C., and D. D. Evans, 1989. Fluid Flow and Solute Transport Modeling Through Three-Dimensional Networks of Variably Saturated Discrete Fractures. NUREG/CR-5239, 193 pp.

Rasmussen, T. C., and D. D. Evans, 1992. Nonisothermal Hydrologic Transport Experimental Plan. NUREG/CR-5880, 41 pp.

Rasmussen, T. C., D. D. Evans, P. J. Sheets, and J. H. Blanford, 1990. Unsaturated Fractured Rock Characterization Methods and Data Sets at the Apache Leap Tuff Site. NUREG/CR-5596, 125 pp.

Rasmussen, T. C., D. D. Evans, P. J. Sheets, and J. H. Blanford, 1993. Permeability of Apache Leap Tuff: Borehole and core measurements using water and air. Water Resour. Res. 29(7): 1997-2006.

Rasmussen, T. C., S. C. Rhodes, A. Guzman, and S. P. Neuman, 1996. Apache Leap Tuff INTRAVAL Experiments: Results and Lessons Learned. NUREG/CR-6096, 94 pp.

Schrauf, T. W., and D. D. Evans, 1984. Relationship Between the Gas Conductivity and Geometry of a Natural Fracture. NUREG/CR-3680, 131 pp.

Tidwell, V. C., T. C. Rasmussen, and D. D. Evans, 1988. Saturated hydraulic conductivity estimates for fractured rocks in the unsaturated zone. In: Proceedings of International Conference and Workshop on the Validation of Flow and Transport Models for the Unsaturated Zone. P. J. Wierenga, ed. New Mexico State University, Las Cruces, N.M.

Vickers, B. C., S. P. Neuman, M. J. Sully, and D. D. Evans, 1992. Reconstruction geostatistical analysis of multiscale fracture apertures in a large block of welded tuff. Geoph. Res. Let. 19(10):1029-1032.

Warren, J. E., and P. J. Root, 1963. The behavior of naturally fractured reservoirs. SPEJ 3:245-255.

Weber, D. S., and D. D. Evans, 1988. Stable Isotopes of Authigenic Minerals in Variably-Saturated Fractured Tuff. NUREG/CR-5255, 70 pp.

Yeh, T. C., T. C. Rasmussen, and D. D. Evans, 1988. Simulation of Liquid and Vapor Movement in Unsaturated Fractured Rock at the Apache Leap Tuff Site: Models and Strategies. NUREG/CR-5097, 73 pp.

Zyvoloski, G. A., Z. V. Dash, and S. Kelkar, 1988. FEHM: Finite element heat and mass transfer. Tech. Rep. LA-11224-MS. Los Alamos National Laboratory.

Zyvoloski, G. A., B. A. Robinson, Z. V. Dash, and L. L. Trease, 1996. Users manual for the FEHMN application. Tech. Rep. LA-UR-94-3788, Los Alamos National Laboratory.

Zyvoloski, G. A., B. A. Robinson, Z. V. Dash, and L. L. Trease, 1997. Summary of the models and methods for the FEHM application: A finite-element heat- and mass-transfer code. Tech. Rep. LA-13307-MS, Los Alamos National Laboratory.

11

Parameterization and Upscaling in Modeling Flow and Transport in the Unsaturated Zone of Yucca Mountain

Gudmundur S. Bodvarsson,[1] Hui Hai Liu,[1] C. Fredrik Ahlers,[1] Yu-Shu Wu,[1] and Eric Sonnenthal[1]

ABSTRACT

Parameterization and upscaling for unsaturated rocks are discussed with application to Yucca Mountain in Nevada, the potential site for a geological repository of high-level nuclear waste. A complex three-dimensional model of the unsaturated zone at Yucca mountain (UZ model) has been developed (Bodvarsson et al., 1997) utilizing all of the available data from the site, including those from numerous surface boreholes and underground tunnels. As the typical grid spacing used in the UZ model is on the order of tens of meters to more than 100 m, the parameterization for the model based on much smaller-scale in situ tests and core measurements must be carefully conducted. A methodology for doing this has been developed and used in the UZ model; this methodology is applicable to most other fractured unsaturated sites. The primary upscaling process used in the UZ model is a direct inversion of the observed data, including saturation, moisture tension, pneumatic, perched water, temperature, and geochemical data. The process starts with one-dimensional inversions and then proceeds to two- and three-dimensional inversions.

INTRODUCTION

The unsaturated zone at Yucca Mountain, Nevada, is being considered by the United States Department of Energy (DOE) as a potential site for the geologic disposal of high-level radioactive waste. The unsaturated zone consists of a series

---

[1] Earth Sciences Division, Lawrence Berkeley National Laboratory

of welded and unwelded volcanic tuff layers with different hydrological characteristics and degree of fracturing. The welded units are heavily fractured and characterized by fracture permeabilities that are many orders of magnitude higher than the corresponding matrix permeabilities. The unwelded or poorly welded units, on the other hand, have significant matrix permeabilities and fewer fractures. Prior to the site characterization activities at Yucca Mountain, little work had been performed in deep unsaturated fractured rocks. Most relevant previous research was performed in shallow unsaturated near-surface soils. Because of this, the site characterization activities at Yucca Mountain have been extremely challenging, with a need to develop relevant theories and hypotheses to explain the observed data from this extremely complex system.

The unsaturated zone at Yucca Mountain is about 600 m thick, and the area of characterization amounts to some 50 km^2. Thus, a rock mass with a volume of 30 km^3, including approximately 10^9 fractures and the corresponding matrix blocks, needs to be characterized. It is clear that only a small fraction of this volume and fractures can be characterized with an economical number of boreholes and underground tunnels. Furthermore, the available measurements of important flow and transport parameters are only possible on spatial scales much smaller than what is needed for a large site-scale model intended to represent global parameters, processes, and conditions. Typically, the grid size for a site-scale model is tens of meters, if not more than 100 m. This creates an enormous challenge in the development of the appropriate parameterization for a site-scale model by taking into account parameter measurements performed in laboratories and the corresponding measurement scales, in-situ dynamic and static observations and their scales, and in situ measurements and the scales they represent. All of this needs careful evaluation of the appropriate parameterization and upscaling practices that are required for the development of a robust, defensible flow and transport model of the unsaturated zone at Yucca Mountain.

In this chapter, we describe the methodologies we are using in the development of the flow and transport model (UZ model) of the unsaturated zone at Yucca Mountain. We will first briefly describe the available data and the basic elements of the conceptual model of the unsaturated zone flow and transport. Then, the development of the hydrological and transport parameters are discussed with reference to many relevant issues such as upscaling, applicability of conventional unsaturated flow theories to fractured rocks, fracture and fault parameters, and others. Parameter estimation and one-dimensional data inversion are described using the methodology we have developed over the last few years (Bodvarsson et al., 1997a; Bandurraga and Bodvarsson, 1997; Liu et al., 1998); this again considers upscaling and other processes. These inversion activities are then augmented by a three-dimensional numerical analysis that folds in and considers multidimensional flow and transport effects, the incorporation of thermal analysis, and the extremely important calibration of data associated with perched-water bodies. Finally, the geochemical data and analysis are described in terms of available data, calibration activities, and their constraining value in the development of the UZ flow and transport model.

CONCEPTUAL MODEL

The evolution of the conceptual model of the unsaturated zone at Yucca Mountain is given in Chapter 2. Here we briefly describe the essential features of the conceptual model for water flow through Yucca Mountain. Plate 5 shows a schematic view of a simplified model for water flow through the mountain. Concentrated infiltration into the Tiva Canyon welded unit (TCw) takes place through narrow zones mostly on ridgetops, where exposed fractures are present (Flint et al., 1996). The infiltration is sporadic with major episodes estimated to occur every few years on average. Episodic fracture flow is believed to be prevalent in the Tiva Canyon, with travel time from the ground surface to the Paintbrush nonwelded unit (PTn) only on the order of one year. These episodic pulses are dampened in the PTn due to its porous medium-type characteristics, with high matrix porosity and permeability. The PTn consists of a series of nonwelded tuff layers with variable porosity and permeability that average about 40 percent and 300 md, respectively (Flint, 1998). Flow through this unit is believed to be primarily vertical (Wu et al., 1999a), although some lateral flow may occur due to its layering structure and associated heterogeneities, and also capillary redistribution. Water flow within the Topopah Springs unit (TSw) is dominated by fracture flow, estimated to range from 50-90 percent of the total flow (Wu et al., 1999a), depending on the fracture and matrix hydrological parameters of the subunits. In general, most of the subunits are heavily fractured with fracture spacings of 20-50 cm (Sweetkind and Williams-Stroud, 1996). However, only a small percentage of these fractures are believed to currently transmit liquid water through the mountain. Pruess (1999) and Pruess et al. (1997) estimated spacings of water flow paths (weeps) in the TSw to be on the order of 10 m, based on various hydrological and geochemical observations. Thus, one expects that there are hundreds of thousands of water flow paths of various sizes within the repository unit at Yucca Mountain.

The main tuff units below the repository are the Calico Hills (CHn) and Prow Pass, Crater Flat (CFu) units. Both of these units have vitric and zeolitic components that differ by the degree of hydrothermal alteration. The zeolitic rocks have low matrix permeabilities, on the order of microdarcies, and some fracture permeabilities. The current conceptual model, primarily based on perched water data, favors a small amount of water flow through the zeolitic units, with most of the water flowing laterally in perched water bodies and then vertically down through highly permeable faults (Wu et al., 1999b; Kwicklis et al., 1999; Flint et al., 1999). The vitric units, on the other hand, possess relatively high matrix permeability (0.1 to 10 Darcies), and therefore porous medium-type flow dominates. Fracture flow is believed to be limited in these units. A much more detailed description of conceptual models for water flow at Yucca Mountain is given by Kwicklis et al. (1999).

HYDROLOGICAL PARAMETERS

Parameterization of the Hydraulic Property Distribution

An important aspect in characterizing a site is to use a number of parameters to represent heterogeneous hydraulic property distributions at the site. These parameters are determined or inferred from the relevant observed data. Heterogeneities at different scales are presented in fractures and matrix. A variety of approaches are available for constructing subsurface heterogeneous hydraulic property distributions, as reviewed by Koltermann and Gorelick (1996). A geology-based, deterministic approach, in which entire model layers are assigned uniform hydraulic properties, has been mainly used for characterizing the unsaturated zone of Yucca Mountain (e.g., Bodvarsson et al., 1997a). The reasons for this are as follows: first, it is generally believed that overall behavior of site-scale flow and transport processes are mainly determined by relatively large-scale heterogeneities associated with the geological structure of the mountain. Second, the complexity of a heterogeneity model needs to be consistent with the availability of relevant data. More complicated models introduce a larger degree of uncertainty in rock property estimations based on inverse modeling, when data are limited. Third, the adopted layered approach is also supported by field observations, such as matrix saturation distributions. For a given geological unit, measured matrix saturation distributions are similar in different boreholes, indicating that large-scale matrix flow behavior and effective fracture and matrix hydraulic parameters should be similar within a unit. Note that matrix saturation distributions are determined by both fracture and matrix properties through fracture-matrix interactions.

Matrix Parameters and Upscaling

Core-scale values for matrix permeability and van Genuchten (1980) parameters are available for each model layer from measurements in several boreholes (Flint, 1998). A practical upscaling approach has been developed to obtain the large-scale effective parameter values for model layers from these small-scale measurements.

It is well known that hydraulic parameters for porous media are generally scale-dependent. For example, large-scale effective permeability can be considerably larger than that measured at a small scale (Neuman, 1990, 1994). A variety of theories have been developed for upscaling parameters such as permeability (Neuman, 1990, Paleologos et al., 1996). However, these theories may not be directly applicable for the matrix in unsaturated fractured rocks due to fracture-matrix interaction. Large fractures can act as capillary barriers for flow between matrix blocks separated by these fractures, even when the matrix is essentially saturated (capillary pressure is close to the air entry value). Therefore, the existence of fractures could reduce the effective permeability of the matrix continuum. Based on these considerations, the currently available relations for upscaling porous medium permeability (Paleologos et al., 1996) are not used for determining effective permeability for model layers. Instead, a two-step approach is employed. We use geometric means of core measurements to determine effective permeabilities for model layers. Then, parameter calibration by data inversion, as discussed in a later section of this paper, is employed to further refine the permeability values.

In general, the upscaling of the water retention curve, characterized by van Genuchten parameters, is a more difficult task. Simply speaking, the main reason behind the upscaling is differences in liquid water distribution mechanisms at different scales. On the core-scale, the liquid water distribution is controlled by the capillary force. Small pores are always filled with liquid water before relatively large pores during a wetting process. At a larger scale, however, the water distribution mechanism is more complicated. As an extreme example, when subgrid fingering occurs for a large grid block, liquid water is distributed based on the fingering pattern that results from gravitational instability and heterogeneity, rather than pore size distribution. Obviously, these differences give rise to very different water retention curves at different scales, if the retention curve can be defined at large scales.

Compared with most natural soils, however, the matrix in these highly fractured rocks has a very small average pore size. In other words, capillary force plays a very important role in the water-flow process within the matrix. It is therefore reasonable to assume that matrix liquid water distribution within a site-scale grid block, similar to that at the core scale, is mainly controlled by capillary force under the steady-state flow condition. In this case, the upscaled retention curve can be simply obtained by averaging the retention curves measured from core samples for a given model layer. To further clarify this point, let us consider an ideal case, in which the capillary force is so strong that there is a uniform capillary pressure distribution within the grid-block while local retention curves are very different. Obviously, the retention curve for the grid block can be estimated as

$$\theta_{block}(P) = \frac{1}{N} \sum_{i=1}^{N} \frac{n_i \theta_i(P)}{\bar{n}}, \tag{11.1}$$

where θ_{block} is the average saturation of the grid block, P is the capillary pressure, N is the number of core samples within the block, n_i and $\bar{n}$ are core sample porosity and average porosity for the grid-block, respectively, and θ_i is the liquid water saturation for sample i at capillary pressure P. Although Equation 11.1 was developed for a grid block, it can be used for a model layer if uniform hydraulic property distributions are assumed within the model layer.

In summary, compared to other subsurface porous media, matrix in fractured rocks is unique in two aspects, the existence of fractures and the generally strong capillarity. These aspects make the currently available upscaling approaches for matrix permeability, developed from stochastic hydrology, not applicable. While the focus of this study is on the development of physically reasonable and practically workable approaches, further study on upscaling of unsaturated properties of matrix in fractured rocks, based on stochastic methodologies, is useful for refining the approaches developed in this study.

Analysis and Development of Fracture Hydrologic Parameters

The purpose of this section is to provide an overview for incorporating measured properties of fractured rock in a large-scale model to capture flow and transport in unsaturated rocks. A conceptual model of unsaturated flow in fractured rock must represent processes within individual fractures, in fracture networks, and the coupling between these systems and the rock matrix. The characterization of flow in fractured rock requires consideration of the orientation, connectivity, and aperture of fractures at different scales. In contrast to saturated systems, additional hydrologic parameters (van Genuchten parameters) are required that cannot be measured or calculated as satisfactorily as permeability or porosity. Furthermore, the equations describing unsaturated flow in fractures are usually loosely borrowed from soil physics, with little experimental work to support their applicability. Thus, we must use any calculated parameters with great caution, utilizing data inversion, to be discussed later on, to refine the values so that the overall system behavior is captured.

Considerable progress has been made recently in improving our understanding of unsaturated flow in fractures (Bodvarsson et al., 1997a). Glass et al. (1996) used lab-scale experiments to demonstrate that the main flow mechanism for a vertical unsaturated fracture is fingering as a result of gravitational instability and aperture heterogeneities, which is further supported by the numerical model studies of Pruess et al. (1997). Tokunaga and Wan (1997) showed that film water flow along the rough fracture surface could be important when the matrix potential is close to or larger than the air entry value. Faybishenko (1999) suggested that unsaturated fracture flow might be described using the nonlinear dynamics (chaos).

Figure 11-1 shows fractures at different scales. It is a very challenging task to develop an approach to describe large-scale fracture flow, which incorporates fracture flow mechanisms observed at smaller scale. While different approaches are available, we believe that a continuum approach is a suitable and robust approach for the large-scale fracture flow in the unsaturated zone of Yucca Mountain. This is mainly based on the following considerations. First, bomb-pulse ^{36}Cl has been found at depth in Yucca Mountain at only a few locations that are associated with geological features (Fabryka-Martin et al., 1996), and there is no

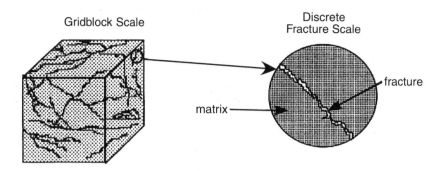

FIGURE 11-1 Schematic of fractures at different scales.

correlation between these locations and enhanced matrix water saturation and potential, suggesting that the related fast-flow paths likely carry a small amount of water. Second, the existence of many dispersed flowing fractures, allowing the continuum approximation, is also supported by many field observations. For example, very similar matrix saturation distributions were observed from different deep boreholes. Calcite coatings, signatures of water flow history, are found in many fractures within the welded units. It may also be important to note that it is difficult, if not impossible, to construct and calibrate a discrete fracture-network model for a site-scale problem. However, we recognize the importance of detailed study on relations between small-scale (say, a single fracture scale) flow processes and those at large scales.

Commonly, there is a positive correlation between fracture permeability and the scale of the system. Measurements of air permeability done at Yucca Mountain over a wide range of scales allow for an examination of this scale dependence (Figure 11-2). Here we show the range in fracture permeability measurements within the welded zones of the Topopah Spring Tuff. Faults are the largest scale features and exhibit the highest permeability along their strike (lateral) for a measurement scale of approximately 500-700 m. Vertical permeability in faults, measured at a scale of some 100-200 meters, is about an order of magnitude less. Air permeability in rocks outside of faults is shown in the hatched and shaded boxes, and by the dashed arrow denoted "niche." The upper permeability limit is similar for all measurement scales, except at the largest scale, with the lower permeability limit generally decreasing at the smaller scales. This relationship is fully expected, because the probability of intersecting a conductive fracture declines as the volume of the region probed diminishes. The unfractured rock matrix has a maximum permeability of about 10^{-16} m² with a minimum value below the detection limit of about 5×10^{-19} m², thus setting the lower bound for any larger-scale permeability measurement.

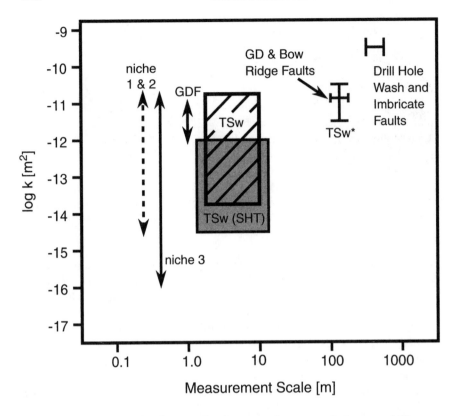

FIGURE 11-2 Schematic diagram showing range in measured air permeability as a function of the measurement scale. TSw* and GD and Bow Ridge faults refer to calibrated permeability based on matching to ambient pneumatic signal (Bandurraga and Bodvarsson, 1997). GDF refers to Ghost Dance Fault (air-k test; LeCain and Patterson, 1997). TSwSHT measurements refer to Single-Heater Test pretest characterization air-k tests (Tsang et al., 1999). TSw refers to in-situ tests in boreholes (LeCain, 1997).

Recently, a number of researchers (Neuman, 1990, 1994; Molz et al., 1997; Liu and Molz, 1997) have indicated that in many cases subsurface heterogeneities can be characterized by stochastic fractals. The air permeability data seem to support this reasoning. According to Neuman (1990, 1994), a fractal-based relation between effective permeability and measurement (support) scale can be given as

$$\ln\left(\frac{k}{k_g}\right) \propto L^{2H} \tag{11.2}$$

where k is the effective permeability, k_g is the geometric mean, L is the measurement scale, and H is the Hurst coefficient. When the fractional Brownian motion (fBm) is applicable for characterizing subsurface heterogeneities, previous studies showed that the H values range from 0.1-0.5, and the average value is about 0.25 (Neuman, 1990, 1994; Molz et al., 1997; Liu and Molz, 1997). Figure 11-3 also shows a curve determined from Equation 11.2 using $H = 0.25$ and $\log(k_g) = -13.09$, as compared with the average $\log(k)$ data. The curve is consistent with the overall trend of the air permeability scale-dependent behavior. One important implication of this result is that there may be some similarities between spatial variations of fracture permeability at different scales. The above discussion was used to demonstrate the complex scale-dependent behavior of fracture permeability. Different fracture permeability values should be used for modeling flow and transport at different scales. The permeability values for the site-scale model were determined by calibrating the model against field observations, including ambient pneumatic signals, while the small-scale (<10 m) measurements were employed as initial guesses for model calibration. A more detailed discussion is given in the section "Parameter Estimation/Data Inversion."

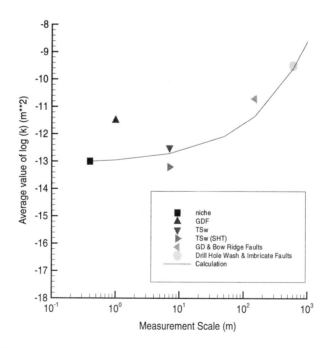

FIGURE 11-3 Comparison of the average $\log(k)$ data obtained from Figure 11-2 and the curve calculated from Equation 11.2.

Once the permeability (k) of the fracture system has been determined, the equivalent hydraulic fracture aperture (b) may be estimated using a relation such as the cubic law, as follows:

$$k = \frac{fb^3}{12},$$

(11.3)

where f is the fracture frequency. Assuming a simplified form (Altman et al., 1996) of the Young-Laplace equation, values of α (Pa^{-1}) may be calculated directly from the fracture aperture b,

$$\alpha = \frac{b\rho g}{2\tau\cos\theta},$$

(11.4)

where ρ is the density of water, g is the acceleration due to gravity, τ is the surface tension of water, and θ is the contact angle (here assumed to be zero). Although an estimate of α may be obtained in this way, it is preferable to consider a range of apertures determined from the spread of measured permeability and fracture frequencies, followed by curve-fitting using, for example, the method of van Genuchten (1980), as done for the Topopah Spring welded tuff (Sonnenthal et al., 1997).

The fracture parameters, developed above, are intended for the fracture continuum. However, it is very important to note that van Genuchten models were developed and have been successfully used for describing water flow in porous media without fingering flow. Unlike the rock matrix, unsaturated flow in fractures is mainly gravity-driven, giving rise to fingering flow at both a single fracture scale and a fracture network scale. In this case, the van Genuchten relations may not be valid. Recently, we developed an "active fracture model," which is based on the hypothesis that not all connected fractures actively conduct liquid water in the unsaturated zone at Yucca Mountain because of fingering (Liu et al., 1998). In this model, van Genuchten relations are generalized by including a new parameter, which is used to characterize the degree of fingering in a connected fracture network. This additional parameter is estimated based on inverse modeling. A detailed description of the active fracture model and a procedure to determine values for the additional parameter are given in Liu et al. (1998).

TRANSPORT PARAMETERS

In this section, we discuss only parameters used to describe diffusion and dispersion processes. A discussion of reactive transport is given in a later section. In unsaturated fractured rock, transport processes occur in both fractures and rock matrix. Within the matrix, molecular diffusion is considered to be much more important than the mechanical dispersion due to high saturation and low pore velocity. Measured diffusion coefficient values for rock matrix at the Yucca Mountain site are summarized by Robinson et al. (1995). The values for saturated rocks are of order of 10^{-10} m^2/s. These lab-scale results can be directly used in the site-scale model, in which the matrix mechanical dispersion is ignored. Note that the diffusion process is generally not subject to upscaling.

Within the fracture continuum, diffusion can be ignored compared with dispersion. We expect the dispersivity for the fracture continuum to be very large, and it may be dependent on travel distance if a dispersivity can be defined for that continuum. However, a large-scale tracer test, which is needed to determine the dispersivity value for the site-scale model, is not considered to be a realistic task for the unsaturated zone of Yucca Mountain considering the temporal scale required for conducting the test. Since fracture dispersivity values from field observations are currently not available, in order to evaluate the importance of fracture dispersivity to the overall chemical transport behavior, we simulate tracer transport along a vertical column extracted from the three-dimensional model for the unsaturated zone of Yucca Mountain (Bodvarsson et al., 1997a). The flow is at steady state, and a constant concentration condition was used at the potential repository. Figure 11-4 shows simulated breakthrough curves at the water table for two different fracture dispersivity values (0 and 100 m). These breakthrough curves were obtained based on combined mass flux from fractures and matrix. A random walk particle method (Pan et al., 1999) is used such that numerical dispersion is essentially eliminated. These two breakthrough curves are very similar, indicating that the overall tracer transport behavior is not very sensitive to the fracture dispersivity value.

This insensitivity of the transport to fracture dispersivity is physically understandable. The large-scale dispersion results from subsurface heterogeneities. For a dual-continua system (matrix and fracture), the largest heterogeneity corresponds to property differences between the two continua. The importance of heterogeneities within each continuum becomes secondary. To further clarify this point, Figure 11-4 also shows a breakthrough curve with a matrix molecular diffusion coefficient value larger than that for the other two curves. A considerable change in the breakthrough curve is observed in this case. This is simply because mass transfer between fracture and matrix continua is directly related to the matrix molecular diffusion. Therefore, parameters to characterize mass transfer between fractures and matrix, such as matrix molecular diffusion coefficients and tortuosity, are the most important for chemical transport. In our current study, effective fracture/matrix interface area values are obtained from inverse modeling studies (see the section "Parameter Estimation/Data Inversion"). Note that this argument cannot be applied to the saturated fractured system, where flow and transport are generally characterized by a single (fracture) continuum while the matrix acts as a source/sink.

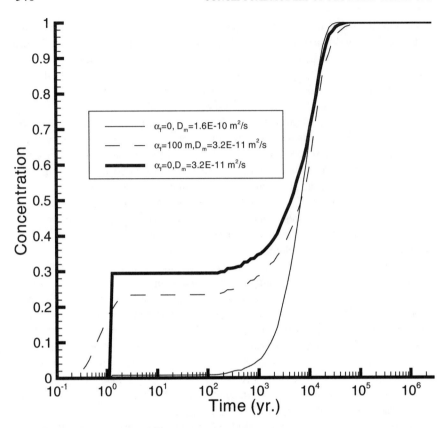

FIGURE 11-4 Breakthrough curves at the water table (α_f is the dispersivity for fracture continuum, and D_m is the matrix molecular coefficient).

PARAMETER ESTIMATION/DATA INVERSION

Approach

In order to further refine parameters for the UZ model, determined based on the procedures given in the section "Hydrological Parameters," data from the site are inverted.

Data inversion involves using numerical models to predict conditions in the unsaturated zone and then comparing them to observations of these conditions from the field (e.g., saturation data and gas pressure data). Model parameters are adjusted (calibrated) so that the difference between the model predictions and the observed data is minimized. This procedure can be carried out by trial and error, analyzing the results of each parameter adjustment by hand, or automatically using a program such as ITOUGH2 (Finsterle, 1999).

Data are inverted using a series of increasingly complex models. Three-dimensional modeling studies indicate that vertical flow of fluids and heat is predominant and therefore one-dimensional models are reasonable approximations for many units. In addition, one-dimensional models are used for initial parameter estimation, because of computational efficiency constraints. For the Yucca Mountain problem, we are trying to estimate approximately 210 parameters, which means that thousands of forward simulations are necessary when using an automated inversion program, such as ITOUGH2 (Finsterle, 1999). Moisture dispersion may limit the validity of the vertical percolation assumption. Perched water and lateral movement on top of low-permeability layers/lenses may also limit the validity of the vertical percolation assumption. However, we believe that the vertical percolation assumption is valid for Yucca Mountain from the surface to the bottom of the TSw. Below the TSw, where low-permeability/altered/zeolitized/perched water zones exist, the vertical percolation assumption may break down.

Using one-dimensional models, data from multiple boreholes are inverted simultaneously to estimate layer average properties. Saturation and moisture tension data inversion is used to estimate the matrix and fracture flow parameters, including permeability, a fracture-matrix interaction parameter, and unsaturated flow parameters (van Genuchten's α and m). Pneumatic pressure (barometric pumping response) data inversion is used to estimate fracture diffusivity, or alternatively either permeability or porosity; in our inversions, we choose permeability. Figure 11-5 shows the flow of one-dimensional inversions to multidimensional inversions that are used for estimating parameters for the UZ model.

Two-dimensional, cross-sectional models are used to refine parameter estimates. Two-dimensional models allow for lateral flow and therefore will provide a better estimate of properties of low-permeability/altered/zeolitized/perched water layers; then also allow for estimation of the properties of faults. In the two-dimensional models, faults are modeled perpendicular to the plane of the model. Again, saturation, water tension, and pneumatic pressure data are inverted.

Computational expense of using two-dimensional models versus one-dimensional models is increased for each forward simulation, but we are able to reduce the number of parameters that are being estimated because some of the parameters are reliably estimated by the one-dimensional inversions. Therefore, the overall computational expense of the automated two-dimensional model inversions does not increase substantially with respect to the automated one-dimensional model inversions.

Three-dimensional models are used for limited trial-and-error calibrations and verification of simulation results against many different data sets, including data sets used for one- and two-dimensional automated inversions. These will be

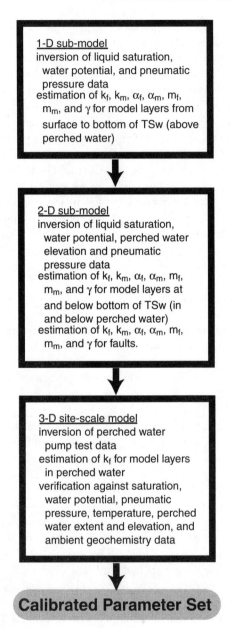

FIGURE 11-5 Flow chart showing inversion/calibration/verification approach. Parameter estimation progresses from one-dimensional inversion of saturation, water tension, and pneumatic pressure data, through two-dimensional inversion of fault and perched water data, to three-dimensional calibration/verification of all data.

discussed further in the section "Multidimensional Model Calibration and Sensitivity Analysis."

One-Dimensional Data Inversion

Saturation and potential data are inverted first. In order to speed up the inversion process, an attempt is made to eliminate some of the parameters being estimated. The sensitivity of all the parameters is evaluated periodically throughout the inversion process, which takes an iterative approach, and only those parameters that exceed a sensitivity cut-off are included in the set of parameters being estimated for the next few iterations. This is a feature of the ITOUGH2 code (Finsterle, 1999) that is used for the automated inversions. The quality of the data must also be considered. ITOUGH2 allows the data to be weighted; better quality data (judged by number of measurements and/or measurement uncertainty) can be more heavily weighted. Even automated inversion requires some human intervention to ensure that calibration is not being trapped in local minima and that the estimated parameters are reasonable. This is especially important when a large number of parameters is being estimated with relatively sparse data, and when nonuniqueness is a problem, as is the case here. Figure 11-6 shows the match between the simulated and observed matrix saturation and

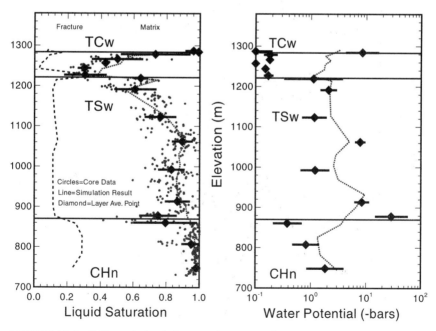

FIGURE 11-6 Calibrated simulation match to saturation and moisture tension data at borehole USW SD-9.

potential values for one of the boreholes used in the simultaneous inversion of several boreholes. Generally, the parameters best constrained by these matrix data are the matrix parameters (i.e., matrix permeability and unsaturated flow parameters).

Time-series pneumatic pressure data (subsurface response to barometric pumping) inversion is the second step in the one-dimensional inversion process. The simulated response to barometric pumping is mainly sensitive to fracture diffusivity, which is proportional to permeability divided by porosity. This is especially true in the densely welded, high-saturation rocks. Yucca Mountain fractures are assumed to be, on average, dry, so that gas-relative permeability of fractures may be assumed to be one or nearly one. Therefore the choice of gas-relative permeability parameters for the fractures is not crucial. Figure 11-7 shows the match between the simulated and observed pneumatic pressure values at one borehole used in the multiple borehole simultaneous inversion. Parameters estimated using inversion of pneumatic data are checked to see that they have not significantly changed saturation and moisture tension data matches.

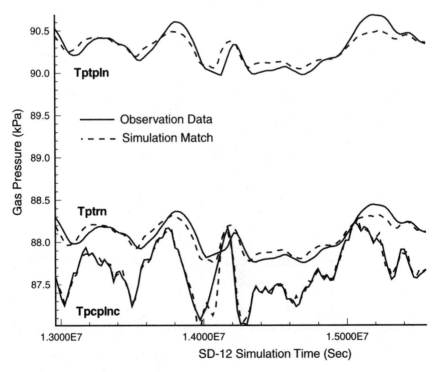

FIGURE 11-7 Calibrated simulation match to pneumatic pressure data at borehole USW SD-12.

Upscaling Issues

Inverting data using mountain-scale models inherently considers upscaling of rock parameters. Matrix parameters change little for many model layers during the inversion process. This is consistent with the previous discussion on the effect of fractures on the effective permeability of matrix in the section on hydrological parameters. Matrix permeability shows little increase, generally less than one-half an order of magnitude, with permeability even being reduced in some layers. Above the bottom of the TSw, where one-dimensional saturation and water potential inversions are reliable, matrix van Genuchten α also shows little variation. Fracture parameters, on the other hand, are expected to upscale because of the difference between the measurement scale and the connected mountain scale. Fracture permeability is measured on the 1- to 10-m air injection testing scale. In the TSw, the calibrated fracture permeability is one order of magnitude, or much larger than the average of the air injection measurements. This is because mountain-scale models and data consider larger fracture connectivity structures than those tested by air injection.

Parameter Uncertainty

ITOUGH2 quantitatively evaluates the uncertainty of the estimated parameters by comparing the match between the simulation and the data (saturation, water potential, etc.) for variations of each of the parameters. Matrix parameters typically have lower uncertainty than the fracture parameters because the saturation and water potential data are measured on the matrix. Where pneumatic data is used to calibrate fracture permeability, the uncertainty of the fracture permeability is much less than the uncertainty of the matrix permeability. This is because there are substantially more pneumatic data points than saturation and water potential data points.

Supplemental information and feedback from downstream users of the calibrated model parameters can also be used to qualitatively judge parameter uncertainty. Modeling of smaller-scale experiments performed at Yucca Mountain using the mountain-scale parameter sets provides invaluable insight into the multiscale applicability of the parameter sets.

MULTIDIMENSIONAL MODEL CALIBRATION AND SENSITIVITY ANALYSIS

Discussion

The three-dimensional nature of fluid flow, heat transfer, and chemical transport at Yucca Mountain makes it a challenge to conduct modeling analysis in such a complicated hydrological system. Even with a significant amount of field data collected from the Yucca Mountain site and good estimates of modeling parameters, as discussed above, a direct input of these parameters into a three-dimensional model may produce physically inconsistent results, as compared with observations. In particular, the model results cannot match certain important observed phenomena well, such as perched water occurrence, temperature profiles, and geochemical data. This is primarily due to limitations in data collected on both spatial and temporal scales of the three-dimensional models, and also because the one-dimensional and two-dimensional simplification and analysis cannot fully capture the complexity of flow and transport processes in the highly heterogeneous fracture-matrix system. In addition, there exist considerable uncertainties associated with data collected from the site. Furthermore, numerical model results depend not only on input rock and fluid parameters, but also on the conceptual hydrogeological models, spatial and temporal scaling and grid discretization, and numerical modeling approaches selected. Therefore, three-dimensional model calibration and sensitivity studies are necessary to obtain a consistent set of modeling parameters using all the available, observed data in combination with inverse and direct (forward) modeling analyses.

The methodology presented in this section consists of the final stage of parameter estimation process, involving the use of one-dimensional and two-dimensional inversion results in a three-dimensional model to perform further calibration and sensitivity analyses. Direct modeling studies are used here to calibrate multidimensional models against field-measured liquid saturation, water potential, perched water, environmental isotopes, geochemical, and temperature data. During this calibration process some rock parameter adjustments may be made to provide a better match between model results and observed field data in order to reduce parameter uncertainties and to evaluate the differences that occur due to changes in dimensionality of the model. This section focuses on model calibration studies using perched water, temperature, and geochemical data.

Lateral Flow and Perched Water Occurrence

Perched water in the vicinity of the potential repository at Yucca Mountain has been observed in several boreholes during the site characterization study (Patterson, 1998; Striffler et al., 1996). The presence of perched water bodies in the unsaturated zone of Yucca Mountain has many implications for assessment of the potential repository. At the same time, it may provide invaluable insights into water movement, flow and transport pathways, or surface infiltration history of the mountain (Wu et al., 1999b). Perched water at Yucca Mountain has been found to be associated with the densely welded basal vitrophyre of the Topopah Spring Tuff or with zeolitic intervals of the Crater Flat and Calico Hills Formations. The presence of perched water indicates that the vertical percolation flux locally exceeds the saturated hydraulic conductivity at those perching layers and suggests that flow paths may not be vertical through the entire thickness of the unsaturated zone to the water table. Instead, water may be diverted laterally to a fault zone or another high-permeability channel that serves to focus flow downward to the water table. As a result, substantial lateral flow may occur at perching layers, which leads to complex flow pathways of groundwater through the unsaturated zone fractured media.

Perched water is observed across Yucca Mountain within the lower TSw and the upper CHn hydrologic units, as indicated in a number of boreholes at Yucca Mountain, including UZ-14, SD-7, SD-9, SD-12, NRG-7a, G-2, and WT-24. The locations of these boreholes are shown in Plate 6 for a plan view. Also shown in the figure are the three-dimensional model domain of the UZ model and the horizontal grid.

The permeability of the zeolitic tuffs within CHn is generally very low (in microdarcies). This low value is converted to as low as 0.5 mm/yr or so of percolation flux in terms of flow capacity through the zeolitized tuffs. Recent studies (Flint et al., 1996) estimate an average net infiltration rate of 5 mm/yr over the UZ model domain. As a result, on average, about 90 percent of the water must bypass the low-permeability zeolitic units and travel through faults, other well-connected fractures, or high-permeability vitric formations to reach the saturated zone. This will affect radionuclide transport from the repository horizon to the water table, which directly impacts geologic repository performance. To capture these phenomena of lateral flow caused by water perching conditions, three-dimensional model calibrations are needed.

The genesis of perched water at Yucca Mountain is much debated among Yucca Mountain Project scientists, and several conceptual models have been discussed (Rousseau et al., 1998; Wu et al., 1999b). The permeability barrier conceptual model for perched-water occurrence has been used in unsaturated-zone flow modeling studies for the Yucca Mountain Project. In this conceptualization, perched-water bodies in the vicinity of the northern part of the model domain (near boreholes UZ-14, SD-9, NRG-7a, G-2, and WT-24) are assumed to lie along the base of the TSw, a zone of altered, more intensely fractured rock, underlain by low-permeability, zeolitized rock. The perched body in this northern area of the repository may be interconnected. The perched-water zones at boreholes SD-7 and SD-12 are considered to be localized, isolated bodies.

The perched-water numerical model used in this work is based on the three-dimensional UZ flow and transport model (Bodvarsson et al., 1997a). The UZ model domain, as shown in Plate 6, covers nearly 43 km^2 and extends from Yucca Wash in the north to the approximate latitude of borehole G-3 in the south. In the east-west direction, the model extends from the Bow Ridge Fault to approximately 1 km west of the Solitario Canyon Fault. The land surface of the mountain (or the tuff/alluvium contact, in areas of significant alluvial cover) is taken as the top model boundary. The water table is treated as the bottom boundary. Both top and bottom boundaries of the models are treated as Dirichlet-type conditions, with specified constant (but areally distributed) gas pressures and constant liquid saturation. A spatially distributed infiltration map (Flint et al., 1996) was used as the top water recharge boundary of the three-dimensional model. This map gives an average net infiltration rate of 4.9 mm/yr of water distributed over the site-scale model domain.

The simulation results presented in this section were obtained using the TOUGH2 code (Pruess, 1991), and the dual-permeability method was used to treat fracture and matrix interactions. The integral finite-difference model grid used consisted of 150,000 elements and 500,000 connections.

Many flow scenarios were investigated to calibrate the perched water model. The simulation results for these modeling scenarios were calibrated against the observed matrix liquid saturation and water potential data, and perching elevations. In general, it was found that the permeability barrier conceptual model of water perching could reproduce the moisture and perched-water conditions beneath Yucca Mountain. Plate 7 shows the extent of perched water, as predicted by the permeability barrier conceptual model. The simulation uses the calibrated perched-water parameters for fractures and matrix in several grid layers near the top of the CHn unit. In this figure, the blue isosurface reflects 100 percent liquid-saturation within fractures, indicating a perched-water body, while the green surface represents the top elevation of the CHn.

Plate 7 shows clearly that several perched bodies develop in the northern part of the model domain, located near the basal vitrophyre of the Topopah Spring Tuff above the top of the CHn unit. These perched bodies match the observed perching elevations from the boreholes in these areas.

The simulated percolation flux at the water table is shown in Plate 8. Comparing the fluxes and their distribution in Plate 8 with that of the surface infiltration map of Flint et al. (1996) indicates the occurrence of significant lateral flow or diversion. Significant lateral flow occurs mainly as water travels from the repository horizon to the water table in perched areas. A large amount of water, when travelling across the CHn, is diverted laterally to the east, along a dip, and intersects faults that focus flow downward to the water table.

The modeling study and sensitivity analysis of this work concludes that it is necessary to conduct three-dimensional model calibrations in order to describe perched water occurrences at Yucca Mountain. Several key factors were identified for creating a perched-water zone in a numerical model. These factors are (1) a water-perching geologic structure with low permeability zones and/or a capillary barrier underlain and surrounding perched-water zones, (2) weak capillary forces under high saturation condition within and near perched-water zones, and (3) sufficient water infiltration rates.

Heat Flow and Geothermal Conditions

In parameter estimation and model calibration using measured temperature data, the ambient geothermal condition at Yucca Mountain is assumed to be at a steady-state condition. The three-dimensional steady-state solution of fluid (water and air) and heat flow were used to match available temperature data measured at 26 boreholes located within the model domain (Wu et al., 1999a). The locations of some of these boreholes are also shown on Plate 6. The objective of the heat flow studies is to confirm if use of the parameters, derived from isothermal calibrations, can predict geothermal conditions at the mountain.

Unlike the above three-dimensional perched-water calibrations, which used a dual-permeability model, fracture-matrix interaction was treated using the effective continuum model (ECM) approach in this section. The ECM assumes a thermodynamic equilibrium between the fracture and the matrix, and adequately approximates steady-state heat transfer, which is dominated by heat conduction.

Measured temperature data (Sass et al., 1988; Rousseau, 1996) were compared with the three-dimensional simulation results for 26 boreholes within and near the study area. In general, temperature data from all the boreholes matched reasonably well with the three-dimensional model (Bodvarsson et al., 1997; Ahlers et al., 1995), in which surface and bottom temperatures were slightly adjusted. We present only one of the 26 comparisons conducted in the model calibration as demonstration examples. The selected borehole is H-5, and it is located in the middle of the model domain, shown in Plate 6. The simulated temperature profile is plotted in Figure 11-8. The simulated temperature profile matches the data very well, as was the case for all the other borehole data/ simulation comparisons.

The three-dimensional temperature calibrations have helped us in constraining percolation fluxes and estimating geothermal conditions for thermal studies (Bodvarsson et al., 1997; Wu et al., 1999a).

Unsaturated Zone Geochemistry

The geochemistry of water (and gas) in the unsaturated zone can be used as an important constraint on net infiltration rates, percolation fluxes, and flow pathways, and as an indicator of the extent of fracture-matrix and water-rock interaction. Mineral paragenesis, distribution, and composition are also useful, yet complex, systems for quantifying flow and transport processes. In contrast to hydrologic parameters, such as water potential or saturation, and possibly temperature, large-scale attainment of steady-state chemical distributions is unlikely, because molecular diffusivity is several orders of magnitude smaller than water or thermal diffusivity. Therefore, climate changes can have a long-lasting effect on the geochemistry of the unsaturated zone, especially in an arid environment such as at Yucca Mountain. There, chemical patterns at depth seem to reflect periods of higher infiltration during the last glacial period, while those close to the surface are consistent with the modern climate (Sonnenthal and Bodvarsson, 1999).

Conservative Chemical Species

The sensitivity of natural conservative tracers such as Cl to infiltration rate is well known, e.g., the chloride mass-balance method. For such constituents the

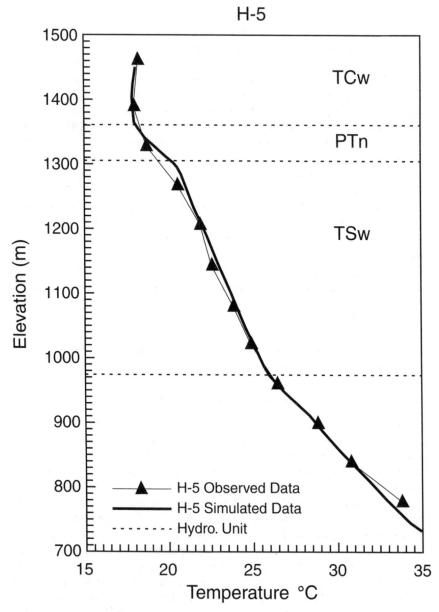

FIGURE 11-8 Comparison between measured and simulated temperatures for borehole H-5.

flux is related to the precipitation rate and additions from other sources such as windblown dust. The extent of evapotranspiration in the shallow soil zone then controls the concentration in infiltrating water. In arid environments, with significant topographic and soil cover thickness variations, the infiltration rate can vary by orders of magnitude over short distances (Flint et al., 1996). Thus, it is expected that spatial distributions of a conservative tracer in the subsurface may also vary by a similar range. In contrast to directly using a method such as chloride mass balance and making assumptions as to the absence of diffusion and lateral flow, limiting the system to one-dimensional piston-flow, it is preferable to incorporate directly such tracers into the calibrated flow model.

An example using chloride as a conservative tracer in UZ flow and transport simulations of Yucca Mountain is presented in Figure 11-9. Chloride fluxes to the

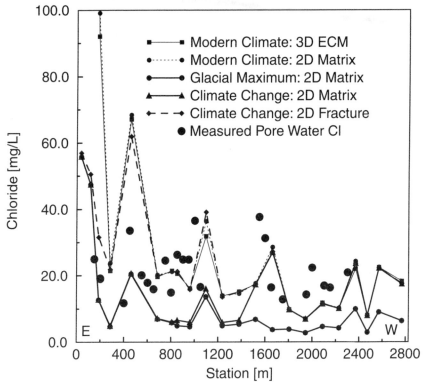

FIGURE 11-9 Prediction of chloride concentrations in pore waters prior to excavation of the ECRB, for two-dimensional dual permeability and three-dimensional ECM simulations, compared to data collected subsequently. The climate change simulation considered a glacial maximum infiltration rate, followed by 10,000 years of modern climate. From Flint et al. (1996).

surface were calculated using precipitation maps of the region (Bodvarsson et al., 2000) and assuming a uniform average effective concentration in precipitation (Fabryka-Martin et al., 1998). Combined with infiltration rates (Flint et al., 1996) the concentrations in infiltrating waters are generated. The simulations were performed as part of a prediction prior to the excavation of a tunnel at the potential repository level (Ritcey et al., 1998), and recent preliminary Cl results (Fabryka-Martin et al., 2000) confirm the trend of lower concentrations under areas of higher predicted net infiltration. Another outcome of the dual-permeability modeling is the remnant disequilibrium between fractures and matrix, resulting from the climate change that took place about 10,000 years ago. Under areas of lower infiltration, matrix pore waters are predicted to have retained their chloride concentrations from the last glacial period. These results are generally consistent with the ^{14}C ages of some perched waters (up to 8,000 to 12,000 years; Yang et al., 1996), and the average ^{14}C age of air at this level (~6,000 years; Yang et al., 1996); however, data from this particular zone will be needed for model validation.

The large-scale continuum model has proven remarkably effective in predicting the range in concentrations and their trends in the unsaturated zone (Sonnenthal and Bodvarsson, 1999). Looking a little closer though, it is clear from numerous chemical analyses of pore waters at Yucca Mountain (Fabryka-Martin et al., 1998), that variability is the norm at nearly every scale. The difficulty in the interpretation of this variability is due to the superposition of short-term transient flow processes (i.e., fast-path flow) on top of longer-term climate effects, the spatial variability induced at the surface, variability in local fracture-matrix interaction, and diffusion-limited fracture-matrix chemical equilibration. Small-scale variability could be due to tempered variations and differences in transport velocities and fracture-matrix interaction, even if over shorter time periods there is spatial redistribution in the PTn.

Reactive Geochemical Systems

Reactive geochemical systems range from the simplest sorption or exchange reactions involving trace ions, to those strongly coupled systems involving multicomponent water-gas-rock interaction with mineral precipitation and dissolution. Analysis of such systems through measurement and modeling yields otherwise unattainable information regarding transport processes at a variety of scales, both spatial and temporal.

Here we present a relatively simple example of a chemical species, strontium (Sr), that shows dominantly conservative behavior in the upper part of the unsaturated zone at Yucca Mountain and strong cation exchange behavior in lower zeolitized units. An important, and previously overlooked, aspect to the behavior of Sr in the unsaturated zone is its relatively high concentration in infiltrating waters, as a result of evapotranspiration, compared to the Sr added through kinetically slow reactions with unzeolitized tuffaceous rocks and that lost to precipitation of calcite in fractures.

Results from a three-dimensional simulation (ECM assumption) at specific locations corresponding to boreholes that intersect perched water are shown in Figure 11-10. Sr is considered to be nearly conservative in unzeolitized rocks (k_D = 0.02 m^2/kg), and extremely reactive in zeolitized rocks (k_D = 1000 m^2/kg). Modeled values are compared to measured Sr concentrations in perched waters (Patterson et al., 1998). Concentrations in waters in unzeolitized rocks (UZ-14 and WT-24) are similar to (but generally higher than) the predicted concentrations (governed mainly by the extent of surface evapotranspiration). Perched waters that are in contact with zeolitized rocks have Sr concentrations one to two orders of magnitude lower. The mismatch between concentrations in the zeolitic units could be due to lateral flow or errors in geologic stratigraphy.

Abundant data on mineral distributions in fractures and their compositions (collected by the USGS and Los Alamos National Laboratory) have been used to infer flow pathways and long-term infiltration rates (Marshall et al., 1998; Vaniman and Chipera, 1996). These data are also important for inferring flow

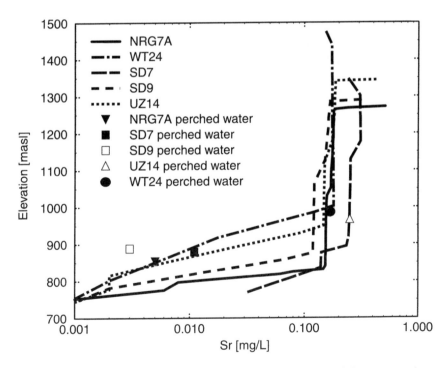

FIGURE 11-10 Modeled (three-dimensional ECM) and measured Sr concentrations (perched water) as a function of elevation.

processes within individual fractures (e.g., fracture mineralization occurs on the footwall of fractures only; Paces et al., 1998). Some of these data are now being evaluated using multicomponent reactive transport modeling for further calibration of the UZ flow and transport model, to assess relations between infiltration rates and mineral precipitation. Such studies reveal information on processes that have taken place over tens of thousands to millions of years, in contrast to other geochemical systems, such as nuclear fallout-generated ^{36}Cl, which records transient processes that have occurred over the past 50 years.

Spatial and Temporal Variability and Uncertainties

The UZ flow and transport model plays an important role in better understanding of flow and transport processes at Yucca Mountain. However, the accuracy and reliability of the model predictions are critically dependent on the accuracy of estimated model properties. Past site investigations have shown that there exists a large variability of the flow and transport parameters over the large spatial and temporal scales of the mountain. Even though considerable progress has been made in this area, uncertainty associated with the UZ model input parameters will continue to be a key issue for future studies. The major uncertainties in model parameters are: (1) accuracy in estimated current, past, and future net infiltration rates over the mountain; (2) quantitative descriptions of the heterogeneity of welded and non-welded tuffs, their flow properties, and detailed spatial distributions within the mountain, especially below the repository; (3) insufficient field studies, especially for fracture properties; and (4) evidence of lateral diversion in the CHn units, where the zeolitic zones may play an important role in diverting moisture laterally. All of these uncertainties have been addressed to a certain extent in past studies; however, a comprehensive study is still needed to reduce these parameter uncertainties further by continuous field, laboratory, and modeling efforts.

SUMMARY AND CONCLUSIONS

The unsaturated zone flow and transport model of Yucca Mountain, Nevada, is a very complex three-dimensional, dual-permeability numerical model. The model considers all relevant geological, hydrological, and geochemical data from the unsaturated zone, and is calibrated to the extent possible using these data. A major challenge in development of the model is the parameterization and upscaling of important fracture and matrix parameters used in the model. The following points represent our current understanding and beliefs regarding parameterization and upscaling in the UZ model.

1. Conventional upscaling methods cannot be used to upscale core-scale (matrix) permeabilities for unsaturated fractured rocks, because fractures may reduce effective matrix permeability at larger scales.

2. No upscaling of fracture permeability is needed, as pneumatic tests have been done at various scales. Fracture permeability data show typical increases with scale and can be represented as stochastic fractals. Other fracture parameters, such as the van Genuchten parameters and fracture porosity, can be derived from the fracture permeability data using the cubic law, but this may yield inaccurate results. For example, in-situ seepage, gas tracer, and thermal test data suggest fracture porosities that are orders of magnitude higher than derived values.

3. Matrix diffusion is more important than dispersion, and neither fracture diffusion or dispersion significantly affects transport at Yucca Mountain. The fracture/matrix interaction factor has large effects on flow and transport; the UZ model utilizes the newly developed "active" fracture concept.

4. The primary upscaling process used in the UZ model is a direct inversion of the observed data, including saturation, moisture tension, pneumatic, perched water, temperature, and geochemical data. A generic approach has been developed that starts with one-dimensional inversions, and then proceeds to two-dimensional and three-dimensional inversions.

5. The one-dimensional inversions are performed using saturation, moisture tension, and pneumatic data. The results obtained show that only limited upscaling is needed for most parameters, with the fracture and matrix α and matrix permeability being the most sensitive parameters.

6. Three-dimensional trial-and-error calibrations are performed primarily for the perched water, temperature, and geochemistry data. The perched water calibrations greatly alter the hydrological parameters below the repository region, but the temperature data match model results reasonably well using the current infiltration maps.

7. Three-dimensional model validation was performed with various chemical species, including Cl, ^{36}Cl, and Sr. Cl modeling gives valuable constraints on infiltration/percolation fluxes and patterns, whereas ^{36}Cl modeling yields the fast component of the flow. Sr reacts strongly in zeolitic rocks, and its modeling yields valuable information, including flow paths and fracture-matrix interaction.

ACKNOWLEDGMENTS

The authors thank Yvonne Tsang, Jianchun Liu, and Dan Hawkes (Lawrence Berkeley National Laboratory) for careful technical and editorial review of this paper. This work was supported by the Director, Office of Civilian Radioactive Waste Management, U.S. Department of Energy, through Memorandum Purchase Order EA9013MC5X between TRW Environmental Safety Systems and the Ernest Orlando Lawrence Berkeley National Laboratory (Berkeley Lab). The support is provided to Berkeley Lab through the U.S. Department of Energy Contract No. DE-AC03-76SF00098.

REFERENCES

Ahlers, C. F., T. M. Bandurraga, G. S. Bodvarsson, G. Chen, S. Finsterle, and Y. S. Wu, 1995. Summary of Model Calibration and Sensitivity Studies Using the LBNL/USGS Three-Dimensional Unsaturated Zone Site-Scale Model. Yucca Mountain Project Milestone 3GLM107M. Berkeley, Calif.: Lawrence Berkeley National Laboratory.

Altman, S. J., B. W. Arnold, R. W. Barnard, G. E. Barr, C. K. Ho, S. A. McKenna, and R. R. Eaton, 1996. Flow Calculations for Yucca Mountain Groundwater Travel Time (GWTT-95). SAND96-0819. Albuquerque, N.M.: Sandia National Laboratories.

Bandurraga, T. M., and G. S. Bodvarsson, 1997. Calibrating matrix and fracture properties using inverse modeling (Chapter 6). In: The Site-Scale Unsaturated Zone Model of Yucca Mountain, Nevada, for the Viability Assessment. G. S. Bodvarsson, T. M. Bandurraga, and Y. S. Wu, eds. LBNL-40376, Lawrence Berkeley National Laboratory, Berkeley, Calif.

Bodvarsson, G. S., C. F. Ahlers, M. Cushey, F. H. Dove, S. A. Finsterle, C. B. Haukwa, J. Hinds, C. K. Ho, J. Houseworth, Q. Hu, H. H. Liu, M. Pendleton, E. L. Sonnenthal, A. J. Unger, J. S. Y. Wang, M. Wilson, and Y.-S. Wu, 2000. Unsaturated Zone Flow and Transport Model Process Model Report. Civilian Radioactive Waste Management System, Management and Operating Contractor, Las Vegas, Nev.

Bodvarsson, G. S., T. M. Bandurraga, and Y. S. Wu (eds.), 1997a. The Site-Scale Unsaturated Zone Model of Yucca Mountain, Nevada, for the Viability Assessment. Rep. LBNL-40376. Berkeley, Calif.: Lawrence Berkeley National Laboratory.

Bodvarsson, G. S., C. Shan, A. Htay, A. Ritcey, and Y. S. Wu, 1997b. Estimation of percolation flux from temperature data. Chapter 11. In: The Site-Scale Unsaturated Zone Model of Yucca Mountain, Nevada, for the Viability Assessment. G. S. Bodvarsson, T. M. Bandurraga, and Y. S. Wu, eds. Report LBNL-40376. Berkeley, Calif.: Lawrence Berkeley National Laboratory.

Fabryka-Matin, J. T., A. Meijer, B. Marshall, L. Neymark, J. Paces, J. Whelan, and A. Yang, 2000. Analysis of Geomechanical Data for the Unsaturated Zone. Civilian Radioactive Waste Management System, Management and Operating Contractor, Las Vegas, Nev.

Fabryka-Martin, J. T., A. V. Wolfsberg, J. L. Roach, S. T. Winters, L. E. Wolfsberg, 1998. Using Chloride to Trace Water Movement in the Unsaturated Zone at Yucca Mountain. In: Proceedings of the Eighth International Conference on High-Level Radioactive Waste Management. American Nuclear Society, May 11-14, 93-96.

Fabryka-Martin, J. T., P. R. Dixon, S. Levy, B. Liu, H. J. Turin, and A. V. Wolfsburg, 1996. Summary Report of Chlorine-36 Studies: Systematic Sampling for Chlorine-36 in the Exploratory Studies Facility. Los Alamos National Laboratory Milestone Report 3783AD. Los Alamos, N.M.: Los Alamos National Laboratory.

Faybishenko, B., 1999. Evidence of Chaotic Behavior in Flow Through Fractured Rocks, and How We Might Use Chaos Theory in Fractured Rock Hydrology. Proceedings of the International Symposium on Dynamics of Fluids in Fractured Rocks in Honor of Paul A. Witherspoon's 80th Birthday. Lawrence Berkeley National Laboratory Report LBNL-42718. Berkeley, Calif.: Lawrence Berkeley National Laboratory.

Finsterle, S., 1999. ITOUGH2 User's Guide. Report LBNL-40040. Berkeley, Calif.: Lawrence Berkeley National Laboratory.

Flint, A. L., J. A. Hevesi, and L. E. Flint, 1996. Conceptual and Numerical Model of Infiltration for the Yucca Mountain Area, Nevada. Milestone 3GU1623M. U.S. Geol. Surv. Water Res. Invest. Rep. Denver, Colo.: U.S. Geological Survey.

Flint, A. L., L. E. Flint, G. S. Bodvarsson, and E. M. Kwicklis, 1999. Evolution of the Conceptual Model of Vadose Zone Hydrology for Yucca Mountain. National Research Council, National Academy of Sciences, U.S. National Committee for Rock Mechanics. Presented at the "Conceptual Models of Flow and Transport in the Fractured Vadose Zone," March 18-19, 1999, Irvine, Calif.

Flint, L. E., 1998. Matrix properties of hydrogeologic units at Yucca Mountain, Nevada. U.S. Geological Survey Report 97-4243. Denver, Colo.: U.S. Geological Survey.

Glass, R. J., M. J. Nicholl, and V. C. Tidwel, 1996. Challenging and Improving Conceptual Models for Isothermal Flow in Unsaturated, Fractured Rocks Through Exploration of Small-Scale Processes. Rep. SAND95-1824. Albuquerque, N.M.: Sandia National Laboratories.

Koltermann, C. E., and S. M. Gorelick, 1996. Heterogeneity in sedimentary deposits: A review of structure imitating, process-imitating, and descriptive approaches. Water Resour. Res. 32: 2617-2658.

Kwicklis, E. M., G. S. Bodvarsson, and A. L. Flint, 1999. A Conceptual Model of the Unsaturated Zone Flow and Transport, Yucca Mountain, Nevada. U.S. Geol. Surv. Water Resour. Rep. Denver, Colo.: U.S. Geological Survey.

LeCain, G. D., 1997. Air-Injection Testing in Vertical Boreholes in Welded and Nonwelded Tuff, Yucca Mountain, Nevada. U.S. Geol. Surv. Water Resour. Invest. Rep. 96-4262. Denver, Colo.: U.S. Geological Survey.

LeCain, G., and G. Patterson, 1997. Technical Analysis and Interpretation, Air-Permeability and Hydrochemistry Data through January 31, 1997. Memo/Report to Robert Craig. Level 4 Yucca Mountain Milestone SPH35EM4. Denver, Colo.: U.S. Geological Survey.

Liu, H. H., and F. J. Molz, 1997, Multifractal analyses of hydraulic conductivity distributions. Water Resour. Res. 33(11): 2483-2488.

Liu, H. H., C. Doughty, and G. S. Bodvarsson, 1998. An active fracture model for unsaturated flow and transport in fractured rocks. Water Resour. Res. 34(10): 2633-2646.

Marshall, B. D., J. B. Paces, L. A. Neymark, J. F. Whelan, Z. E. Peterman, 1998. Secondary minerals record past percolation flux at Yucca Mountain, Nevada. In: Proceedings of the Eighth Annual International Conference, High Level Radioactive Waste Management, May 11-14. Las Vegas, Nev.: American Nuclear Society.

Molz, F. J., H. H. Liu, and J. Szulga, 1997. Fractional Brownian motion and fractional Gaussian noise in subsurface hydrology: A review, presentation of fundamental properties, and extensions. Water Resour. Res. 33(10): 2273-2286.

Neuman, S. P., 1990. Universal scaling of hydraulic conductivities and dispersivities in geologic media. Water Resour. Res. 26(8): 1749-1758.

Neuman, S. P., 1994. Generalized scaling of permeabilities: Validation and effect of support scale. Geophy. Res. Lett. 30(21): 349-352.

Paces, J. B., B. D. Marshall, L. A. Neymark, J. F. Whelan, Z. E. Peterman, 1998. Inferences for Yucca Mountain unsaturated zone hydrology from secondary minerals. In: Proceedings of the Eighth Annual International Conference, High Level Radioactive Waste Management, May 11-14, 1998. Las Vegas, Nev.: American Nuclear Society, Pp. 36-39.

Paleologos, E. K., S. P. Neuman, and D. Tatakovsky, 1996. Effective hydraulic conductivity of bounded, strongly heterogeneous porous media. Water Resour. Res. 32: 1333-1341.

Pan, L., H. H. Liu, M. Cushey, and G. S. Bodvarsson, 1999. A New Random Walk Particle Tracker for Dual-Continua. LBNL-42928, Lawrence Berkeley National Laboratory, Calif.

Patterson, G., 1998. Occurrences of Perched Water in the Vicinity of the Exploratory Studies Facility North Ramp. Section 4.2.4 (including sections 4.2.4.1-4.2.4.5). In: J. P Rousseau, E. M. Kwicklis, and D. C. Gillies, eds. Hydrogeology of the Unsaturated Zone, North Ramp Area of the Exploratory Studies Facility, Yucca Mountain, Nevada. Yucca Mountain Project Milestone Report 3GUP667M. U.S. Geological Survey Water Resources Investigations Report 98-4050. Denver, Colo.: U.S. Geological Survey.

Patterson, G. L., Z. E. Peterman, and J. B. Paces, 1998. Hydrochemical evidence for the existence of perched water at USW WT-24, Yucca Mountain, Nevada. In: Proceedings of the Eighth Annual International Conference, High Level Radioactive Waste Management, May 11-14. Las Vegas, Nev.: American Nuclear Society, pp. 277-278.

Pruess K., 1991. TOUGH2: A General Purpose Numerical Simulator for Multiphase Fluid and Heat Flow. Lawrence Berkeley National Laboratory Report LBNL-29400. Berkeley, Calif.: Lawrence Berkeley National Laboratory.

Pruess, K., 1999. A mechanistic model for water seepage through thick unsaturated zones in fractured rocks of low matrix permeability. Water Resour. Res. 35(4): 1039-1051.

Pruess, K., B. Faybishenko, and G. S. Bodvarsson, 1997. Alternative Concepts and Approaches for Modeling Unsaturated Flow and Transport in Fractured Rocks. Chapter 24 of The Site-Scale Unsaturated Zone Model of Yucca Mountain, Nevada, for the Viability Assessment. G. S. Bodvarsson, T. M. Bandurraga, and Y. S. Wu, eds. Rep. LBNL-40376. Berkeley, Calif.: Lawrence Berkeley National Laboratory.

Ritcey, A. C., E. L. Sonnenthal, Y. S. Wu, C. Haukwa, and G. S. Bodvarsson, 1998. Final Prediction of Ambient Conditions along the East-West Cross Drift Using the 3-D UZ Site-Scale Model. Yucca Mountain Project Milstone SP22ABM4. Berkeley, Calif.: Lawrence Berkeley National Laboratory.

Robinson, B. A., A. V. Wolfsberg, G. A. Zyvoloski, and C. W. Gable, 1995. An Unsaturated Zone Flow and Transport Model of Yucca Mountain, Milestone 3468. Los Alamos, N.M.: Los Alamos National Laboratory.

Rousseau, J. P., 1996. Data Transmittal of Pneumatic Pressure Records from 10/24/94 through 3/31/ 96 for Boreholes UE-25 UZ#4, UE-25 UZ#5, USW NRG-6, USW NRG-7a, USW SD-12, and USW UZ-7a. U.S. Geological Survey Preliminary Data from Yucca Mountain, Nevada. Denver, Colo.: U.S. Geological Survey.

Rousseau, J. P., E. M. Kwicklis, and D. C. Gillies, eds., 1998. Hydrogeology of the Unsaturated Zone, North Ramp Area of the Exploratory Studies Facility, Yucca Mountain, Nevada. USGS-WRIR-98-4050 Yucca Mountain Project Milestone 3GUP667M (formerly 3GUP431M). Denver, Colo.: U.S. Geological Survey.

Sass J. H., A. H. Lachenbruch, W. W. Dudley Jr., S. S. Priest, and R. J. Munroe, 1988. Temperature, thermal conductivity, and heat flow near Yucca Mountain, Nevada: Some tectonic and hydrologic implications. USGS OFR-87-649. U.S. Geological Survey Open File Rep. 87-649. DE89 002697. Denver, Colo.: U.S. Geological Survey.

Sonnenthal, E. L., and G. S. Bodvarsson, 1999. Constraints on the hydrology of the unsaturated zone at Yucca Mountain, Nevada, from Three-Dimensional Models of Chloride and Strontium Geochemistry. Journal of Contaminant Hydrology 38(1-3): 107-156.

Sonnenthal, E. L., C. F. Ahlers, and G. S. Bodvarsson, 1997. Fracture and Fault Properties for the UZ Site-Scale Flow Model. In: G. S. Bodvarsson, T. M. Bandurraga, and Y. S. Wu, eds. The Site-Scale Unsaturated Zone Model of Yucca Mountain, Nevada, for the Viability Assessment. Chapter 7. Yucca Mountain Site Characterization. Lawrence Berkeley National Laboratory Report LBNL-40376. Berkeley, Calif.: Lawrence Berkeley National Laboratory.

Striffler, P., G. M. O'Brien, T. Oliver, and P. Burger, 1996 (Unpublished). Perched Water Characteristics and Occurrences, Yucca Mountain, Nevada. Yucca Mountain Milestone 3LGUS600M. Submitted for publication as a USGS Water Resources Investigation Report. Denver, Colo.: U.S. Geological Survey.

Sweetkind, D. S., and S. C. Williams-Stroud, 1996. Characteristics of Fractures at Yucca Mountain, Nevada. YMP Milestone 3GGF205M. Denver, Colo.: U.S. Geological Survey.

Tokunaga, T. K., and J. Wan, 1997. Water film flow along fracture surfaces of porous rock. Water Resour. Res. 33(6): 1287-1295.

Tsang, Y. W., J. A. Apps, J. T. Birkholzer, B. Freifeld, J. Peterson, M. Q. Hu, E. Sonnenthal, and N. Spycher, 1999. Single Heater Test Final TDIF Submittal and Final Report. Lawrence Berkeley National Laboratory Report LBNL-42537. Berkeley, Calif.: Lawrence Berkeley National Laboratory.

Van Genuchten, M., 1980. A closed-form equation for predicting the hydraulic conductivity of unsaturated soils. Soil Sci. Soc. Amer. J 44(5): 892-898.

Vaniman, D. T., and S. J. Chipera, 1996. Paleotransport of lanthanides and strontium recorded in calcite compositions from Tuffs at Yucca Mountain, Nevada, USA. Geochimica et Cosmochimica Acta 60(22): 4417-4433.

Wu, Y. S., C. Haukwa, and G. S. Bodvarsson, 1999a. A site-scale model for fluid and heat flow in the unsaturated zone of Yucca Mountain, Nevada. Journal of Contaminant Hydrology 38(1-3): 185-217.

Wu, Y. S., A. C. Ritcey, and G. S. Bodvarsson, 1999b. A modeling study of perched water phenomena in the unsaturated zone at Yucca Mountain. Journal of Contaminant Hydrology 38(1-3): 157-184.

Yang, I. C., G. W. Rattray, and P. Yu, 1996. Interpretation of Chemical and Isotopic Data from Boreholes in the Unsaturated Zone at Yucca Mountain. U.S. Geol. Surv. Water Resour. Invest. Rep. 96-4058. Denver, Colo.: U.S. Geological Survey.

Appendixes

Appendix A

Workshop Attendees

Panel Members

Paul A. Hsieh, *Chair*, U.S. Geological Survey, Menlo Park, California
Jean M. Bahr, University of Wisconsin, Madison
Thomas W. Doe, Golder Associates, Inc., Redmond, Washington
Alan L. Flint, U.S. Geological Survey, Sacramento, California
Glendon Gee, Battelle Pacific Northwest Laboratory, Richland, Washington
Lynn W. Gelhar, Massachusetts Institute of Technology, Cambridge
D. Kip Solomon, University of Utah, Salt Lake City
Rien van Genuchten, U.S. Salinity Laboratory, USDA, ARS, Riverside, California
Stephen W. Wheatcraft, University of Nevada, Reno

Guests

Brian Berkowitz, Weizmann Institute of Science, Rehovot, Israel
G. S. (Bo) Bodvarsson, E. O. Lawrence Berkeley National Laboratory, California
William Boyle, DOE, Yucca Mt. Project, Las Vegas, Nevada
John D. Bredehoeft, The Hydrodynamics Group, La Honda, California
Ralph Cady, Nuclear Regulatory Commission, Washington, D.C.
Jeffrey Ciocco, Nuclear Regulatory Commission, Washington, D.C.
June Fabryka-Martin, Los Alamos National Laboratory, New Mexico
Boris Faybishenko, E. O. Lawrence Berkeley National Laboratory, California

Randy Fedors, Southwest Research Institute, San Antonio, Texas
Jan Hendrickx, New Mexico Institute of Mining and Technology, Socorro
Philip M. Jardine, Oak Ridge National Laboratory, Tennessee
Nicholas Jarvis, Swedish University of Agricultural Sciences, Uppsala, Sweden
Vivek Kapoor, Georgia Institute of Technology, Atlanta
Edward Kwicklis, Los Alamos National Laboratory
Jane C. S. Long, University of Nevada, Reno
Larry D. McKay, University of Tennessee, Knoxville
James W. Mercer, HSIGeoTrans, Inc., Sterling, Virginia
Philip D. Meyer, Pacific Northwest National Laboratory, Portland, Oregon
Bimal Mukhopadhyay, DOE, Yucca Mt. Project, Las Vegas, Nevada
Shlomo P. Neuman, University of Arizona, Tucson
Thomas Nicholson, Nuclear Regulatory Commission, Washington, D.C.
Russ Patterson, DOE, Yucca Mt. Project, Las Vegas, Nevada
Jean-Yves Parlange, Cornell University, Ithaca, New York
Fred Phillips, New Mexico Institute of Mining and Technology, Socorro
Harihar Rajaram, University of Colorado, Boulder
Bruce A. Robinson, Los Alamos National Laboratory, New Mexico
Bridget R. Scanlon, Texas Bureau of Economic Geology, Austin, Texas
J. Leslie Smith, University of British Columbia, Vancouver
Ed A. Sudicky, University of Waterloo, Ontario
Tetsu K. Tokunaga, E. O. Lawrence Berkeley National Laboratory, California
Chin-Fu Tsang, E. O. Lawrence Berkeley National Laboratory, California
Joseph S. Wang, E. O. Lawrence Berkeley National Laboratory, California
Ed Weeks, U. S. Geological Survey, Lakewood, Colorado
Paul A. Witherspoon, Berkeley, California
Tian-Chyi Jim Yeh, The University of Arizona, Tucson

NRC Staff

Thomas Usselman, *Study Director*

Appendix B

Invited Presentations

General Overview of Conceptual Models, *Leslie Smith*

Evolution of Conceptual Model for Yucca Mountain, *Alan Flint*

Conceptual Model of Vadose-Zone Transport in Fractured Weathered Shale, *Philip M. Jardine, G. V. Wilson, and J. P. Gwo*

Evaluation of Conceptual and Quantitative Models of Fluid Flow and Chemical Transport in Fractured Media, *Brian Berkowitz, Ronit Nativ, and Eilon Adar*

Mechanisms of Preferential Flow in the Subsurface, *Jan M. H. Hendrickx and Markus Flury*

Modeling Macropore Flow in Soils: Field Validation and Use for Management Purposes, *Nicholas Jarvis and Martin Larsson*

Mechanisms of Fracture Flow and Fracture-Matrix Interactions, *Joe Wang*

Free-Surface Flow, *Maria Ines Dragila and Stephen W. Wheatcraft*

Fracture Network Geometry, *Tom Doe*

Investigating Flow and Transport in the Fractured Vadose Zone Using Environmental Tracers, *Fred M. Phillips*

Lessons from Field Studies at the Apache Leap Research Site in Arizona, *Shlomo P. Neuman, Walter A. Illman, Velimir V. Vesselinov, Dick L. Thompson, A. Guzman*

Parameterization and Upscaling, *Bo Bodvarsson*

Appendix C

Panel Biographies

PAUL A. HSIEH, is a research hydrologist with the U.S. Geological Survey in Menlo Park, California. Dr. Hsieh is also an adjunct professor in the Department of Geological and Environmental Sciences at Stanford University. He currently serves on the U.S. National Committee for Rock Mechanics and was a member of the Committee on Fracture Characterization and Fluid Flow (Rock Fractures and Fluid Flow: Contemporary Understanding and Applications, NRC, 1996). He conducts theoretical and field research on fluid flow and solute transport in fractured rocks, development of computer simulation models, and laboratory investigations of hydraulic properties of low-permeability rocks. He is a member of the American Geophysical Union, American Society of Civil Engineers, and the National Ground Water Association.

JEAN M. BAHR is a professor in the Department of Geology and Geophysics at the University of Wisconsin-Madison. Dr. Bahr specializes in field monitoring and tracer studies of the effects of these transport processes in real time, core and outcrop studies to provide data on hydrologic and geochemical properties of aquifer materials, and numerical modeling of groundwater transport systems. She served on several NRC activities: (1) as a member of the Board on Radioactive Waste Management (1992-1997); (2) as vice-chair of the NRC Committee on Yucca Mountain Peer Review: Surface Characteristics, Preclosure Hydrology, and Erosion; and (3) as a member of the Committee on Technical Bases for Yucca Mountain Standards. She was awarded a Ph.D. by Stanford University. She is a member of the American Geophysical Union and an Associate Editor of AGU's journal "Water Resources Research." She is also a member of the American Geological Institute's editorial board of Geotimes.

THOMAS W. DOE is a principal of Golder Associates, Inc., located in Seattle, Washington. He is manager of the FracMan Technology Group, which develops software for analysis and modeling of heterogeneous and fractured rock masses containing discrete features such as faults, fractures, paleochannels, karsts, and stratigraphic contacts. Dr. Doe is a pioneer in the development of fractional dimension type-curve methods for understanding connectivity patterns, permeability, and storativity. He has also worked extensively on interpretation of well tests in fractured rock to determine hydraulic and geometric characteristics of fractures. Dr. Doe earned his Ph.D. from the University of Wisconsin-Madison. He is a member of the American Geophysical Union, and is currently serving as President of the American Rock Mechanics Association.

ALAN L. FLINT is a research hydrologist with the U.S. Geological Survey in the California District, Sacramento. Dr. Flint earned his Ph.D. from Oregon State University in soil physics. He has primarily been involved in modeling and field characterization of unsaturated zone infiltration and deep percolation through the vadose zone; much of his recent work has been related to hydrological conditions at Yucca Mountain, Nevada. Dr. Flint is currently conducting regional research on rainfall, runoff, infiltration, and recharge in the arid southwest. He is a member of the American Geophysical Union and the Soil Science Society of America.

GLENDON GEE is a Senior Staff Scientist in the Environmental Technology Division of the Batelle Pacific Northwest Laboratories. Dr. Gee specializes in the study of recharge in arid regions and the physical and chemical observations of fluid flow and transport under field conditions.

LYNN W. GELHAR is a Professor in the Department of Civil and Environmental Engineering at the Massachusetts Institute of Technology. Dr. Gelhar investigates the effects of natural heterogeneity in the underground aqueous systems using mathematical theory to stochastically represent the variability of flow and transport properties, and evaluates the applicability of the theoretical results through controlled field experimentation. He is a fellow of the American Geophysical Union and a recipient of the Meinzer Award by the Geological Society of America. He is the author of the advanced textbook, *Stochastic Subsurface Hydrology*, and has authored over 140 research publications. He has more than two decades of experience on aspects of subsurface hydrology relating to problems of radioactive waste disposal, and has served on multidisciplinary groups reviewing environmental aspects of the Hanford site in Washington, the WIPP radioactive waste disposal site in New Mexico, and the Nevada Test Site and the Yucca Mountain site in Nevada.

D. KIP SOLOMON is a chemical hydrogeologist with the Department of Geology and Geophysics at the University of Utah. He specializes in fluid flow in soils and shallow aquifers, emphasizing the fate and transport of contaminants. Dr. Solomon has also worked on techniques for determining the age of shallow groundwater using tritium and helium isotopes and chlorofluorocarbons and using these tools to examine mass transport in fractured, clay-rich systems. He earned a Ph.D. from the University of Waterloo. He is a member of the American Geophysical Union and the Geological Society of America.

MARTINUS TH. VAN GENUCHTEN is a soil scientist and research leader in the Soil Physics and Pesticide Research Unit of the Department of Agriculture's U.S. Salinity Laboratory, Riverside, California. Dr. van Genuchten is also an adjunct professor in the Department of Soil and Environmental Sciences at the University of California, Riverside. He is an expert in water flow and solute transport in the vadose zone, analytical and numerical methods for simulating water and solute movement in the subsurface, preferential flow of water and solutes in structured media, and description and measurement of unsaturated soil hydraulic properties. He is a fellow of both the American Geophysical Union and the Soil Science Society of America. Dr. van Genuchten earned his bachelors and masters degrees from the Agricultural University (Wageningen, The Netherlands) and a Ph.D. in soil physics from New Mexico State University. From 1975 to 1978, he was with the Water Resources Program at Princeton University prior to his career with the U.S. Salinity Laboratory.

STEPHEN W. WHEATCRAFT is a Professor in the Department of Geological Sciences and Associate Dean of the Mackay School of Mines at the University of Nevada, Reno. His research ranges from classical problems such, as sea water intrusion, to the use of chaos to describe flow and transport characteristics in unsaturated fracture flow. Most recently, he has used fractional calculus to develop a new transport theory in saturated heterogeneous aquifers. Dr. Wheatcraft earned a Ph.D. from the University of Hawaii. He is a member of the American Geophysical Union, the Association of Groundwater Scientists and Engineers, and a fellow of the Geological Society of America.